T0137229

Computational Intelligence for Network Structure Analytics

Maoguo Gong · Qing Cai
Lijia Ma · Shanfeng Wang
Yu Lei

Computational Intelligence for Network Structure Analytics

Maoguo Gong
Xidian University
Xi'an, Shaanxi
China

Shanfeng Wang
Xidian University
Xi'an, Shaanxi
China

Qing Cai
Hong Kong Baptist University
Hong Kong
China

Yu Lei
Northwestern Polytechnical University
Xi'an, Shaanxi
China

Lijia Ma
Hong Kong Baptist University
Hong Kong
China

ISBN 978-981-13-5167-9 ISBN 978-981-10-4558-5 (eBook)
DOI 10.1007/978-981-10-4558-5

Printed on acid-free paper

This Springer imprint is published by Springer Nature
The registered company is Springer Nature Singapore Pte Ltd.
The registered company address is: 152 Beach Road, #21-01/04 Gateway East, Singapore 189721, Singapore

Preface

Many real-world problems are actually optimization problems and it is our long standing quest to address these real-world optimization problems by means of techniques from diverse fields. Theories and techniques for canonical mathematical optimization have been well studied for centuries and play a leading role in solving optimization problems. Nevertheless, most if not all network issues are essentially hard optimization problems which very often cannot be well solved by traditional mathematical optimization techniques. The performances of canonical methods deteriorate rapidly especially when the real problems involve many optimization objectives and the number of decision variables is large.

In order to remedy the drawbacks of canonical optimization techniques, computational intelligence, a class of artificial intelligence techniques, has come into being and is widely recognized as a promising computing paradigm. Evolutionary computation, an important branch of computational intelligence, emerges and tremendous efforts have been done to design many kinds of efficient evolutionary algorithms for solving diverse hard optimization problems. Apart from evolutionary computation, other computational intelligence techniques such as swarm intelligence, meta-heuristics, and artificial neural networks have all find their niche in the optimization field.

This book makes efforts to delineate in detail the existing state-of-the-art methods based on computational intelligence for addressing issues related to complex network structures. Using computational intelligence techniques to address network issues may facilitate smart decisions making by providing multiple options to choose from, while conventional methods can only offer a decision maker a single suggestion.

Meanwhile, evolutionary computation provides a promising outlet toward network issues and in turn network structure patterns may provide novel inspiration toward the design of next-generation computational intelligence techniques.

As a comprehensive text, the contents of the whole book cover most emerging topics of both network structures analytics and evolutionary computation, including theories, models, algorithm design, and experimental exhibitions. This book summarizes the researches achievements on the topics by the authors, their postgraduate students and their alumni ever since 2008. Offering a rich blend of theories and practices, the book is suitable for students, researchers, and practitioners interested in network analytics and computational intelligence, both as a textbook and as a reference work.

We would like to take this great opportunity to extend our sincere thanks to the editors at Springer Press and the anonymous reviewers for their helpful comments for improving the quality of this book. Finally, we would like to thank all the members with the Collaborative Innovation Center for Computational Intelligence (OMEGA) at Xidian University.

This research was supported by the National Natural Science Foundation of China (Grant nos. 61273317, 61422209, and 61473215), the National Program for Support of Top-notch Young Professionals of China, and the National key research and development program of China (Grant no. 2017YFB0802200).

Xi'an, China Maoguo Gong
Hong Kong, China Qing Cai
Hong Kong, China Lijia Ma
Xi'an, China Shanfeng Wang
Xi'an, China Yu Lei

Contents

Chapter 1
Introduction

Abstract Complex network structure analytics contribute greatly to the understanding of complex systems, such as Internet, social network, and biological network. Many issues in network structure analytics, for example, community detection, structure balance, and influence maximization, can be formulated as optimization problems. These problems usually are NP-hard and nonconvex, and generally cannot be well solved by canonical optimization techniques. Computational intelligence-based algorithms have been proved to be effective and efficient for network structure analytics. This chapter gives a holistic overview of complex networks and the emerging topics concerning network structure analytics as well as some basic optimization models for network issues.

1.1 Network Structure Analytics with Computational Intelligence

Recent years have witnessed an enormous interest in complex network, such as, social networks and traffic networks. It is a challenge to understand and mine knowledge from these systems. Then network structure analytics give an insight into complex systems. Network-related topics, such as community detection, structure balance, network robustness, and influence maximization are attracting researchers from all over the world.

Many issuess related to network structure analytics can be modeled as optimization problems, and these optimization problems usually are nonconvex and NP-hard. Then computational intelligence-based algorithms, especially evolutionary algorithms, have been proved to be promising for solving these optimization problems efficiently.

In this chapter, some network structure analytics and their optimization models will be introduced. First, the concepts of networks are introduced. Afterward, several networks analytics topics are described and their optimization models are presented in detail. The introduced topics include community detection, structural balance, network robustness, etc.

M. Gong et al., *Computational Intelligence for Network Structure Analytics*,
DOI 10.1007/978-981-10-4558-5_1

1.1.1 *Concepts of Networks*

Our daily life is intertwined with complex systems. Internet helps us to freely communicate with each other from all over the world. Traffic networks facilitate our daily traveling. The power grid provides us with electricity from power stations. To better understand these complex systems, they are commonly represented as graphs. Individuals in complex systems are represented by nodes of graphs and nodes who have interactions with each other are connected by edges in graphs.

1.1.1.1 Representation of Networks

Any complex graphs consist of two components: node and edge. From the perspectives of mathematics, we can model a network by $G = (V, E)$, where V is the set of nodes and E is the set of edges. The number of nodes is usually denoted by $n = |V|$ and the number of edges is $m = |E|$.

A network is usually represented as an adjacency matrix A. Each element a_{ij} is denoted as follows.

$$a_{ij} = \begin{cases} w_{ij} & if\ (i, j) \in E \\ 0 & if\ (i, j) \notin E \end{cases} \tag{1.1}$$

where w_{ij} represents the weight of the edge between nodes i and j. $(i, j) \in V$ denotes an edge between node i and j.

According to the weight of the edge between nodes i and j, networks are classed into different categories. When $w_{ij} = 1$, the network is an unsigned and unweighted one. When $w_{ij} \neq 1$ and $w_{ij} > 0$, the network is an unsigned and weighted one. In reality, a special case is that individuals are with friendly/hostile relationships. One can use signed edges to represent these relationships. Positive edges between two nodes represent friendships and hostile relationships are represented by negative edges. In this way, w_{ij} is positive or negative and such a network is called a signed network.

Figure 1.1 gives an illustration of unsigned network and signed network in real areas. Figure 1.1a is an unsigned network and Fig. 1.1b is a signed network.

1.1.1.2 Network Data Sets

This section will list the network data sets commonly used in the literature for testing purpose. The data sets contain two types, artificial benchmark networks and real-world networks. Benchmark networks have controlled topologies. They are used to mimic real-world networks. Different real-world networks may have different properties. Hence, real-world networks are still needed for testing purpose.

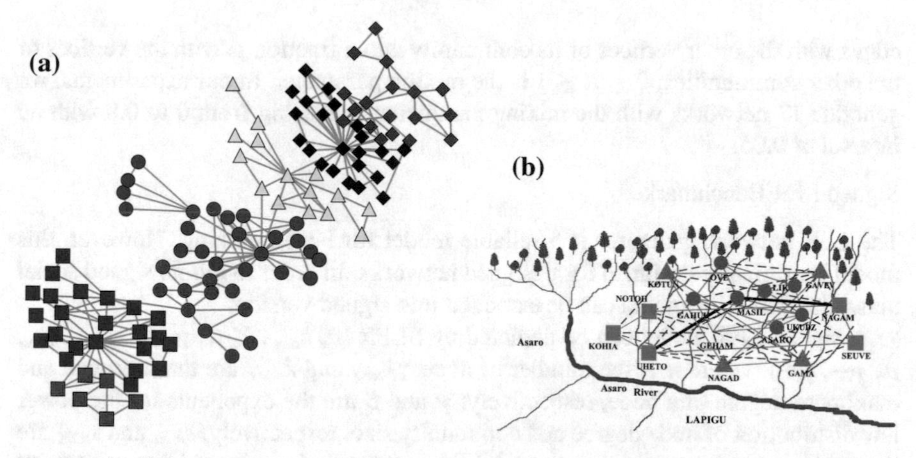

Fig. 1.1 An illustration of signed networks. **a** An unsigned network. **b** A signed network

GN Benchmark and Its Extended Version

Girvan and Newan (GN) in [30] put forward a benchmark network generator which is normally recognized as the GN benchmark. For a GN benchmark network, it was constructed with 128 vertices divided into four communities of 32 vertices each. Edges were placed between vertex pairs independently at random, with probability P_{in} for vertices belonging to the same community and P_{out} for vertices in different communities, with $P_{out} < P_{in}$. The probabilities were chosen so as to keep the average degree z of a vertex equal to 16.

An extended version of the GN model was introduced in [53]. The extended benchmark network also consists of 128 nodes divided into four communities of 32 nodes each. Every node has an average degree of 16 and shares a fraction μ of links with the rest in its community, and $1 - \mu$ with the other nodes of the network. Here, μ is called the mixing parameter. When $\mu < 0.5$, the neighbors of a vertex inside its community are more than the neighbors belonging to the rest groups.

LFR Benchmark

Standard benchmarks, like the GN benchmark or its extended version, do not account for important features in graph representations of real systems, like the fat-tailed distributions of node degree and community size, since on those benchmark networks, all vertices have approximately the same degree, moreover, all communities have exactly the same size by construction.

To overcome these drawbacks, a new class of benchmark graphs have been proposed by Lancichinetti, Fortunato, and Radicchi (LFR) in [53], in which the distributions of node degree and community size are both power laws with tunable exponents. They assume that the distributions of degree and community size are power laws, with exponents τ_1 and τ_2, respectively. Each vertex shares a fraction $1 - \mu$ of its edges with the other vertices of its community and a fraction μ with the vertices of the other communities; $0 \leq \mu \leq 1$ is the mixing parameter. In our experiments, we generate 17 networks with the mixing parameter increasing from 0 to 0.8 with an interval of 0.05.

Signed LFR Benchmark

The LFR network generator is a reliable model for benchmarking. However, this model is originally designed for unsigned networks. In order to mimic signed social networks, The LFR model can be extended into signed version.

A signed LFR model can be depicted by SLFR (n, k_{arg}, k_{max}, γ, β, s_{min}, s_{max}, μ, $p-$, $p+$), where n is the number of nodes; k_{arg} and k_{max} are the averaged and maximum degree of a node, respectively; γ and β are the exponents for the power law distribution of node degree and community size, respectively; s_{min} and s_{max} are the minimum and maximum community size, respectively. μ is a mixing parameter. Each node shares a fraction $1 - \mu$ of its links with the other nodes of its community and a fraction μ with the other nodes of the network. $p-$ is the fraction of negative edges within communities, and $p+$ is the fraction of positive edges between different communities.

Real-World Networks

Tables 1.1 and 1.2 list the parameters of commonly tested unsigned and signed networks. In the Tables, m^+ and m^- denote the numbers of positive and negative edges, respectively.

Table 1.1 Properties of unsigned networks

Network	# Node	# Edge	# Clusters	References
Karate	34	78	2	[112]
Journal	40	189	4	[92]
Dolphin	62	159	2	[70]
Football	115	613	12	[30]
Polbooks	105	441	Unknown	[79]
SFI	118	200	Unknown	[30]
VAST	400	9834	Unknown	[37]
Jazz	198	2742	Unknown	[31]
Elegans	453	2025	Unknown	[18]
Email	1133	5451	Unknown	[41]
Netscience	1589	2742	Unknown	[79]
Power	4941	6594	Unknown	[108]
Ca_grqc	5242	14490	Unknown	[58]
Geom	7743	11898	Unknown	[72]
Pgp	10680	24340	Unknown	[7]
Internet	22963	48436	Unknown	[72]

Table 1.2 Properties of signed networks

Network	# Node	# Edge	m^+	m^-	References
SPP	10	45	18	27	[51]
GGS	16	58	29	29	[89]
SC	28	42	30	12	[111]
War	216	1433	1295	138	[29]
CP	119	686	528	158	
EGFR	329	779	515	264	[82]
Macrophage	678	1,425	947	478	[81]
Yeast	690	1080	860	220	[75]
Ecoli	1461	3215	1879	1336	[94]
WikiElec	7114	100321	78792	21529	[57]
Wiki-rfa	11276	170973	132988	37985	[56]
Slashdot	22936	288638	217563	71075	[56]
Slashdot-large	77357	466666	352890	113776	[56]

1.1.2 Community Structure and Its Detection in Complex Networks

1.1.2.1 Definition of Community Structure

Community is an important property of network. Broadly speaking, a community is a set of individuals with common interests or characteristics. Communities have received great attention in many fields [24]. For example, users with similar interests are clustered into the same community and then recommendation can be generated based on community detection results. A community is also called a cluster which is a subset of the network. Nodes in the same community are more likely to have commonly properties and features.

There is no uniform definition of community structure. One generally recognized definition of community for unsigned network was given by Radicchi et al. [88]. This definition is based on the hypothesis that the connections within communities of a network are denser than those between communities. Given a network $G = (V, E)$, where V is the set of nodes in network G, and E is the set of edges in this network. Let d_i denote the degree of node i and A represent the adjacency matrix. Suppose C is a community of network G and $k_i^{in} = \sum_{i,j \in C} A_{ij}$ and $k_i^{out} = \sum_{i \in C, j \notin C} A_{ij}$ are the internal and external degree of node i, then C is a community in a strong sense if

$$\forall i \in C, k_i^{in} > k_i^{out} \tag{1.2}$$

C is a community in a weak sense if

$$\sum_{i \in C} k_i^{in} > \sum_{i \in C} k_i^{out} \tag{1.3}$$

Later, the definition of community is extended to signed networks. In signed networks, nodes linked by positive edges have the same interests and may be divided into the same community. Nodes linked by negative edges are hostile and they should be divided into different communities. Based on this cognition, Gong et al. proposed a definition of community in signed networks [40]. Given a subgraph C, $(k_i^+)^{in} = \sum_{j \in C, w_{ij}>0} A_{ij}$ and $(k_i^-)^{in} = \sum_{j \in C, w_{ij}<0} A_{ij}$ are positive and negative internal degree of node i, respectively. C is a community in a strong sense if

$$\forall i \in C, (k_i^+)^{in} > (k_i^-)^{in} \tag{1.4}$$

Suppose $(k_i^+)^{out} = \sum_{j \notin C, w_{ij}>0} A_{ij}$ and $(k_i^-)^{out} = \sum_{j \notin C, w_{ij}<0} |A_{ij}|$ denote the positive and negative external degree of node j, respectively. C is a community in a weak sense if

$$\begin{cases} \sum_{i \in C} (k_i^+)^{in} > \sum_{i \in C} (k_i^+)^{out} \\ \sum_{i \in C} (k_i^-)^{out} > \sum_{i \in C} (k_i^+)^{in} \end{cases} \tag{1.5}$$

Network community detection or community discovery is potentially essential to analyze and understand the topological structure of complex networks [24]. Formally, for a graph G, community detection aims to find a set of communities C_i.

1.1.2.2 Introduction to Optimization Models of Community Detection

Though definitions of community detection are given, it is still a challenge to evaluate whether community detection results are right or not. In this section, evaluation functions of community detection and their optimization models are summarized.

Single-Objective Optimization Models for Community Detection

- *Modularity-based model*
  Considering single-objective optimization, the widely used fitness function for community detection is the modularity (normally denoted by Q) originally introduced by Newman and Girvan [78]. The fitness function is as follows:

$$Q = \frac{1}{2m} \sum_{i,j}^{n} (A_{ij} - \frac{k_i \cdot k_j}{2m}) \delta(i, j) \tag{1.6}$$

  where n and m represent the number of nodes and edges in the network, respectively. $\delta(i, j) = 1$, if node i and j are in the same cluster; otherwise, $\delta(i, j) = 0$. By assumption, the larger the value of Q is, the better the partition is.
  Many nature-inspired algorithms combined with network-specific knowledge have been applied to detect the community structure by optimizing Q value [13, 15, 26, 32, 44, 47–49, 59, 61–63, 66, 67, 77, 96, 100, 105, 110]. However, Q has several drawbacks. First, the optimization of modularity is a NP-hard problem

[8]. Second, there exists resolution limit which may lead to unsatisfying results [25]. To resolve this dilemma, some new evaluation criteria have been designed, such as extended modularity [5, 87, 91], multi-resolution index [60], and so forth. For the analysis of networks that are weighted, signed, and have self-loops, a reformulation of $Q(Q_{sw})$ was proposed [33]. Based on the Q_{sw} metric, the authors in [11] suggested a discrete particle swarm optimization (DPSO) algorithm to solve community detection problem of signed networks. Q_{sw} can also be utilized to handle directed, weighted graphs with moderate modifications [6, 55, 93]. Under the circumstances that a node may belong to more than one community, Q metric has been extended in the literature to identify overlapping community structure [80, 99, 115]. In this line, memetic algorithms for overlapping community detection can be found in [10, 52, 54, 65, 68, 85, 90, 114, 119]. There are also some other extended criteria, such as the local modularity [74], the triangle modularity [4] and the bipartite modularity [42].

- *Multi-resolution model*
  To avoid the resolution limit problem, many multi-resolution models have been proposed recently. The author in [84] employed a memetic algorithm to discover community structure. In [87], the authors put forward a novel fitness function, named as modularity density D.
  Given $G = (V, E)$ with $|V| = n$ vertexes and $|E| = m$ edges. The adjacency matrix is A. If V_1 and V_2 are two disjoint subsets of V, $L(V_1, V_2) = \sum_{i \in V_1, j \in V_2} A_{ij}$ and $L(V_1, \overline{V_1}) = \sum_{i \in V_1, j \in \overline{V_2}} A_{ij}$. Given a partition $S = (V_1, V_2, \cdots, V_k)$ of the graph, where V_i is the vertex set of subgraph G_i for $i = 1, 2, \cdots, k$, the modularity density is defined as

$$D = \sum_{i=1}^{k} \frac{L(V_i, V_i) - L(V_i, \overline{V_i})}{|V_i|} \tag{1.7}$$

  Based on modularity density, a number of algorithms have been developed [12, 17, 28, 34, 35, 43, 64, 76, 118].

Multi-Objective Optimization Models for Community Detection

Many real-world optimization problems involve multiple objectives, optimizing only one objective often leads to low performance in one or more of the other objectives. As discussed earlier, community detection also could be formulated as a multiobjective optimization problem. A number of multiobjective optimization methods have been applied to solve community detection problems.

- *General model*
  With the objective that maximizes the intra-connections in each community and minimizes the inter-connections between different communities, multiobjective memetic algorithms to discover community structure called MOGA-Net were proposed [83, 86]. In these methods, the community detection problem was modeled as a multiobjective optimization problem which optimizes the community score and community fitness. The fast elitist non-dominated sorting genetic algorithm (NSGA-II) framework is utilized to solve this problem. An extended version

based on MOGA-Net can be found in [9]. Following this, the hybrid EA based on harmony search algorithm [27, 53], non-dominated neighbor immune algorithm [36, 54], and multiobjective enhanced firefly algorithm [3] have been employed to optimize the model above. In [37], Gong et al. proposed a multi-objective memetic algorithm for community detection. The highlight of this work is the newly designed multiobjective model which optimizes two objectives termed as negative ratio association (NRA) and ratio cut (RC). Similar multiobjective optimization models can be found in [39, 97]. Some other optimization models like optimizing the combinations of Q and CS can be found in [1], maximizing the two parts of the Q can be found in [102]. A three objectives model can be found in [98]. Surveys on the selection of objective functions in multiobjective community detection can be found in [101, 103].

- *Signed model*
  A network with positive and negative links between vertices is called a signed network. To detect the signed network communities, in [40], the authors put forward a new discrete framework of PSO algorithm which optimizes two objectives named as signed ratio association and signed ratio cut. Another signed optimization model that employs the NSGA-II was developed for community detection [2]. Recently, a signed optimization model based on novel nodes similarity was presented [69].
- *Overlapping model*
  It is noted that in plenty of real-world networks, some nodes might belong to not only one community. To find overlapping communities in networks, Liu et al. [68] designed a three objective memetic algorithm for detecting overlapping community structure, in which one objective can control the partition quality, and another two objectives can get a tradeoff between separating and overlapping communities.
- *Dynamical model*
  Dynamical networks are networks those evolve over time. Analyzing the community structures of dynamical networks will be beneficial to predict the probable change tendency. Many memetic approaches have been proposed to analyze the community structures of dynamical networks [16, 21–23, 38, 50].

Individual Representation

For the community detection problem, each individual represents a network partition. In general, two typical representative individual representation schemas can be found in the existing researches, which are the locus-based [16, 23, 28, 37, 39, 84, 86, 102] and the string-based [12, 32, 35, 63, 64, 69, 72, 76, 96, 97, 109].

The locus-based coding schema explicitly presents the networks linkage relations, but makes the corresponding evolutionary operators like the crossover and mutation operations hard to design. The string-based coding schema takes the advantage of both individual coding and evolutionary operators. One should select a suitable coding schema according to the optimization problem.

Individual Reproduction

Crossover and mutation are two widely used operators to produce offsprings. With regard to locus-based coding schema, uniform crossover and two-point crossover are the most representative choices for community detection problem. Both uniform and two-point crossover can guarantee the maintenance of effective connections of nodes in the network. Uniform crossover has been utilized in [16, 23, 36, 37, 49, 50, 83, 84, 86], and two-point crossover has been adopted in [37, 39, 102]. As to string-based coding schema, one-way crossover and two-way crossover are the most popular choices for community detection problem. The two-way crossover takes the advantage of generating offsprings those combine the common features from their parents. The one-way crossover is employed in [63, 64, 69, 76, 109] and the two-way crossover is used in [12, 35, 72, 96, 97].

The most common mutation operator for both two code schemas is the one-point neighbor-based mutation, which can avoid an unnecessary exploration of the search space effectively [12, 16, 23, 34–37, 39, 43, 64, 72, 76, 77, 83, 84, 86, 102, 109]. Many approaches either use these methods or just make some modifications based on specific problems to them.

Individual Local Search

In order to improve the abilities of community detection methods, the local search operators are adopted to improve the effectiveness of genetic algorithms.

In [35], the authors designed a hill-climbing strategy as the local search method. The hill-climbing strategy is actually a greedy algorithm since it makes a minor change of the solution to get biggest fitness increment. A simulated annealing based local search was proposed in [34]. Local search operators as introduced in [35] and [34] are time-consuming that may cause useless exploitation. The authors in [12] proposed a local search by the motivation that two nodes with no connection are less likely to be divided in the same community. In this local search, a node is assigned to one of its neighbor communities that can get largest fitness increment. For the purpose of improving the search ability, a new local search strategy was designed [72]. A three-level learning strategy is designed to enhance both the population diversity and global searching ability of genetic algorithm. In [117], Zhou et al. designed a multiobjective local search. In their work, two objective-wise local searches are designed. The first one can accelerate convergence and the second one can improve the diversity.

1.1.3 Structure Balance and Its Transformation in Complex Networks

Structural balance is one of the important notions of signed networks [116]. It has received great attention from researchers in various fields, such as psychologist, economist, and ecologist. Analyzing the structural balance allows scholars to study the interactions among different entities in depth and understands the evolving dynamics of social systems from unbalanced status to the unbalanced one. There are various applications of structural balance in different fields, such as international relationships and political elections. For example, different nations express different states of equilibrium and disequilibrium in the international relationships [19].

1.1.3.1 Definition of Structure Balance

Structure balance theory investigates the balance of three interconnected individuals. Assuming that an edge labeled + (−) indicates a friendly (hostile) relationship between the corresponding two individuals, there are the following four types of social relations [45]:

- $+++$: "the friend of my friend is my friend";
- $++-$: "the friend of my friend is my enemy";
- $--+$: "the enemy of my enemy is my friend";
- $---$: "the enemy of my enemy is my enemy".

Structural balance theory in its strong definition claims that the social relations $+++$ and $--+$ are balanced while the relations $++-$ and $---$ are unbalanced. Structural balance theory in its weak definition illustrates that the social relations $+++$, $--+$ and $---$ are balanced while the relation $++-$ is unbalanced. Figure 1.2 illustrates the differences between the strong and the weak definitions of structural balance.

Structural balance theory states that a complete signed network is balanced if and only if all its signed triads are balanced. For a noncomplete signed network, it is balanced if it can be filled into a complete network by adding edges in such a way that the resulting complete network is balanced [19]. Wasserman gives an equivalent theorem from the perspective of clustering as follows [107].

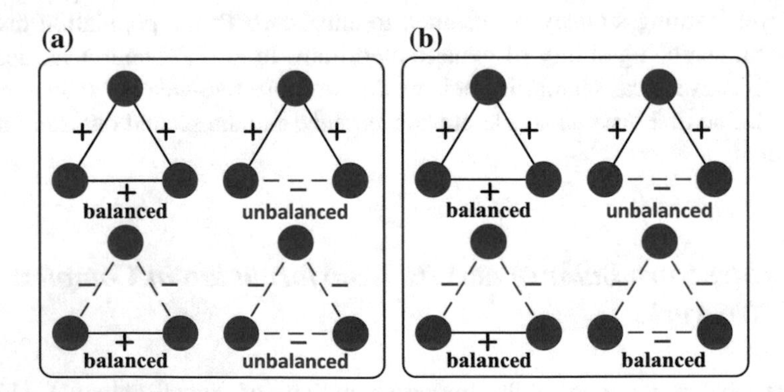

Fig. 1.2 Illustration of the structural balance theorem in **a** the strong definition and **b** the weak definition. A *solid line* represents a positive relationship (+) and a *dash line* indicates a negative relationship (−)

The strong (weak) definition of structural balance: "A signed network G is completely balanced, if and only if its nodes can be divided into $k = 2$ ($k \geq 2$) clusters such that the nodes are positively linked within each cluster whereas negatively linked between the clusters" [107].

According to the equivalent balance theorem, the social relation $- - -$ is unbalanced (balanced) when its individuals are divided into $k = 2$ ($k = 3$) clusters.

1.1.3.2 Introduction to Optimization Models of Structure Balance

Optimization Models of Structure Balance

In the analysis of structural balance, there are two challenges: how unbalanced a signed network is and how unbalanced networks evolve to the balanced one.

Facchetti et al. proposed an energy function to compute the global balance of networks [20]. Energy function is defined as follows.

$$H(s) = \sum_{(i,j)} (1 - J_{ij} s_i s_j)/2 \tag{1.8}$$

where $J_{ij} \in \{\pm 1\}$ represents the sign of edge between nodes v_i and v_j. $s_i \in \{\pm 1\}$ represents the clustering identifier of node v_i. In this function, the summation runs over all adjacent pairs of nodes.

It has proved that computing the energy function is a NP-hard problem. This energy function is not suited for the case of weak balance. Many authors have designed extended energy function. In reality, there is a certain cost bias towards flipping positive or flipping negative signs in the transformation of structurally unbalanced networks to balanced one. An extended energy function is designed by introducing the cost of edge change [106]. Optimizing both of the proposed objective functions above is to find the minimum optimal solutions by clustering networks into two clusters. Cai et al. proposed an extended energy function based on weak balance [14]. Ma et al. proposed an extended energy function based on weak balance. In this objective function, they also introduced the cost of sign change [73].

Individual Representation

For the structural balance problem, string-based individual representation is adopted. To optimize energy function, Sun et al. adopt a chromosome consisting of $+1$ and -1 to encode the clustering of each node [104]. In the chromosome, nodes with the same genes are clustered into the same cluster. In [106], Wang et al. adopt a string-based representation to encode the clustering of each node and the sign change of each edge. Ma et al. dealing with weak balance adopt integer encoding [73]. All of chromosome representations are easily performed in a low complexity. These representations are well-suited for the design of genetic operators.

Individual Reproduction

In [106] and [104], two-point crossover operator is adopted, since it is effective and easy. Ma et al. employed a two-way crossover because offspring chromosomes can inherit the communities from their parents in two-way crossover [73].

To reduce useless explorations of search space, in [73, 104, 106], the authors employed neighbor-based mutation procedures.

Individual Local Search

In [104], Sun et al. employ a neighbor-based mutation, in which neighbor is defined based on the knowledge that two nodes linked by a positive edge are to be clustered into the same cluster. In [106], a greedy local search is proposed. This kind of local search is effective but it is time-consuming. To better exploit structural information from networks, a multi-level learning memetic algorithm is proposed in [73]. This multi-level learning memetic algorithm can effectively accelerate the convergence.

1.1.4 Network Robustness and Its Optimization in Complex Networks

1.1.4.1 Definition of Network Robustness

One vital characteristic of complex networks is its ability to resist failures and fluctuations, namely the robustness. The structural integrity of the networks would be damaged when abrupt failures or attack occurred on them. Thus, the robustness of networks is of great importance to guarantee the security of network systems.

There are also many definitions of robustness of networks. Commonly, a network is robust if the network functionality is not affected under different attacks. Holme et al. proposed a criterion to measure the functionality of network [46]. Network functionality is measured by the average inverse geodesic length and the size of the largest connected subgraph.

1.1.4.2 Introduction to Optimization Models of Network Robustness

Recently, Schneider et al. [95] proposed a measure, node robustness (R_n), to evaluate the robustness of networks under node attacks. When nodes are gradually damaged due to random failures or targeted attacks, a network may be split into several unconnected parts. The node robustness (R_n) considers the size of the largest connected component during all possible node attacks

$$R_n = \frac{1}{N}\sum_{q=1}^{N} s(q) \tag{1.9}$$

where N is the number of nodes in the network and $s(q)$ is the integrity of nodes in the largest connected part after removing q nodes [95]. The normalization factor $1/N$ makes it possible to make a comparison of the node robustness between networks with different sizes. Generally, the larger the value of R_n, the more robust the network is. Zhou and Liu proposed a memetic algorithm to enhance the robustness of scale-free networks against malicious attacks by optimizing node robustness (R_n). Zeng and Liu [113] proposed a link-robustness index (R_l) to evaluate the robustness of networks under link attacks. They also proposed a new algorithm to improve the node robustness and link robustness simultaneously by optimizing R_n and R_l.

Community is an important topic in network structure analytics. When a network suffers from attacks, the community integrity would be damaged. Ma et al. modeled the attack on the network as a two-level targeted one and a community robustness index was proposed to evaluate the integrality of the community structure [71]. Based on the proposed community robustness index, a greedy algorithm is devised to mitigate the network attack. Gong et al. proposed a memetic algorithm with two-level learning strategies to optimize community robustness index.

Another important issue in the study of robustness of networks is enhancing the robustness of coupled network under targeted recoveries. In [71], the cascading failures of coupled networks during recoveries was analyzed. Then, a recovery robustness index was presented to evaluate the resilience of coupled networks to cascading failures in the recovery processes.

For a coupling network $G = (C, D, E_{CD})$, its functionality depends on not only the remaining functionality of network C, but also that of network D. Therefore, it is necessary for the recovery robustness R_{rc} of the coupling network to consider the functionality integrity of both networks C and D during the recoveries. R_{rc} is computed as

$$\begin{aligned} R_{rc} &= \frac{1}{N} \sum_{p=1/N}^{1} f_{rc}(p) \\ &= \frac{1}{N} \sum_{p=1/N}^{1} [(p \cdot S_C^r(\psi^r(p)))^{\lambda} \cdot (p \cdot S_D^r(\eta^r(p)))^{1-\lambda}] \\ &= \frac{1}{N} \sum_{p=1/N}^{1} [p \cdot (S_C^r(\psi^r(p)))^{\lambda} \cdot (S_D^r(\eta^r(p)))^{1-\lambda}] \end{aligned} \tag{1.10}$$

where $f_{rc}(p)$ is the remaining functionality integrity of the coupled network after the fraction of nodes p in network C has been recovered, and $\psi^r(p)$ is the fraction of nodes in the largest linked parts of network $C(D)$. λ is a mixing parameter ranging from 0 to 1. When $\lambda = 0$ or $\lambda = 1$, the robustness of the coupled network system is determined by that of D or C, respectively.

To enhance the robustness of coupled network, Gong et al. [71] designed a strategy based on the protection of several influential nodes to enhance the robustness Eq. (1.10) of coupled networks under their recoveries.

1.2 Book Structure

The contents of this book cover most emerging topics of network structures analytics. The network issues have been modeled as either single or multiple objective optimization problems. Afterward, efficient evolutionary algorithms are designed to solve these problems. The organizations of the book are as follows:

In this chapter, we give a brief introduction to network structure analytics. Network issues mainly include network community discovery, network structure balance, and network robustness.

For the network community discovery problem, Chap. 2 formulates it as a single optimization problem and a basic memetic algorithm with a hill-climbing strategy as the local search procedure was introduced. In order to improve its ability for handling large- scale networks, an improved memetic algorithm with multi-level learning was introduced. From the perspective of swarm intelligence, we finally presented an efficient discrete greedy particle swarm optimization algorithm for large-scale social network community discovery.

To overcome the resolution limit, Chap. 3 formulates network community discovery as a multiobjective optimization problem. Three representative evolutionary multiobjective algorithms, i.e., MOEA/D, NNIA, and a multiobjective discrete particle swarm optimization algorithm were delineated to solve multi-resolution community detection. Finally, a NNIA-based algorithm was introduced for multi-resolution community discovery from dynamic networks.

Chapter 4 mainly studies the network structure balance problem. For a real-world network, there may exist both friendly and antagonistic relationships between its vertices and the network is often structurally unbalanced. In order to find a way with minimum costs to change a structurally unbalanced network into a balanced one, Chap. 4 presents in detail both single and multiple objective optimization-based algorithms.

Chapter 5 discusses the robustness properties of complex networks. When analyzing the robustness of a network subject to perturbations, meta-heuristic method was introduced to aromatically discover network structures with higher robustness.

Chapter 6 presents our attempts on solving other network-based issues such as network-based recommendation, network influence maximization and global biological network alignment. These network-related issues were modeled as optimization problems and efficient evolutionary algorithms were specially designed to solve the modeled problems.

Besides the above mentioned network issues, Chap. 7 will list available and promising research directions about computational intelligence for network structure analytics.

References

1. Agrawal, R.: Bi-objective community detection (bocd) in networks using genetic algorithm. In: Contemporary Computing, pp. 5–15. Springer (2011)
2. Amelio, A., Pizzuti, C.: Community mining in signed networks: a multiobjective approach. In: Proceedings of the 2013 IEEE/ACM International Conference on Advances in Social Networks Analysis and Mining, pp. 95–99. ACM (2013)
3. Amiri, B., Hossain, L., Crawford, J.W., Wigand, R.T.: Community detection in complex networks: multi-objective enhanced firefly algorithm. Knowl. Based Syst. **46**, 1–11 (2013)
4. Arenas, A., Duch, J., Fernández, A., Gómez, S.: Size reduction of complex networks preserving modularity. New J. Phs. **9**(6), 176 (2007)
5. Arenas, A., Fernandez, A., Fortunato, S., Gomez, S.: Motif-based communities in complex networks. J. Phs. A: Math. Theor. **41**(22), 224,001 (2008)
6. Arenas, A., Fernández, A., Gomez, S.: Analysis of the structure of complex networks at different resolution levels. New J. Phys. **10**(5), 053,039 176 (2008)
7. Boguñá, M., Pastor-Satorras, R., Díaz-Guilera, A., Arenas, A.: Models of social networks based on social distance attachment. Phys. Rev. E **70**(5), 056,122 (2004)
8. Brandes, U., Delling, D., Gaertler, M., Görke, R., Hoefer, M., Nikoloski, Z., Wagner, D.: Maximizing modularity is hard. arXiv:physics/0608255 (2006)
9. Butun, E., Kaya, M.: A multi-objective genetic algorithm for community discovery. In: 2013 IEEE 7th International Conference on Intelligent Data Acquisition and Advanced Computing Systems (IDAACS), vol. 1, pp. 287–292. IEEE (2013)
10. Cai, Q., Gong, M., Ma, L., Jiao, L.: A novel clonal selection algorithm for community detection in complex networks. Comput Intell. **31**(3), 442–464 (2015)
11. Cai, Q., Gong, M., Shen, B., Ma, L., Jiao, L.: Discrete particle swarm optimization for identifying community structures in signed social networks. Neural Netw. **58**, 4–13 (2014)
12. Cai, Q., Gong, M., Ma, L., Jiao, L.: A novel clonal selection algorithm for community detection in complex networks. Comput. Intell. **31**(3), 442–464 (2015)
13. Cai, Q., Gong, M., Ma, L., Ruan, S., Yuan, F., Jiao, L.: Greedy discrete particle swarm optimization for large-scale social network clustering. Inf. Sci. **316**, 503–516 (2015)
14. Cai, Q., Gong, M., Ruan, S., Miao, Q., Du, H.: Network structural balance based on evolutionary multiobjective optimization: a two-step approach. IEEE Trans. Evol. Comput. **19**(6), 903–916 (2015)
15. Cao, C., Ni, Q., Zhai, Y.: A novel community detection method based on discrete particle swarm optimization algorithms in complex networks. In: 2015 IEEE Congress on Evolutionary Computation (CEC), pp. 171–178. IEEE (2015)
16. Chen, G., Wang, Y., Wei, J.: A new multiobjective evolutionary algorithm for community detection in dynamic complex networks. Math. Probl. Eng. **2013** (2013)
17. Chen, G., Wang, Y., Yang, Y.: Community detection in complex networks using immune clone selection algorithm. Int. J. Digital Content Technol. Appl. **5**(6), 182–189 (2011)
18. Duch, J., Arenas, A.: Community detection in complex networks using extremal optimization. Phys. Rev. E **72**(2), 027,104 (2005)
19. Easley, D., Kleinberg, J.: Networks, Crowds, and Markets: Reasoning About a Highly Connected World. Cambridge University Press (2010)
20. Facchetti, G., Iacono, G., Altafini, C.: Computing global structural balance in large-scale signed social networks. Proc. Natl. Acad. Sci. **108**(52), 20953–20958 (2011)
21. Folino, F., Pizzuti, C.: A multiobjective and evolutionary clustering method for dynamic networks. In: 2010 International Conference on Advances in Social Networks Analysis and Mining (ASONAM), pp. 256–263. IEEE (2010)
22. Folino, F., Pizzuti, C.: Multiobjective evolutionary community detection for dynamic networks. In: Proceedings of the 12th Annual Conference on Genetic and Evolutionary Computation, pp. 535–536. ACM (2010)
23. Folino, F., Pizzuti, C.: An evolutionary multiobjective approach for community discovery in dynamic networks. IEEE Trans. Knowl. Data Eng. **26**(8), 1838–1852 (2014)

24. Fortunato, S.: Community detection in graphs. Phys. Rep. **486**(3), 75–174 (2010)
25. Fortunato, S., Barthelemy, M.: Resolution limit in community detection. Proc. Natl. Acad. Sci. **104**(1), 36–41 (2007)
26. Gach, O., Hao, J.K.: A memetic algorithm for community detection in complex networks. In: International Conference on Parallel Problem Solving from Nature, pp. 327–336. Springer (2012)
27. Geem, Z.W., Kim, J.H., Loganathan, G.: A new heuristic optimization algorithm: harmony search. Simulation **76**(2), 60–68 (2001)
28. Ghorbanian, A., Shaqaqi, B.: A genetic algorithm for modularity density optimization in community detection (2015)
29. Ghosn, F., Palmer, G., Bremer, S.A.: The mid3 data set, 1993–2001: procedures, coding rules, and description. Conflict Manag. Peace Sci. **21**(2), 133–154 (2004)
30. Girvan, M., Newman, M.E.: Community structure in social and biological networks. Proc. Natl. Acad. Sci. **99**(12), 7821–7826 (2002)
31. Gleiser, P.M., Danon, L.: Community structure in jazz. Adv. Complex Sys. **6**(04), 565–573 (2003)
32. Gog, A., Dumitrescu, D., Hirsbrunner, B.: Community detection in complex networks using collaborative evolutionary algorithms. In: European Conference on Artificial Life, pp. 886–894. Springer (2007)
33. Gómez, S., Jensen, P., Arenas, A.: Analysis of community structure in networks of correlated data. Phys. Rev. E **80**(1), 016,114 (2009)
34. Gong, M., Cai, Q., Chen, X., Ma, L.: Complex network clustering by multiobjective discrete particle swarm optimization based on decomposition. IEEE Trans. Evol. Comput. **18**(1), 82–97 (2014)
35. Gong, M., Cai, Q., Li, Y., Ma, J.: An improved memetic algorithm for community detection in complex networks. In: 2012 IEEE Congress on Evolutionary Computation (CEC), pp. 1–8. IEEE (2012)
36. Gong, M., Chen, X., Ma, L., Zhang, Q., Jiao, L.: Identification of multi-resolution network structures with multi-objective immune algorithm. Appl. Soft Comp. **13**(4), 1705–1717 (2013)
37. Gong, M., Ma, L., Zhang, Q., Jiao, L.: Community detection in networks by using multi-objective evolutionary algorithm with decomposition. Phys. A: Stat. Mech. Appl. **391**(15), 4050–4060 (2012)
38. Gong, M.G., Zhang, L.J., Ma, J.J., Jiao, L.C.: Community detection in dynamic social networks based on multiobjective immune algorithm. J. Comput. Sci. Technol. **27**(3), 455–467 (2012)
39. Gong, M., Fu, B., Jiao, L., Du, H.: Memetic algorithm for community detection in networks. Phys. Rev. E **84**(5), 056,101 (2011)
40. Gong, M., Hou, T., Fu, B., Jiao, L.: A non-dominated neighbor immune algorithm for community detection in networks. In: Proceedings of the 13th Annual Conference on Genetic and Evolutionary Computation, pp. 1627–1634. ACM (2011)
41. Guimera, R., Danon, L., Diaz-Guilera, A., Giralt, F., Arenas, A.: Self-similar community structure in a network of human interactions. Phys. Rev. E **68**(6), 065,103 (2003)
42. Guimerà, R., Sales-Pardo, M., Amaral, L.A.N.: Module identification in bipartite and directed networks. Phys. Rev. E **76**(3), 036,102 (2007)
43. Guoqiang, C., Xiaofang, G.: A genetic algorithm based on modularity density for detecting community structure in complex networks. In: 2010 International Conference on Computational Intelligence and Security (CIS), pp. 151–154. IEEE (2010)
44. He, D., Wang, Z., Yang, B., Zhou, C.: Genetic algorithm with ensemble learning for detecting community structure in complex networks. In: Fourth International Conference on Computer Sciences and Convergence Information Technology, 2009. ICCIT'09, pp. 702–707. IEEE (2009)
45. Heider, F.: Attitudes and cognitive organization. J. Psychol. **21**(1), 107–112 (1946)
46. Holme, P., Kim, B.J., Yoon, C.N., Han, S.K.: Attack vulnerability of complex networks. Phys. Rev. E **65**(5), 056,109 (2002)

47. Huang, Q., White, T., Jia, G., Musolesi, M., Turan, N., Tang, K., He, S., Heath, J.K., Yao, X.: Community detection using cooperative co-evolutionary differential evolution. In: International Conference on Parallel Problem Solving from Nature, pp. 235–244. Springer (2012)
48. Jia, G., Cai, Z., Musolesi, M., Wang, Y., Tennant, D.A., Weber, R.J., Heath, J.K., He, S.: Community detection in social and biological networks using differential evolution. In: Learning and Intelligent Optimization, pp. 71–85. Springer (2012)
49. Jin, D., He, D., Liu, D., Baquero, C.: Genetic algorithm with local search for community mining in complex networks. In: 22nd IEEE International Conference on Tools with Artificial Intelligence (ICTAI), 2010, vol. 1, pp. 105–112. IEEE (2010)
50. Kim, K., McKay, R.I., Moon, B.R.: Multiobjective evolutionary algorithms for dynamic social network clustering. In: Proceedings of the 12th Annual Conference on Genetic and Evolutionary Computation, pp. 1179–1186. ACM (2010)
51. Kropivnik, S., Mrvar, A.: An analysis of the slovene parliamentary parties network. In: Ferligoj, A. Kramberger, A. (eds.) Developments in Statistics and Methodology, pp. 209–216 (1996)
52. Lancichinetti, A., Fortunato, S., Kertész, J.: Detecting the overlapping and hierarchical community structure in complex networks. New J. Phys. **11**(3), 033,015 (2009)
53. Lancichinetti, A., Fortunato, S., Radicchi, F.: Benchmark graphs for testing community detection algorithms. Phys. Rev. E **78**(4), 046,110 (2008)
54. Lázár, A., Ábel, D., Vicsek, T.: Modularity measure of networks with overlapping communities. EPL(Europhysics Letters) **90**(1), 18,001 (2010)
55. Leicht, E.A., Newman, M.E.: Community structure in directed networks. Phys. Rev. Lett. **100**(11), 118,703 (2008)
56. Leskovec, J., Huttenlocher, D., Kleinberg, J.: Signed networks in social media. In: Proceedings of the SIGCHI Conference on Human Factors in Computing Systems, pp. 1361–1370 (2010)
57. Leskovec, J., Kleinberg, J., Faloutsos, C.: Graph evolution: Densification and shrinking diameters. ACM Trans. Knowl. Discov. Data **1**(1), 2 (2007)
58. Leskovec, J., Krevl, A.: SNAP Datasets: Stanford large network dataset collection. http://snap.stanford.edu/data (2014)
59. Li, J., Song, Y.: Community detection in complex networks using extended compact genetic algorithm. Soft Comput. **17**(6), 925–937 (2013)
60. Li, S., Chen, Y., Du, H., Feldman, M.W.: A genetic algorithm with local search strategy for improved detection of community structure. Complexity **15**(4), 53–60 (2010)
61. Li, X., Gao, C.: A novel community detection algorithm based on clonal selection. J. Comput. Inf. Sys. **9**(5), 1899–1906 (2013)
62. Li, Y., Liu, G., Lao, S.y.: Complex network community detection algorithm based on genetic algorithm. In: The 19th International Conference on Industrial Engineering and Engineering Management, pp. 257–267. Springer (2013)
63. Li, Z., Zhang, S., Wang, R.S., Zhang, X.S., Chen, L.: Quantitative function for community detection. Phys. Rev. E 77(3), 036,109 (Mar. 2008)
64. Li, Y., Liu, J., Liu, C.: A comparative analysis of evolutionary and memetic algorithms for community detection from signed social networks. Soft Comput. **18**(2), 329–348 (2014)
65. Lin, C.C., Liu, W.Y., Deng, D.J.: A genetic algorithm approach for detecting hierarchical and overlapping community structure in dynamic social networks. In: 2013 IEEE Wireless Communications and Networking Conference (WCNC), pp. 4469–4474. IEEE (2013)
66. Lipczak, M., Milios, E.: Agglomerative genetic algorithm for clustering in social networks. In: Proceedings of the 11th Annual Conference on Genetic and Evolutionary Computation, pp. 1243–1250. ACM (2009)
67. Liu, C., Liu, J., Jiang, Z.: A multiobjective evolutionary algorithm based on similarity for community detection from signed social networks. IEEE Trans. Cybernet. **44**(12), 2274–2287 (2014)
68. Liu, J., Zhong, W., Abbass, H.A., Green, D.G.: Separated and overlapping community detection in complex networks using multiobjective evolutionary algorithms. In: 2010 IEEE Congress on Evolutionary Computation (CEC 2010), pp. 1–7. IEEE (2010)

69. Liu, X., Li, D., Wang, S., Tao, Z.: Effective algorithm for detecting community structure in complex networks based on ga and clustering. In: International Conference on Computational Science, pp. 657–664. Springer (2007)
70. Lusseau, D., Schneider, K., Boisseau, O.J., Haase, P., Slooten, E., Dawson, S.M.: The bottlenose dolphin community of doubtful sound features a large proportion of long-lasting associations. Behav. Ecol. Sociobiol. **54**(4), 396–405 (2003)
71. Ma, L., Gong, M., Cai, Q., Jiao, L.: Enhancing community integrity of networks against multilevel targeted attacks. Phys. Rev. E **88**(2), 022,810 (2013)
72. Ma, L., Gong, M., Liu, J., Cai, Q., Jiao, L.: Multi-level learning based memetic algorithm for community detection. Appl. Soft Comput. **19**, 121–133 (2014)
73. Ma, L., Gong, M., Du, H., Shen, B., Jiao, L.: A memetic algorithm for computing and transforming structural balance in signed networks. Knowl. Based Sys. **85**, 196–209 (2015)
74. Massen, C.P., Doye, J.P.: Identifying communities within energy landscapes. Phys. Rev. E **71**(4), 046,101 (2005)
75. Milo, R., Shen-Orr, S., Itzkovitz, S., Kashtan, N., Chklovskii, D., Alon, U.: Network motifs: simple building blocks of complex networks. Science **298**(5594), 824–827 (2002)
76. Mu, C.H., Xie, J., Liu, Y., Chen, F., Liu, Y., Jiao, L.C.: Memetic algorithm with simulated annealing strategy and tightness greedy optimization for community detection in networks. Appl. Soft Comput. **34**, 485–501 (2015)
77. Naeni, L.M., Berretta, R., Moscato, P.: Ma-net: A reliable memetic algorithm for community detection by modularity optimization. In: Proceedings of the 18th Asia Pacific Symposium on Intelligent and Evolutionary Systems, vol. 1, pp. 311–323. Springer (2015)
78. Newman, M.E.: Modularity and community structure in networks. Proc. Natl. Acad. Sci. **103**(23), 8577–8582 (2006)
79. Newman, M.E.: Modularity and community structure in networks. Proc. Natl. Acad. Sci. **103**(23), 8577–8582 (2006)
80. Nicosia, V., Mangioni, G., Carchiolo, V., Malgeri, M.: Extending the definition of modularity to directed graphs with overlapping communities. J. Stat. Mech.: Theory Exp. **2009**(03), P03,024 (2009)
81. Oda, K., Kimura, T., Matsuoka, Y., Funahashi, A., Muramatsu, M., Kitano, H.: Molecular interaction map of a macrophage. AfCS Res. Rep. **2**(14), 1–12 (2004)
82. Oda, K., Matsuoka, Y., Funahashi, A., Kitano, H.: A comprehensive pathway map of epidermal growth factor receptor signaling. Mol. Syst. Biol. **1**(1), 2005 (2005)
83. Pizzuti, C.: Ga-net: a genetic algorithm for community detection in social networks. In: Parallel Problem Solving from Nature (PPSN), vol. 5199, pp. 1081–1090. Springer (2008)
84. Pizzuti, C.: A multi-objective genetic algorithm for community detection in networks. In: 21st International Conference on Tools with Artificial Intelligence, 2009. ICTAI'09, pp. 379–386. IEEE (2009)
85. Pizzuti, C.: Overlapped community detection in complex networks. In: Proceedings of the 11th Annual Conference on Genetic and Evolutionary Computation, pp. 859–866. ACM (2009)
86. Pizzuti, C.: A multiobjective genetic algorithm to find communities in complex networks. IEEE Trans. Evol. Comput. **16**(3), 418–430 (2012)
87. Pons, P., Latapy, M.: Post-processing hierarchical community structures: quality improvements and multi-scale view. Theor. Comput. Sci. **412**(8), 892–900 (2011)
88. Radicchi, F., Castellano, C., Cecconi, F., Loreto, V., Parisi, D.: Defining and identifying communities in networks. Proc. Natil. Acad. Sci. USA **101**(9), 2658–2663 (2004)
89. Read, K.E.: Cultures of the central highlands, New Guinea. Southwestern J. Anthropol. pp. 1–43 (1954)
90. Rees, B.S., Gallagher, K.B.: Overlapping community detection using a community optimized graph swarm. Soc. Netw. Anal. Min. **2**(4), 405–417 (2012)
91. Reichardt, J., Bornholdt, S.: Statistical mechanics of community detection. Phys. Rev. E **74**(1), 016,110 (2006)
92. Rosvall, M., Bergstrom, C.T.: An information-theoretic framework for resolving community structure in complex networks. Proc. Natl. Acad. Sci. **104**(18), 7327–7331 (2007)

93. Rosvall, M., Bergstrom, C.T.: Maps of random walks on complex networks reveal community structure. Proc. Natl. Acad. Sci. **105**(4), 1118–1123 (2008)
94. Salgado, H., Gama-Castro, S., Peralta-Gil, M., Díaz-Peredo, E., Sánchez-Solano, F., Santos-Zavaleta, A., Martínez-Flores, I., Jiménez-Jacinto, V., Bonavides-Martínez, C., Segura-Salazar, J., et al.: Regulondb (version 5.0): Escherichia coli k-12 transcriptional regulatory network, operon organization, and growth conditions. Nucleic Acids Res. **34**(suppl 1), D394–D397 (2006)
95. Schneider, C.M., Moreira, A.A., Andrade, J.S., Havlin, S., Herrmann, H.J.: Mitigation of malicious attacks on networks. Proc. Natl. Acad. Sci. **108**(10), 3838–3841 (2011)
96. Shang, R., Bai, J., Jiao, L., Jin, C.: Community detection based on modularity and an improved genetic algorithm. Phys. A: Stat. Mech. Appl. **392**(5), 1215–1231 (2013)
97. Shang, R., Luo, S., Zhang, W., Stolkin, R., Jiao, L.: A multiobjective evolutionary algorithm to find community structures based on affinity propagation. Phys. A: Stat. Mech. Appl. **453**, 203–227 (2016)
98. Shelokar, P., Quirin, A., Cordón, Ó.: Three-objective subgraph mining using multiobjective evolutionary programming. J. Comput. Syst. Sci. **80**(1), 16–26 (2014)
99. Shen, H., Cheng, X., Cai, K., Hu, M.B.: Detect overlapping and hierarchical community structure in networks. Phys. A: Stat. Mech. Appl. **388**(8), 1706–1712 (2009)
100. Shi, C., Yan, Z., Cai, Y., Wu, B.: Multi-objective community detection in complex networks. Appl. Soft Comput. **12**(2), 850–859 (2012)
101. Shi, C., Yu, P.S., Cai, Y., Yan, Z., Wu, B.: On selection of objective functions in multi-objective community detection. In: Proceedings of the 20th ACM International Conference on Information and Knowledge Management, pp. 2301–2304. ACM (2011)
102. Shi, Z., Liu, Y., Liang, J.: Pso-based community detection in complex networks. In: Second International Symposium on Knowledge Acquisition and Modeling, 2009. KAM'09, vol. 3, pp. 114–119. IEEE (2009)
103. Shi, C., Yu, P.S., Yan, Z., Huang, Y., Wang, B.: Comparison and selection of objective functions in multiobjective community detection. Comput. Intell. **30**(3), 562–582 (2014)
104. Sun, Y., Du, H., Gong, M., Ma, L., Wang, S.: Fast computing global structural balance in signed networks based on memetic algorithm. Phys. A: Stat. Mech. Appl. **415**, 261–272 (2014)
105. Tasgin, M., Herdagdelen, A., Bingol, H.: Community detection in complex networks using genetic algorithms. arXiv:0711.0491 (2007)
106. Wang, S., Gong, M., Du, H., Ma, L., Miao, Q., Du, W.: Optimizing dynamical changes of structural balance in signed network based on memetic algorithm. Social Netw. **44**, 64–73 (2016)
107. Wasserman, S., Faust, K.: Social Network Analysis: Methods and Applications, vol. 8. Cambridge University Press (1994)
108. Watts, D.J., Strogatz, S.H.: Collective dynamics of small-worldnetworks. Nature **393**(6684), 440–442 (1998)
109. Wu, P., Pan, L.: Multi-objective community detection based on memetic algorithm. PLoS one **10**(5), e0126,845 (2015)
110. Xiaodong, D., Cunrui, W., Xiangdong, L., Yanping, L.: Web community detection model using particle swarm optimization. In: IEEE Congress on Evolutionary Computation, 2008. CEC 2008, pp. 1074–1079. IEEE (2008)
111. Yang, B., Cheung, W.K., Liu, J.: Community mining from signed social networks. IEEE Trans. Knowl. Data Eng. **19**(10), 1333–1348 (2007)
112. Zachary, W.W.: An information flow model for conflict and fission in small groups. J. Anthropol. Res. **33**(4), 452–473 (1977)
113. Zeng, A., Liu, W.: Enhancing network robustness against malicious attacks. Phys. Rev. E **85**(6), 066,130 (2012)
114. Zhan, W., Guan, J., Chen, H., Niu, J., Jin, G.: Identifying overlapping communities in networks using evolutionary method. Phys. A: Stat. Mech. Appl. **442**, 182–192 (2016)

115. Zhang, S., Wang, R.S., Zhang, X.S.: Identification of overlapping community structure in complex networks using fuzzy c-means clustering. Phys. A: Stat. Mech. Appl. **374**(1), 483–490 (2007)
116. Zheng, X., Zeng, D., Wang, F.Y.: Social balance in signed networks. Inf. Syst. Frontiers **17**(5), 1077–1095 (2015)
117. Zhou, D., Wang, X.: A neighborhood-impact based community detection algorithm via discrete pso. Math. Probl. Eng. **2016** (2016)
118. Zhou, X., Liu, Y., Zhang, J., Liu, T., Zhang, D.: An ant colony based algorithm for overlapping community detection in complex networks. Phys. A: Stat. Mech. Appl. **427**, 289–301 (2015)
119. Zhou, Y., Wang, J., Luo, N., Zhang, Z.: Multiobjective local search for community detection in networks. Soft Comput. pp. 1–10 (May 2015)

Chapter 2
Network Community Discovery with Evolutionary Single-Objective Optimization

Abstract Network community detection is one of the most fundamental problems in network structure analytics. With the modularity and modularity density being put forward, network community detection is formulated as a single-objective optimization problem and then communities of network can be discovered by optimizing modularity or modularity density. However, the community detection by optimizing modularity or modularity density is NP-hard. The computational intelligence algorithm, especially for evolutionary single-objective algorithms, have been effectively applied to discover communities from networks. This chapter focuses on evolutionary single-objective algorithms for solving network community discovery. First this chapter reviews evolutionary single-objective algorithm for network community discovery. Then three representative algorithms and their performances of discovering communities are introduced in detail.

2.1 Review of the State of the Art

Complex systems can be modeled by complex networks. Entities of complex systems can be represented by nodes of networks and their relationships are represented by edges of networks. It is found that community structure is an important property of networks. As shown in Chap. 1, a community is a subset of nodes which have dense connections within them and sparse connections between them and other nodes in the network. Community detection is to divide a network into several clusters. Criteria are required to evaluate the partitions of network.

One of the most famous and used criteria is modularity, which is originally introduced by Newman and Girvan [19]. High values of modularity are associated with subjectively good partitions. As a consequence, the partition that has the maximum modularity is expected to be the subjectively best partition, or at least a very good one. However, Fortunato and Barthélemy [9] found that modularity may have the resolution limit and fail to identify modules smaller than a scale. To alleviate the resolution limit, Li et al. [17] introduced a quality function, called modularity density. The authors demonstrated that this quantitative function is superior to the widely used

M. Gong et al., *Computational Intelligence for Network Structure Analytics*,
DOI 10.1007/978-981-10-4558-5_2

modularity. A tunable parameter is introduced into a general version of modularity density which can explore the network at different resolutions.

With the introductions of modularity and modularity density, network community discovery can be formulated as a single-objective optimization problem. Several studies have shown the potential of nature-inspired optimization algorithms for community discovery by optimizing modularity or modularity density [10, 12, 13, 25]. For instance, in [13], Gong et al. proposed a memetic algorithm, named as Meme-Net, to uncover communities at different hierarchical levels. Meme-Net shows its effectiveness. However, its high computational complexity makes it impossible to search communities on slightly large networks. Shang et al. [25] and Gong et al. [12] try to adopt simulated annealing as an individual learning procedure to decrease the computational complexity of Meme-Net. However, their computational complexities are still very high relative to classical modularity-based community detection algorithms. The algorithm in [10] adopted the technique in [3] as the local search. Meanwhile, it takes a large amount of time and energy on generating initial population as it directly uses the algorithm in [3] to initialize a population of solutions. Therefore, the algorithms in [10, 12, 13, 25] are difficult to apply to real-world problems or large-scale networks.

To discover communities in large-scale networks, different strategies are adopted, for example, multi-level local learning strategies and greedy local search [5, 18]. Ma et al. [18] proposed a fast memetic algorithm with multi-level learning strategies for community detection by optimizing modularity. The multi-level learning strategies are devised based on the potential knowledge of the node, community and partition structures of networks, and they work on the network at nodes, communities and network partitions levels, respectively. Cai et al. [5] designed a novel discrete particle swarm optimization algorithm with a greedy local search for community discovery.

This chapter is organized into four sections. Section 2.2 introduces a basic memetic algorithm for community discovery. A memetic algorithm with multi-learning strategies is presented in Sect. 2.3. A greedy discrete particle swarm optimization algorithm is introduced in Sect. 2.4.

2.2 A Node Learning-Based Memetic Algorithm for Community Discovery in Small-Scale Networks

As shown in Sect. 2.1, the discovery of community structure is formulated as a single optimization problem. Discovering community can be solved by maximizing modularity or modularity density. This section is based on maximizing modularity density.

Memetic algorithms (MAs) are hybrid global–local heuristic search methodologies [20]. The global heuristic search is usually a form of population-based method, while the local search is generally considered as an individual learning procedure for accelerating the convergence to an optimum [20]. In general, the population-based global search has the advantage of exploring the promising search space and providing a reliable estimate of the global optimum [4]. However, the population-based global search is difficult to discover an optimal solution around the explored search space in a short time. The local search is usually designed for accelerating the search and finding the best solutions on the explored search space. Therefore, this hybridization, which synthesizes the complementary advantages of the population-based global search and the individual-based local search, can effectively produce better solutions. MAs have been proved to be of practical success in NP-complete problem.

This section introduces a basic memetic algorithm for community discovery, named Meme-Net. The advantages of Meme-Net are as follows. Meme-Net combines GAs and a hill-climbing strategy as the local search procedure. The hill-climbing strategy is designed based on node learning. The algorithm does not require the number of communities present in the graph in advance. In addition, by tuning the parameter in the quality function, it is able to explore the network at different resolutions and may reveal the hierarchical structure of the network.

2.2.1 Memetic Algorithm with Node Learning for Community Discovery

Here, the framework of Meme-Net is given as Algorithm 1.

Algorithm 1 The algorithm framework of Meme-Net

Input: Maximum number of generations: G_{max}; Population size: S_{pop}; Size of mating pool: S_{pool}; Tournament size: S_{tour}; Crossover probability: P_c; Mutation probability: P_m.
Output: Convert the fittest chromosome in **P** into a partition solution and output.

```
P ←GenerateInitialPopulation(S_pop);
repeat
    P_parent ←Selection(P,S_pool,S_tour);
    P_child ←GeneticOperation(P_parent,P_c,P_m);
    P_new ←LocalSearch(P_child);
    P ←UpdatePopulation(P,P_new);
until TerminationCriterion(G_max)
```

In this framework, the GenerateInitialPopulation() procedure is used to create the initial population. The Selection() function is used to select parental population for mating. Here, the deterministic tournament selection is used. The GeneticOperation() function is used to perform crossover and mutation operation. The UpdatePopulation() procedure is used to reconstruct the current population using the population **P** and $\mathbf{P}_{new}$. Here, the current population is constructed taken the best S_{pop} individuals from $\mathbf{P} \cup \mathbf{P}_{new}$. The TerminationCriterion() function is the terminate condition.

In what follows, more detailed descriptions about the initialization procedure, genetic operators and the local search procedure will be given.

2.2.2 Problem Formation

Li et al. [17] designed a function, named modularity density (Eq. 1.7). The authors also proved the equivalence of modularity density and kernel k means, and proposed a more general modularity density measure:

$$D_\lambda = \sum_{i=1}^{m} \frac{2\lambda L(V_i, V_i) - 2(1-\lambda)L(V_i, \overline{V_i})}{|V_i|}. \tag{2.1}$$

When $\lambda = 1$, D_λ is equivalent to the ratio association; when $\lambda = 0$, D_λ is equivalent to the ratio cut; when $\lambda = 0.5$, D_λ is equivalent to the modularity density D. So the general modularity density D_λ can be viewed as a combination of the ratio association and the ratio cut. Generally, optimization of the ratio association algorithm often divides a network into small communities, while optimization of the ratio cut often divides a network into large communities. This general modularity density D_λ, which is a convex combination of these two indexes, can avoid the resolution limits. In other words, by varying the λ value, we can use this general function to analyze the topological structure and uncover more detailed and hierarchical organization of the complex network.

2.2.3 Representation and Initialization

Each chromosome is encoded as an integer string

$$x = \{x^1\, x^2 \cdots x^n\}. \tag{2.2}$$

where, n is the number of the vertices in the graph, and x^i is the integer cluster identifier of vertex v_i, which can be any integer number between 1 and n. The vertices having the same cluster identifier are partitioned into the same community. This representation does not need the number of clusters of in the graph. A graph of n vertices can be partitioned into n clusters at most, and in this case, each cluster contains only one vertex, which can be denoted as $\{1\ 2 \cdots n\}$. However, it should be noted that there are many different representations corresponding to the same partition. For instance, given a graph of 4 vertices, $\{3\,1\,2\,3\}$ and $\{1\,2\,3\,1\}$ represent the same partition $\{\{1, 4\}, \{2\}, \{3\}\}$ of the graph.

The population initialization procedure is given as Algorithm 2. Initially each vertex is put in a different cluster for all chromosomes, that is, each chromosome in the population is $\{1\ 2 \cdots n\}$. However, this initialised population lacks of diversity and each solution is of low quality. To speed up the convergence, here a simple heuristic algorithm is employed. For each chromosome, a vertex is selected randomly and assign its cluster identifier to all of its neighbors. This operation is repeated $\alpha \cdot n$ times for each chromosome in the initial population where α is a parameter and $\alpha = 0.2$ is used. This operation is very fast and results in local small communities, but the resulting clusterings are still far away from being optimal.

Algorithm 2 The Population Initialization Procedure

Input: Population size: S_{pop}.
Output: Population **P**.
1: Generate a population **P**, each chromosome $\mathbf{x}_k$ of which is set to $\{1\ 2 \cdots n\}$, where $k = 1, 2, \cdots, S_{pop}$.
2: **for** each chromosome $\mathbf{x}_k$ **do**
3: $t_{counter} \leftarrow 0$;
4: **repeat**
5: randomly select a vertex v_i;
6: $x_k^j \leftarrow x_k^i$ whenever $(v_i, v_j \in E)$;
7: $t_{counter} \leftarrow t_{counter} + 1$;
8: **until** $t_{counter} = \alpha \cdot n$
9: **end for**

2.2.4 Genetic Operators

The genetic operators, including crossover and mutation, are important to change the genetic composing of the chosen solutions. The crossover operator is to generate offspring chromosomes which can inherit their parent chromosomes' communities, and the mutation operator is to generate spontaneous random changes in chromosomes. The crossover and mutation operators can effectively prevent the proposed algorithm from getting into a local optimal network partition.

Crossover. In Meme-Net, a two-way crossover operation is employed. The crossover procedure is defined as follows. The two selected chromosomes are called x_a and x_b, respectively. We pick a vertex v_i at random, determine its cluster (i.e., x_a^i) in the chromosome x_a and make sure that all the vertices in this cluster of x_a are also assigned to the same cluster in the chromosome x_b (i.e. $x_b^k \leftarrow x_a^i, \forall k \in \{k \mid x_a^k = x_a^i\}$). Simultaneously, we also determine the cluster of the vertex v_i in x_b, and make sure that all the vertices in this cluster of x_b are also assigned to the same cluster in x_a (i.e., $x_a^k \leftarrow x_b^i, \forall k \in \{k \mid x_b^k = x_b^i\}$). This procedure returns two new chromosomes x_c and x_d. An example of two-way crossover is given in Table 2.1. This two-way crossing over operation can generate descendants carrying features common to the parents, which represents the exploitative side of the crossover operator; on the other hand, the crossover operation is exploratory, which means it can generate descendants carrying combinations of features taken from the parents. For instance, as shown in one of the descendants $\mathbf{x}_c$ in Table 2.1, v_4 becomes in the same community with v_3. These properties make the two-way crossover operation suitable for our algorithm.

Mutation. In Meme-Net, one-point mutation is employed on this chromosome: a vertex is picked randomly on the chromosome, then the cluster of the vertex is randomly changed to the cluster of one of its neighbors. This operation is repeated n times on the chromosome. The specialized mutation operator that only considers the neighbors of the vertex can decrease useless exploration and reduce the search space.

Table 2.1 Two-way crossover when v_3 is selected

v		$\mathbf{x}_a$		$\mathbf{x}_b$		$\mathbf{x}_c$	$\mathbf{x}_d$		$\mathbf{x}_a$		$\mathbf{x}_b$		v
1		⑤	→	2	→	⑤	5		5		2		1
2		3		6		6	⑥	←	3	←	⑥		2
③	→	⑤	→	6	→	⑤	⑥	←	5	←	⑥	←	③
4		7		5		5	7		7		5		4
5		2		6		6	⑥	←	2	←	⑥		5
6		⑤	→	3	→	⑤	5		5		3		6
7		3		2		2	3		3		2		7

2.2.5 *The Local Search Procedure*

First the neighbors of a partition are defined. Given a partition $\Omega = \{V_1, V_2, \ldots, V_m\}$ $(2 \leq m \leq n)$ of a graph G, where m is the number of clusters of this partition and n is the number of vertices in the graph, a single vertex, chose from a cluster $V_i (i \in 1, \ldots, m)$, is reassigned into another cluster $V_j (j \in 1, \ldots, m$ and $j \neq i)$. The new partition Ω_{nbr} after this reassignment is called a neighbor of the partition Ω. In particular, when $m = 1$, $\Omega = \{V_1\}$, a neighbor of Ω is defined as $\Omega_{nbr} = \{V_1 - v, \{v\}\}$, where v is a vertex from V_1, and $\{v\}$ is a cluster including the single vertex v (a single-vertex cluster).

Then N_Ω, the number of all possible neighbors of the partition Ω can be figured out. If $m = 1$, $N_\Omega = n$; if $2 \leq m \leq n$, $N_\Omega = n(m-1) - \frac{1}{2}s(s-1)$, where s is the number of single-vertex clusters in this partition. For example, when $m = n$, the value of s must be n, so $N_\Omega = n(n-1) - \frac{1}{2}n(n-1) = \frac{1}{2}n(n-1)$.

Algorithm 3 The Local Search Procedure

```
Input: P_child.
Output: P_child.
 1: n_current ←FindBest(P_child);
 2: y_islocal ← FALSE;
 3: repeat
 4:     L ←FindNeighbors(n_current);
 5:     n_next ←FindBest(L);
 6:     if Eval(n_next)>Eval(n_current) then
 7:         n_current ← n_next;
 8:     else
 9:         y_islocal ← TRUE;
10:     end if
11: until y_islocal is TRUE
```

The local search procedure used in Meme-Net is a hill-climbing strategy, and the implementation is given as Algorithm 3. In Algorithm 3, the FindBest() function is responsible for evaluating the fitness of each chromosome in the input population, and return the chromosome having maximum fitness, on which the local search procedure will be performed. The Eval() function is used to evaluate the fitness of a solution. The FindNeighbors() function is responsible for finding the neighbors of a partition, which can be done easily according to the definition of the neighbors of a partition.

In Meme-Net, the local search is applied on $\mathbf{P}_{child}$, which is the population after crossover and mutation. Not all of the chromosomes in this population are undergoing the local search procedure. Only the fittest chromosome in $\mathbf{P}_{child}$ is found and performed local search on it, until no improvement can be made.

2.2.6 Experimental Results

In this section, the Meme-Net algorithm is tested on artificial generated networks and 4 real-world networks. To show the performance of Meme-Net, Fast Modularity algorithm is taken as a comparison. Some parameters and their values in Meme-Net are summarized in Table 2.2.

A similarity measure *Normalized Mutual Information* (NMI) [8] is used for estimating the similarity between the true partitions and the detected ones. Given two partitions A and B of a network in communities, let C be the confusion matrix whose element C_{ij} is the number of nodes of community i of the partition A that are also in the community j of the partition B. The normalized mutual information $I(A, B)$ is defined as

$$I(A,B)=\frac{-2\sum_{i=1}^{c_A}\sum_{j=1}^{c_B}C_{ij}\log(C_{ij}N/C_{i.}C_{.j})}{\sum_{i=1}^{c_A}C_{i.}\log(C_{i.}/N)+\sum_{j=1}^{c_B}C_{.j}\log(C_{.j}/N)} \tag{2.3}$$

where $c_A(c_B)$ is the number of groups in the partition $A(B)$, $C_{i.}(C_{.j})$ is the sum of elements of C in row i (column j), and N is the number of nodes (note that, some denominations here are different from the ones in previous sections just for convenience). If $A = B$, then $I(A, B) = 1$; if A and B are completely different, then $I(A, B) = 0$.

Table 2.2 Parameter settings for the algorithm

Parameter	Meaning	Value
G_{max}	The number of iterations	50
S_{pop}	Population size	450
S_{pool}	Size of the mating pool	$\frac{S_{pop}}{2}$
S_{tour}	Tournament size	2
P_c	Crossover probability	0.8
P_m	Mutation probability	0.2

2.2.6.1 Results on Artificially Generated Network

The artificially generated network used is the extension of GN benchmark benchmark network. 11 different networks are generated for the value of mixing parameter μ ranging from 0 to 0.5 and NMI is used to measure the similarity between the true partitions and the detected ones. For each network, the average NMI is computed over 10 independent runs. Figure 2.1 shows the average NMI, for different values of λ (the parameter in the objective function D_λ), when the mixing parameter increases from 0 to 0.5. As shown in Fig. 2.1, for $\lambda = 0.5$, when the value of mixing parameter μ is small ($\mu \leq 0.25$), which means the fuzziness of the community in the network is low, our algorithm find the true partition correctly (NMI equals 1). When the mixing parameter increases, it is more difficult to detect the true partition, but the detected partition is also close to the true one (NMI is about 0.9 when $\mu = 0.35$). Only when $\mu \geq 0.4$, there is no community structure detected. However, the higher value of λ could help to discover smaller communities. For instance, for $\lambda = 0.7$, when the mixing parameter $\mu = 0.45$, Meme-Net is able to detect about 80% of the community structure information. On the other hand, lower value of λ could help to discover larger communities. For example, for $\lambda = 0.3$, when $\mu \geq 0.3$, Meme-Net tends to consider the whole network as a large single community (NMI equals 0). Notice that, when $\mu = 0.5$, each node has half of the links inside the community and the other half with the rest the network. This means that the community structure is very fuzzy, and any algorithm can hardly find the true partition of the network.

The local search procedure plays a very important role in Meme-Net. A genetic algorithm (GA) version of Meme-Net algorithm is developed by removing the LocalSearch function. Figure 2.2 shows the average NMI obtained by Meme-Net and this GA version algorithm, for $\lambda = 0.5$, when the mixing parameter increases from 0 to 0.5. It clearly shows that, with the local search procedure, when $\mu \leq 0.25$,

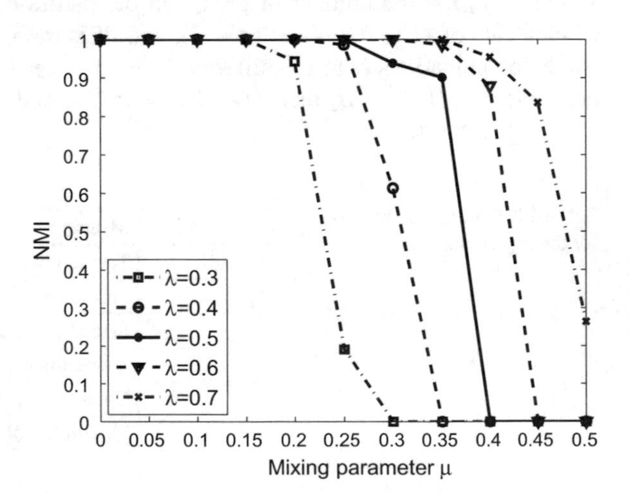

Fig. 2.1 Average NMI versus mixing parameter μ, for different values of λ

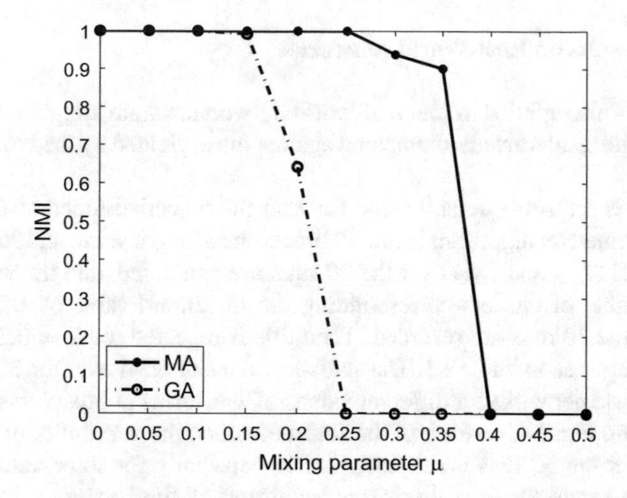

Fig. 2.2 Average NMI versus mixing parameter μ, for $\lambda = 0.5$

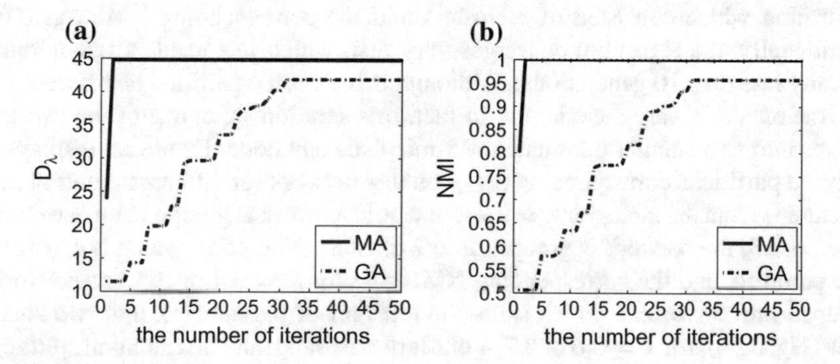

Fig. 2.3 **a** The values of D_λ, (**b**) the corresponding NMI versus the number of iterations, for $\mu = 0.15$ and $\lambda = 0.5$

Meme-Net is able to detect the true partition of the network; however, without the local search procedure, it becomes harder for the algorithm to detect the true partition when $\mu > 0.15$.

The local search procedure also speeds up the convergence of Meme-Net. Figure 2.3 displays the values of D_λ and the corresponding NMI obtained by Meme-Net and the GA version algorithm in one run, for $\mu = 0.15$ and $\lambda = 0.5$, when the number of generations increases from 1 to 50. The figure shows that with the local search procedure, Meme-Net finds a maximum objective function value of 44.75 in just two generations, and the corresponding NMI is 1, which means the true partition is found. However, without the local search procedure, after 50 generations, the algorithm does not find the true partition of the network.

2.2.6.2 Results on Real-World Networks

Meme-Net is also applied to four real-world networks, karate, dolphin, football, and polbooks. The results are also compared against those yielded by the Fast Modularity Algorithm.

$\lambda = 0.5$ is set to the default value for λ in the objective function D_λ. For each network, Meme-Net algorithm is run 30 times, the average value and standard deviation of NMI (I_{avg} and I_{std}) over the 30 runs are computed, and the value of NMI and the number of clusters corresponding the maximum value of D_λ (I_{maxD} and $N_{cluster}$) in the 30 runs are recorded. Then this is repeated for $\lambda = 0.2$ to 0.8. The results are reported in Table 2.3. The statistic values of NMI over the 30 runs on the four real-world networks for different values of λ in terms of box plots are shown in Fig. 2.4. From Fig. 2.4, on each of the four networks, the variability of NMI values obtained over the 30 runs is relatively small, especially for some values of λ. For instance, On karate network, this is true for almost all the λ values.

On karate network, for $\lambda = 0.5$, in fact in all the 30 runs Meme-Net found a partition which consisted of 3 clusters and the corresponding NMI was 0.699. Additionally, the algorithm converges very fast, which just needs a few iterations (usually less than 10 generations). Although this detected partition is different from the true one, it is very meaningful. In fact, this solution splits one of the two large groups into two smaller ones and never misplace any node. Figure 2.5 displays the detected partitions corresponding I_{maxD} on this network for different values of λ. As we can see from the table, for $\lambda = 0.2$, the whole network is grouped into one cluster; for $\lambda = 0.3$, the network is grouped into 2 clusters (Fig. 2.5a), which is exactly the true partition, and the corresponding NMI is 1; for $\lambda = 0.4$ or 0.5, the network is grouped into 3 clusters, which splits the left part of the network into two smaller ones (Fig. 2.5b); for $\lambda = 0.6$ or 0.7, 4 clusters are found, and this solution splits each of the two large groups into two smaller ones (Fig. 2.5c); for $\lambda = 0.8$, the network is grouped into 5 clusters, which further splits the right part into 3 clusters (Fig. 2.5d). If λ is set to 0.9 or larger, many small clusters containing only two or three vertices are detected. We did not display this network partition in Fig. 2.5.

On dolphin network, for $\lambda = 0.3$, 2 clusters are found and the corresponding NMI is 1. This means that the detected partition is exactly the true one, which is shown in Fig. 2.6. For $\lambda = 0.4$, the network is grouped into 3 clusters and the corresponding NMI is 0.756. This partition splits the lower right group of the network into two smaller ones and never misplaces a vertex. For $\lambda = 0.5$, 5 clusters are found and the corresponding NMI is 0.586, which splits the upper left group of the network into 2 smaller ones, and the lower right group into 3 smaller ones. For $\lambda = 0.6$ or larger, more smaller clusters are found. The experiments show that, by tuning the parameter λ, the network could be explored at different resolutions. In general, the larger λ value is, the smaller communities Meme-Net tends to find.

Because of the complexity of the networks themselves, It could not find the "true" partition on the football network (Fig. 2.7a) and the polbooks network (Fig. 2.7b). However, the detected ones are very close to the true partitions. For instance, on the

Table 2.3 The results of 30 runs of Meme-Net on four real-world networks for different values of λ

Network	λ	I_{avg}	I_{std}	I_{maxD}	$N_{cluster}$
Zachary's karate club	0.2	0	0	0	1
	0.3	1	0	1	2
	0.4	0.740	0.104	0.699	3
	0.5	0.699	0	0.699	3
	0.6	0.690	0.007	0.687	4
	0.7	0.688	0.004	0.687	4
	0.8	0.651	0.038	0.628	5
Dolphin social network	0.2	0.889	0	0.889	2
	0.3	1	0	1	2
	0.4	0.787	0.073	0.756	3
	0.5	0.569	0.035	0.586	5
	0.6	0.467	0.048	0.477	9
	0.7	0.400	0.034	0.385	11
	0.8	0.346	0.017	0.334	14
American college football	0.2	0.595	0.118	0.359	2
	0.3	0.723	0.078	0.523	3
	0.4	0.851	0.032	0.824	8
	0.5	0.890	0.033	0.911	11
	0.6	0.904	0.022	0.924	12
	0.7	0.912	0.011	0.911	13
	0.8	0.911	0.010	0.911	13
Books about US politics	0.2	0.583	0.015	0.598	2
	0.3	0.576	0.008	0.574	3
	0.4	0.588	0.011	0.590	4
	0.5	0.481	0.029	0.455	7
	0.6	0.449	0.018	0.434	9
	0.7	0.423	0.013	0.402	11
	0.8	0.394	0.014	0.378	14

football network, for $\lambda = 0.5$, 11 clusters are found and the corresponding NMI is 0.911, only a few vertices are misplaced. On the polbooks network, the results are also competitive with other popular community detection algorithms.

The results obtained by Fast Modularity algorithm are given in Table 2.4. Now the results obtained by Meme-Net for $\lambda = 0.5$ are considered. On karate network, the Fast Modularity algorithm found a solution with $NMI = 0.693$, while in fact for all the 30 runs Meme-Net found a solution with a NMI value of 0.699. On the

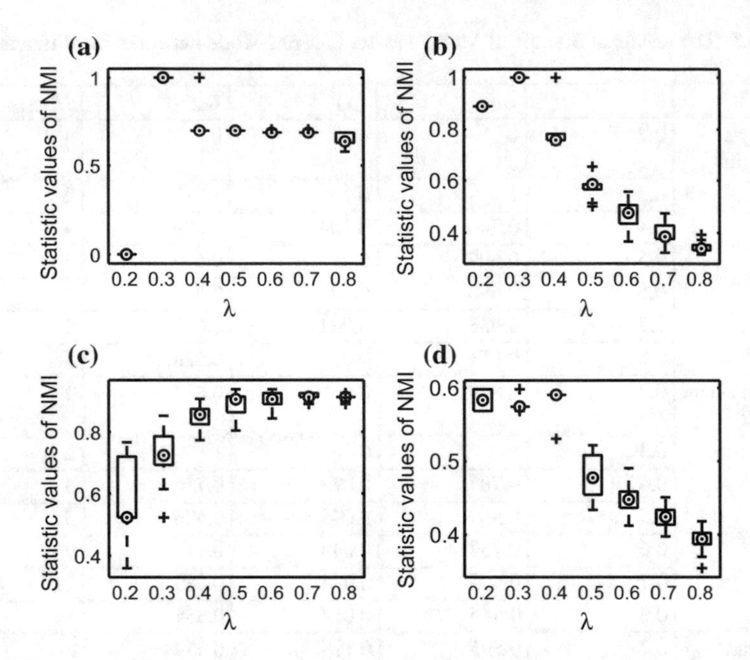

Fig. 2.4 The statistic values of NMI over the 30 runs on (**a**) Zachary's karate club, (**b**) Dolphin social network, (**c**) American college football, (**d**) Books about US politics for different values of λ. Here, box plots are used to illustrate the distribution of these samples. On each box, the central mark $\odot$ is the median, the edges of the box are the 25th and 75th percentiles, the whiskers extend to the most extreme datapoints the algorithm considers to be not outliers, and the outliers are plotted individually. Symbol + denotes outliers

football network, the average NMI value found by Meme-Net is 0.890, while Fast Modularity algorithm found the NMI value of 0.762. On these two networks the results obtained by Meme-Net are better than those of Fast Modularity algorithm. On the dolphin network and polbooks network, the average values of NMI found by Meme-Net are 0.569 and 0.481 respectively, while the Fast Modularity algorithm found the NMI values of 0.573 and 0.531, which are slightly better than Meme-Net. However, by tuning the parameter λ, we can also get better results. For example, on the dolphin network, when $\lambda = 0.3$, the solution found by Meme-Net is exactly the true partition. On polbooks network, when $\lambda = 0.4$, the solutions with the average NMI value of 0.588 found by Meme-Net are more closer to the true partition. This comparison clearly shows the very good performance of Meme-Net with respect to the Fast Modularity algorithm.

In practice, $\lambda = 0.5$ is set as the default value for λ in the objective function D_λ, because this value indeed produces good results as shown in the experiments.

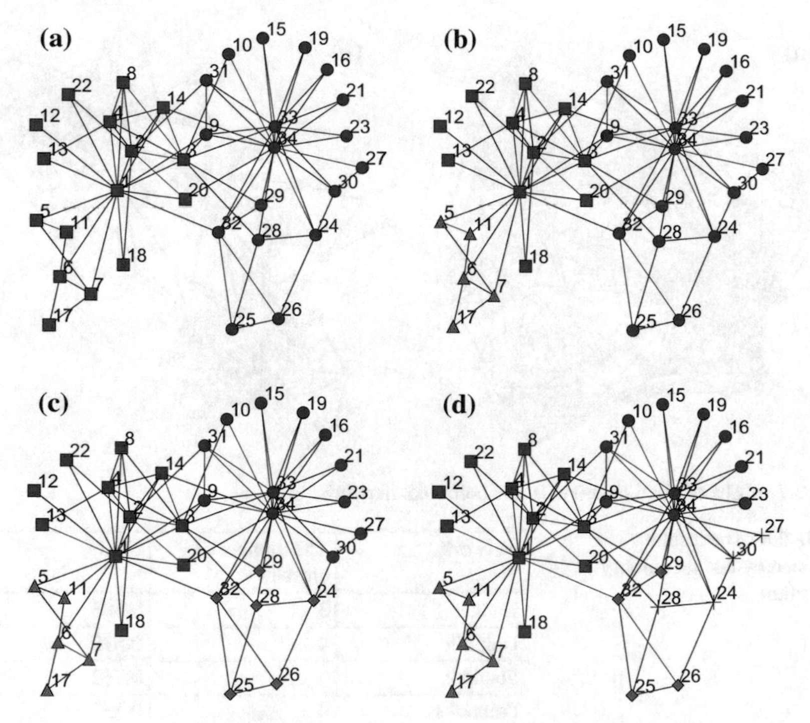

Fig. 2.5 The detected partitions on karate network for **(a)** $\lambda = 0.3$, **(b)** $\lambda = 0.4$ or 0.5, **(c)** $\lambda = 0.6$ or 0.7, **(d)** $\lambda = 0.8$

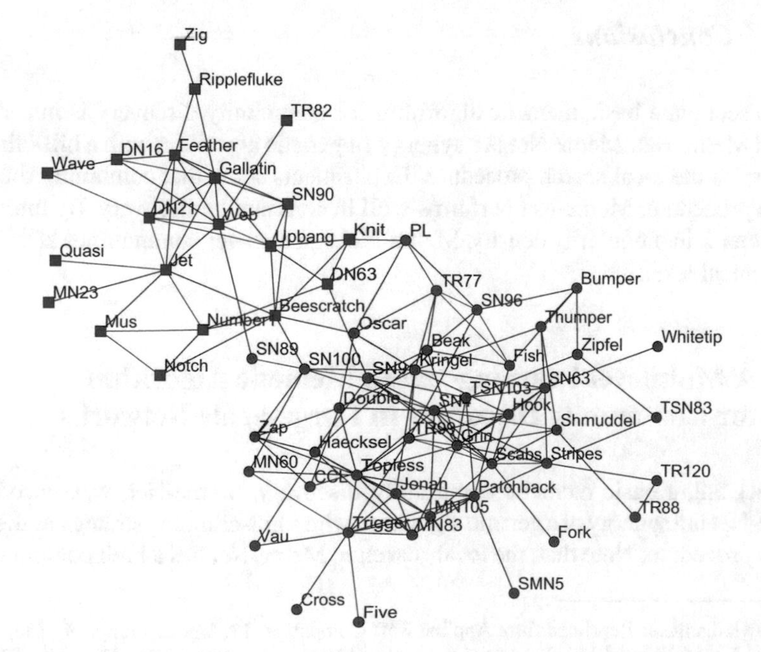

Fig. 2.6 The Dolphin social network

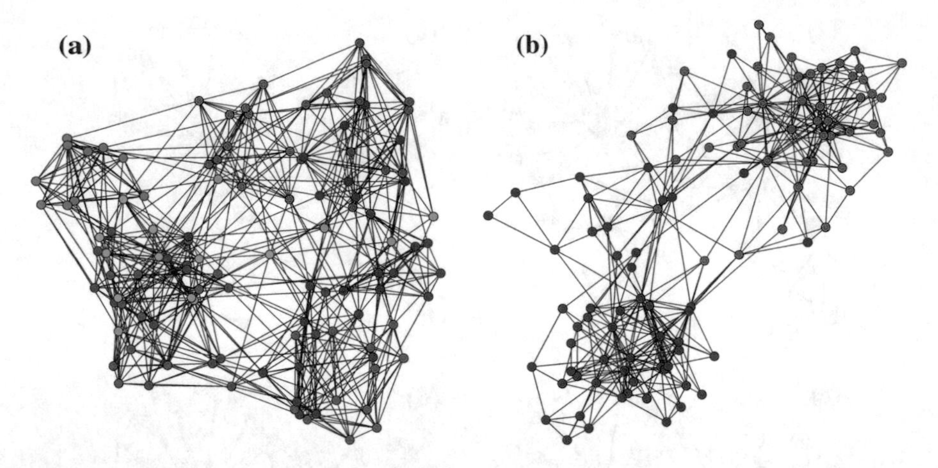

Fig. 2.7 **a** The football network, **b** The polbooks network

Table 2.4 The results obtained by fast modularity algorithm

Network	The number of clusters	NMI
Karate	3	0.693
Dolphin	4	0.573
Football	7	0.762
Polbooks	4	0.531

2.2.7 Conclusions

In this section, a basic memetic algorithm for community discovery is introduced, named Meme-Net. Meme-Net is a synergy of genetic algorithm with a hill-climbing strategy as the local search procedure. Experiments show that combining the local search procedure, Meme-Net performs well in community discovery. By tuning the parameter λ in modularity density, Meme-Net can uncover communities at different hierarchical levels.

2.3 A Multilevel Learning-Based Memetic Algorithm for Community Discovery in Large-Scale Networks

[1]In Sect. 2.2, a basic memetic community discovery, Meme-Net, was introduced. Meme-Net is a synergy of a genetic algorithm with a hill-climbing strategy as the local search procedure. Note that, the local search in Meme-Net has a high computational

[1]Acknowledgement: Reprinted from Applied Soft Computing, 19, Ma, L., Gong, M., Liu, J., Cai, Q., Jiao, L., Multi-level learning based memetic algorithm for community detection, 121–133, Copyright (2014), with permission from Elsevier.

complexity, which makes it impossible to discover communities on large scale of networks.

This section introduces a memetic algorithm with multi-level learning strategies (MLCD) to detect communities. MLCD aims to maximize modularity. The advantages of MLCD are as follows. The proposed multi-level learning strategies work on the network at node, cluster, and network partition levels, respectively. By iteratively executing GA and multi-level learning strategies, a network partition with high modularity can be accurately and stably obtained. A modularity-specific label propagation rule is employed to update the cluster identifier of each node at each operation. The simple update rule guarantees the rapidity of the proposed algorithm.

2.3.1 Memetic Algorithm with Multi-level Learning for Community Discovery

The framework of MLCD is similar to that of Meme-Net, as shown in Algorithm 1. In what follows, more detailed descriptions about the initialization procedure, genetic operators and the multi-level learning strategies will be given.

2.3.2 Representation and Initialization

Like Meme-Net, the string-based representation is also adopted in MLCD, but the initialization is different. The initialization process of MLCD can be described as follows: firstly, each gene in chromosome is put in a different cluster (i.e., $x_a^i \leftarrow i$, $1 \leq i \leq n, 1 \leq a \leq N_p$). Then, generate a random sequence. Next, for each vertex v_i chosen in that sequence, update its cluster identifier with one of its neighbors' (i.e., $x_a^i \leftarrow x_a^j, \exists j \in \{j|A_{ij} = 1\}$). The above processes independently run N_p times, and a population of solutions are generated. The solution, which has maximal modularity in the population, is chosen as x_g.

2.3.3 Genetic Operators

Crossover. The traditional crossover operations, including the uniform crossover [22] and the two-point crossover [14], have great randomness and large uncertainty, and thus the resulting descendants can hardly inherit useful communities from their parents. In MLCD, a two-way crossover operation is employed. The crossover procedure is given as Algorithm 4. First, randomly choose two chromosomes x_a and x_b from X_C. Then, randomly choose a vertex v_i and identify the cluster identifiers of node v_i in x_a and x_b (x_a^i and x_b^i, respectively). Next, generate a random value in the range of 0–1. If the generated value is larger than the crossover probability p_c, then assign the cluster identifier of the corresponding vertices which have the same cluster identifier as x_a^i in x_a with x_a^i in x_b (i.e., $x_b^j \leftarrow x_a^i, \forall j \in \{j \mid x_a^j = x_a^i\}$). At the same time, assign the cluster identifier of the corresponding vertices which have the same cluster identifier as x_b^i in x_b as x_b^i in x_a (i.e., $x_a^j \leftarrow x_b^i, \forall j \in \{j \mid x_b^j = x_b^i\}$). The above processes repeatedly operate $\lfloor N_c/2 \rfloor$ times.

This two-way crossover operation can generate descendants which inherit most communities from their parents. As shown in Fig. 2.8, a toy network G has two communities $s_1 = \{v_1, v_2, v_3, v_4\}$ and $s_2 = \{v_5, v_6, v_7\}$. The parent chromosomes $x_a = \{1, 1, 1, 1, 2, 3, 4\}$ and $x_b = \{1, 2, 3, 4, 5, 5, 5\}$ have the community structure s_1 and s_2, respectively. Assuming that the vertex v_5 is chosen, in the following operations, the nodes v_5, v_6 and v_7 in x_a are assigned with the cluster identifier of the corresponding nodes in x_b, and thus produce a descendant $x_c = \{1, 1, 1, 1, 5, 5, 5\}$.

Algorithm 4 Two-way crossover

Input: X_C and the crossover probability p_c.
Output: the offspring chromosome.

1: $r \leftarrow 1$
2: **while** $r \leq \lfloor N_c/2 \rfloor$ **do**
3: Randomly select two chromosomes x_a and x_b ($a \neq b$) from X_C and randomly select a vertex v_i.
4: Randomly generate a value f within [0 1].
5: **if** $f \leq p_c$ **then**
6: $x_b^j \leftarrow x_a^i, \forall j \in \{j \mid x_a^j = x_a^i\}$.
7: $x_a^j \leftarrow x_b^i, \forall j \in \{j \mid x_b^j = x_b^i\}$.
8: **end if**
9: $r \leftarrow r + 1$
10: **end while**

However, as for the uniform and two-point crossover operations, they can hardly inherit the communities from their parents.

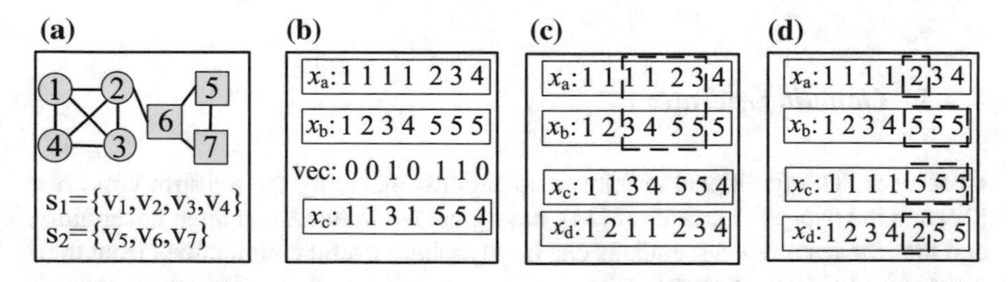

Fig. 2.8 Illustration of the crossover operations. **a** a toy network with two communities s_1 and s_2. **b** the Uniform crossover. The resulting solution x_c is generated by selecting the genes where the binary vector vec is 0 from x_a, and the genes where the binary vector vec is 1 from x_b. **c** Two-point Crossover. The resulting solutions x_c and x_d are generated by swapping the cluster identifier between the two chosen nodes (eg. v_3 and v_6) in x_a and x_b. (**d**) Two-way crossover

Mutation. In order to decrease useless exploration, a neighbor-based mutation operator is used which considers the effective connections among nodes. The neighbor-based mutation procedure works on the generated population X_C, and it works as follows. For each node v_i in a chromosome x_c ($x_c \in X_C$), first, generate a random value in the range of 0–1. If the generated value is smaller than the mutation probability p_m, then the cluster identifier of the node v_i is randomly mutated to one of its neighbors' (i.e., $x_c^i \leftarrow x_c^j, \exists j \in \{j|A_{ij} = 1\}$). Among the generated offspring chromosomes, the chromosome, which has the maximal modularity value, is chosen as x_l.

2.3.4 Multi-level Learning Strategies

In MLCD, a multi-level learning based local search is proposed to refine the individual x_l. The multilevel learning based local search is proposed based on the potential knowledge of the node, community and partition structures of networks, and it is composed of three learning techniques from low level to high level. Each level learning technique can converge to a local optimal solution rapidly and the higher level learning techniques have the ability to escape from the local optima obtained by the lower level learning techniques. In the realization of the proposed multi-level learning techniques, in order to avoid generating unnecessary divisions containing unconnected nodes and communities and to find the best solution as soon as possible, the cluster identifier of nodes and communities with their neighbors' are updated when the increment of modularity ΔQ is maximal. Moreover, a modularity-specific label propagation rule is employed to update the cluster identifier of each node and community at each operation. This simple update rule guarantees the rapidity of the proposed multi-level learning techniques. In addition, in order to reduce the sensitivity of the proposed learning techniques to the optimized order, the cluster identifiers of nodes and communities are updated in a random order.

2.3.4.1 The First-Level Learning

A node-level learning strategy, which is similar to the algorithm LPAm [2], is chosen as the first-level learning technique, and it takes x_l as the initial solution. The node-level learning strategy works on each node of the network and its processes is shown in Algorithm 5. Given a network G with n nodes, the first-level learning strategy works as follows. Firstly, generate a random sequence (i.e.,$\{r_1, r_2, \ldots, r_n\}$). Then, for each vertex v_{r_i} chosen in that generated sequence, update its cluster identifier $x_l^{r_i}$ with one of its neighbors' using the modularity-specific label propagation rule described in Sect. 2.3.4.4. The above processes end when the cluster identifier of every node has no change. The first-level learning strategy can help GA quickly converge to an optimal solution around one of its search space.

Algorithm 5 Node-level learning strategy.

Input: x_l and G. **Output**: x_l.
1: **repeat**
2: Generate a random sequence (i.e.,$\{r_1, r_2, \ldots, r_n\}$).
3: **for** each node v_{r_i} of G **do**
4: Update $x_l^{r_i}$ using the modularity-specific label propagation rule.
5: **end for**
6: **until** the cluster identifier of each node is not changed.

Note that, the first-level learning strategy tends to uncover a network partition with similar communities in total degree, and thus it is easy to fall into a local optimum. For example, as is shown in Fig. 2.9, for a toy network G_1 with 16 nodes, the intuitive community division divides the network into two communities (drawn in circle and square shapes respectively) with $Q = 0.4127$. The first-level learning strategy tends to divide this network into four communities with $Q = 0.3988$, as is shown in Fig. 2.9b. The modularity value of this network partition is not increased by removing any node from its community and placing it in that of its neighbors'. Evidently, the community division obtained by the first-level learning strategy falls into a local optimal solution. Therefore, it is necessary to take measures to help the first-level learning strategy escape from its local optimum.

Actually, for small size networks, the hybrid technique, which iteratively employs GA and the first-level learning strategy, can escape from the local optimum obtained by the first-level learning strategy. A schematic illustration of how GA helps the first-level learning strategy escape from its local optimum is shown in Fig. 2.10. Assuming that the local optimal solution obtained by the first-level learning strategy is x_1 and there is one solution like x_2 in the population, the offspring chromosome $x_1^{'}$ which corresponds to the true community division of the toy network G_1 is generated after employing the two-way crossover operation on the two chromosomes x_1 and x_2. Therefore, the hybrid algorithm, termed as M-LPAm, which iteratively performs GA and the first-level learning strategy, has the ability to help the first-level learning strategy escape from its local optimum. However, for the larger scale networks, M-LPAm is difficulty to solve this problem. This is because the probability of the existence of the chromosome like x_2 in the population becomes smaller. S community-level learning strategy is devised to help the first-level learning strategy further escape from its local optimum.

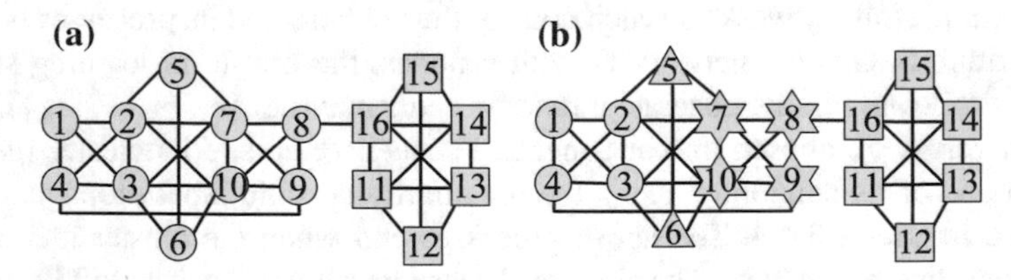

Fig. 2.9 A schematic description of the partitions of a toy network G_1. **a** The best community division with $Q = 0.4127$ divides the network into two communities. **b** The first-level learning strategy divides this network into four communities with $Q = 0.3988$

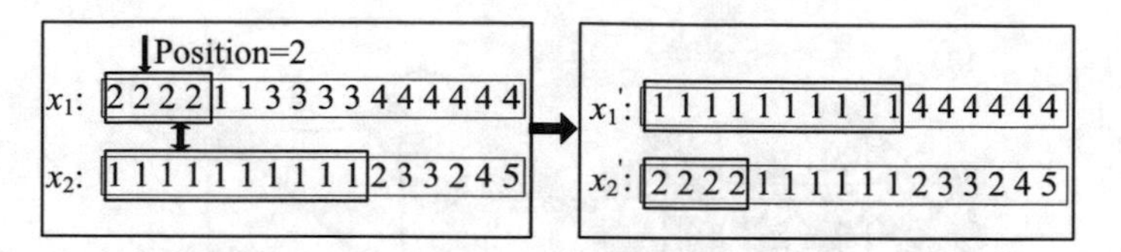

Fig. 2.10 A schematic illustration of how the two-way crossover operation helps the first-level learning strategy escape from its local optimum on a toy network G_1. Chromosome x_1 is obtained by the first-level learning strategy and chromosome x_2 comes from the chosen population. After employing the two-way crossover x_1 and x_2, an offspring chromosome $x_1^{'}$ which is the same as the true community division of the toy network is generated

2.3.4.2 The Second-Level Learning

A community-level learning technique which is similar to the second phase of the BGLL algorithm [3] is devised as the second-level learning strategy, and it is shown in Algorithm 6. The second-level learning strategy works on each community of the solution x_l generated by the first-level learning strategy. Assuming that the chromosome x_l has k communities (e.g., $s_1, s_2, \ldots, s_k$), the second-level learning strategy works as follows. Firstly, a new network, whose nodes are now the communities of x_l, is generated. In this case, the links between the new nodes $v_i^{'}$ and $v_j^{'}$ are the total links between the nodes in the corresponding communities s_i and s_j of x_l. A schematic illustration of the above operation is given in Fig. 2.11a with a toy network G_1. The original partition of the toy network G_1 is composed of 4 communities, which have 4, 2, 4, and 6 nodes respectively. In the generated new network, there are four nodes and each node corresponds to one community of the original network. The weight of links between the new nodes is the total weight of links between nodes in the corresponding communities. Second, assign each new node with a unique cluster identifier, i.e., $X^i = i$, $1 \le i \le k$, where X^i represents the cluster identifier of the new node $v_i^{'}$. Thirdly, employ the first-level learning strategy on X to find a better partition of the new network $G^{'}$. In Algorithm 6, the function of NodeLearning() is used to represent the node-level learning technique. Finally, decode the partition of new network $G^{'}$ to the community division of the original network G. The second-level learning procedure returns a new chromosome x_e. Compared with the second phase of the BGLL algorithm, the community-level learning technique has different strategies in terms of the optimization order and the cluster identifier update rule.

Algorithm 6 Community-level learning strategy.

Input: x_l and G.
Output: x_e.
1: Generate a new network $G^{'} = (V^{'}, E^{'})$, where $V^{'} = \{v_1^{'}, v_2^{'}, \ldots, v_k^{'}\}$. $v_i^{'}$ is the corresponding community s_i in x_l, $1 \le i \le k$. The connections between the new nodes can be represented as an adjacency matrix $A^{'}$, whose elements $A_{ij}^{'}$ is computed as $A_{ij}^{'} \leftarrow \sum_{i \in s_i} \sum_{j \in s_j} A_{ij}$.
2: Set $X^i = i$, $1 \le i \le k$.
3: $X \leftarrow$ NodeLearning(X,$G^{'}$).
4: **while** $i \le k$ **do**
5: $\quad x_e^j \leftarrow X^i, \forall j \in \{j | v_j \in s_i\}$.
6: **end while**

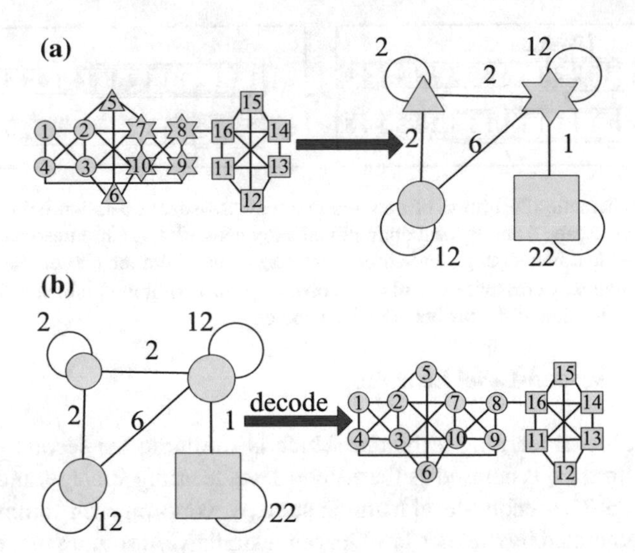

Fig. 2.11 A schematic illustration of the second-level learning technique. **a** Generate a new network. *Left* the original network G_1. *Right* the new network. **b** Generate a new partition of the toy network G_1 by employing the second-level learning strategy. The new partition is the same as the true one

The second-level learning strategy can help the first-level learning strategy jump out of its local maximum. As is shown in Fig. 2.11b, after merging the upper left three communities, the generated partition is the same as the true one with $Q = 0.4238$.

Note that, the first two level learning strategies are also easy to fall into local optimal solutions. This is because the merged communities can hardly be separated again. Therefore, the new partition may not always be good enough (although it is better than the previous local maximum). The above view is confirmed by Rosvall and Axelsoon [23]. For example, as is shown in Fig. 2.12a, the new toy network G_2 with 16 nodes is intuitively divided into three communities (drawn in circle, triangle, and square shapes respectively), with $Q = 0.4298$. After performing the first-level learning and the second-level learning strategies, the obtained partitions tend to

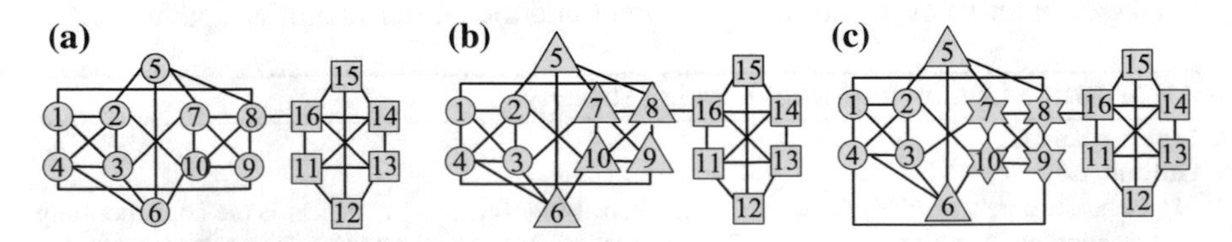

Fig. 2.12 A schematic description of the partitions of a new toy network G_2. **a** The best network partition with modularity $Q = 0.4298$ divides the network in three communities. **b** The first-level learning strategy is easy to a local maximum where the network is divided into 4 communities with $Q = 0.4094$. **c** The technique, which combines the first-level learning strategy and the second-level learning strategy, is easy to get into a local maximum where the network is divided into 2 communities with $Q = 0.4127$

divide the toy network into four communities with $Q = 0.4094$ (Fig. 2.12b) and two communities with $Q = 0.4127$ (Fig. 2.12c), respectively. Evidently, these divisions correspond to local optima in the modularity space. As is shown in Fig. 2.12c, the left large community can hardly be separated again by performing the first two level learning strategies.

In this study, a structural learning strategy, which works on the network partition layer, is proposed to help the first two level learning strategies escape from their local optima.

2.3.4.3 The Third-Level Learning

The third-level learning strategy is the proposed partition-level learning procedure, and it is shown in Algorithm 7. It works on two chromosomes, $x_g = \{g_1, g_2, \dots g_{k_1}\}$ and $x_e = \{s_1, s_2, \dots s_{k_2}\}$, where k_1 and k_2 are the number of clusters of x_g and x_e, respectively. The partition-level learning strategy consists of two phases, and its first phase works as follows. For each cluster g_i, $1 \le i \le k_1$, of x_g, the corresponding nodes in g_i are divided into a set of clusters according to the following two principles: (1) If the nodes are in the same cluster in both x_g and x_e, they are divided into the same cluster; (2) If the nodes are in the same cluster in x_g, while they are in different clusters in x_e, they are divided into different clusters. The first phase returns a basic consensual network partition $x_f = \{g_1^{'}, g_2^{'}, \dots, g_{k_3^{'}}\}$, where k_3 is the number of clusters of x_f and its value depends on the difference between x_g and x_e. The second phase is employing the first two level learning strategies on x_f to find a better community division of networks. In Algorithm 7, the functions of NodeLearning() and CommunityLearning() are used to represent the node-level learning and the community-level learning techniques, respectively. The partition-level learning procedure returns a new chromosome $x_{f'}$.

A schematic illustration of the partition-level learning strategy on two individuals x_g and x_e is given in Fig. 2.13c. As is shown in Fig. 2.13c, individual x_g has two communities $g_1 = \{v_1, v_2, v_3, v_4\}$ and $g_2 = \{v_5, v_6, v_7, v_8, v_9, v_{10}, v_{11}, v_{12}, v_{13}, v_{14}, v_{15}, v_{16}\}$, and individual x_e has two communities $s_1 = \{v_1, v_2, v_3, v_4, v_5, v_6, v_7, v_8, v_9, v_{10}\}$ and $s_2 = \{v_{11}, v_{12}, v_{13}, v_{14}, v_{15}, v_{16}\}$. For the nodes of the first

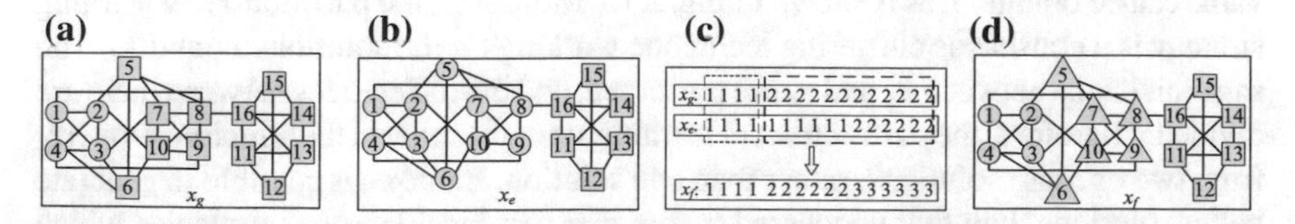

Fig. 2.13 A schematic illustration of the partition-level learning technique. **a** and **b** are the corresponding network partitions of the individuals x_g and x_e, respectively. **c** Generate an individual x_f by performing the partition-level learning technique on x_g and x_e. **d** is the corresponding network partition of the generated individual x_f

cluster g_1 of x_g, they are in the same cluster in x_e. Therefore, in the generated individual x_f, the nodes v_1, v_2, v_3 and v_4 form a cluster. For the nodes of the second community g_2 of x_g, the nodes v_5, v_6, v_7, v_8, v_9 and v_{10} have the same cluster identifier 1 in x_e, and the nodes v_{11}, v_{12}, v_{13}, v_{14}, v_{15} and v_{16} have the same cluster identifier 2 in x_e. Therefore, in the generated individual x_f, the nodes v_5, v_6, v_7, v_8, v_9 and v_{10} form a cluster and the nodes v_{11}, v_{12}, v_{13}, v_{14}, v_{15} and v_{16} form a new cluster.

The partition-level learning strategy can help the first two level learning strategies to escape from their local maxima. As is shown in Fig. 2.13, after performing the partition-level learning strategy, the large community s_1 of x_e has been broken into two communities and the true partition has been uncovered.

Algorithm 7 Partition-level learning strategy

Input: G, $x_g = \{g_1, g_2, \ldots, g_{k_1}\}$ and $x_e = \{s_1, s_2, \ldots, s_{k_2}\}$. **Output**: $x_{l'}$.
1: Set $x_f^i \leftarrow 0, i = 1, 2, \ldots, n$, and $C \leftarrow 0$;
2: **for** each cluster g_i of x_g **do**
3: Identify the nodes in g_i and set them as $g_i = \{v_{i_1}, v_{i_2}, \ldots, v_{i_{|g_i|}}\}$, where $|g_i|$ is the size of g_i.
4: **for** each node v_{i_j} of g_i **do**
5: **if** $x_f^{i_j} = 0$ **then**
6: $C \leftarrow C + 1$;
7: $x_f^q \leftarrow C$, where $\forall q \in \{q | v_q \in g_i \& x_e^{i_j} = x_e^q\}$;
8: **end if**
9: **end for**
10: **end for**
11: $x_f \leftarrow$ NodeLearning(x_f,G).
12: $x_{l'} \leftarrow$ CommunityLearning(x_f,G).

The reasons why the partition-level learning strategy can improve the accuracy of the first two level learning strategies are as follows. First, the first two level learning strategies tend to get into local optimal network partitions, mainly because it can hardly be separated again after merging two or more nodes and communities together and forming a single one. The partition-level learning strategy makes it possible for the merged nodes and communities to be separated again. After reconstructing the separated nodes and communities, a good network partition with high modularity value can be obtained, as is shown in Fig. 2.13. Moreover, the partition-level learning strategy is a consensus clustering technique working on the solutions x_g and x_e. The same divisions between x_g and x_e are preserved, and the differences between them are divided. Therefore, the partition-level learning strategy collects the structure property from two or more solutions, rather than one solution. It makes it possible to generate better solutions than that uncovered by the first two level learning strategies which just collect the structure property from one possible solution. Finally, the structural learning strategy can generate new solutions which are different from x_g and x_e. Therefore, it can enhance both the population diversity and the global searching capability of GA.

2.3.4.4 Modularity-Specific Label Propagation Update Rule

In MLCD, in order to decrease the computation complexity, a simple way is adopted to calculate the ΔQ obtained by moving an isolated node v_i into its neighbor cluster s_j. The ΔQ can easily be computed by:

$$\begin{aligned}\Delta Q &= \left[\frac{l_{s_j} + l_{i,s_j}}{m} - \left(\frac{k_{s_j} + k_i}{2m}\right)^2\right] - \left[\frac{l_{in}}{m} - \left(\frac{k_{s_j}}{2m}\right)^2 - \left(\frac{k_i}{2m}\right)^2\right] \\ &= \frac{l_{i,s_j}}{m} - \frac{k_i k_{s_j}}{2m^2}\end{aligned} \tag{2.4}$$

where l_{s_j} is the number of links inside cluster s_j, l_{i,s_j} is the number of links from node v_i to nodes in s_j, k_i represents the degree of the node v_i and k_{s_j} is the total degree of nodes in s_j. Let x^i and C_i denote the cluster identifier of the node v_i and the cluster s_i, respectively, the above equation also can be written in terms of the adjacency matrix A as follow

$$\Delta Q = \frac{1}{m}\Big(\sum_{q \neq i} A_{iq}\delta(x^q, C_j) - \frac{1}{2m}k_i \sum_{q \neq i} k_q \delta(x^q, C_j)\Big), \tag{2.5}$$

where δ is the Kronecker delta. In this study, the maximum ΔQ is chosen as the update rule for updating the cluster identifier of each isolated node v_i. Therefore, the new cluster identifier X^i of the node v_i can be written as

$$X^i = \arg\underset{C}{max}\Big(\sum_{q \neq i} A_{iq}\delta(x^q, C) - \frac{1}{2m}k_i \sum_{q \neq i} k_q \delta(x^q, C)\Big). \tag{2.6}$$

The rule of Eq. (2.6) is the modularity-specific label propagation update rule. As is known from Eq. (2.6), when $X^i = x^i$, it indicates that updating the cluster identifier of v_i with any one of its neighbors' cannot increase the value of modularity. When $X^i \neq x^i$, it indicates that updating the cluster identifier of v_i with X^i produces the maximal increment of modularity.

2.3.5 *Complexity Analysis of MLCD*

Here, the time complexity of the proposed algorithm MLCD is analyzed. Given a network with n nodes and m edges, at each generation, firstly, the crossover operator is performed $\lfloor N_c/2 \rfloor$ times and the mutation operator N_c times at most, where N_c is the size of the chosen population. The time complexity of the calculation of modularity is $O(m)$. Therefore, the time complexity of the genetic operator is $O(N_c(n+m))$. Then, the multi-level learning strategies are performed. The time complexity of the first-level learning strategy is $O(rm)$, where r is the number of steps required to reach a local maximum in modularity space. The second-level learning algorithm needs to merge pairs of communities which requires $O(m \log n)$ basic operations. Therefore, the time complexity of the second-level learning algorithm is $O(hm \log n)$, where h is the number of steps required to reach a local maximum in modularity space. The time complexity of the first step and the second step of the partition-level learning strategy are $O(n)$ and $O(rm + hm \log n)$, respectively. Generally, $r = \log n$ and $h \approx \log n$. Therefore, the overall time cost for the multi-level learning strategies at each generation is $O(m(\log n)^2)$. Finally, the population is updated, which needs $O(N_p + N_c)$ basic operations. In practical applications, $N_p + N_c \leq N_c(n + m) \leq m(\log n)^2$. Therefore, the time complexity of the proposed MLCD algorithm at each generation is $O(m(\log n)^2)$.

2.3.6 Comparisons Between MLCD and Meme-Net

By comparing MLCD with Meme-Net, it can be seen that both of them model the community detection in networks into an optimization problem, and use the framework of memetic algorithm to solve the modeled optimization problem. Moreover, both of them have excellent performance in discovering the potential community structure of networks. However, they are different in the following aspects.

First, the motivation of Meme-Net is to find a set of hierarchical community structures on small-scale networks by optimizing an objective with a resolution parameter λ. To reveal a community structure at a certain hierarchical level, it is necessary to set the value of λ in advance. The motivations of MLCD are to find the best community division of networks quickly and stability without knowing the number of clusters in advance, and to extend Meme-Net to handle larger scale networks.

Secondly, the modeled optimization problems are different. Meme-Net and MLCD model the community detection of networks into an optimization problem based on modularity density and modularity, respectively. The optimization of modularity density tends to reveal a community structure of a network which has greater internal link density than external link density of communities, while the optimization of modularity tends to discover a community structure which has more intra-community links than inter-community links.

Moreover, Meme-Net adopts a hill-climbing strategy based local search which works on a network at nodes level, while MLCD uses the proposed multilevel learning techniques based local search which works on a network at nodes, communities, and network partitions levels, respectively. Compared with the proposed multilevel learning techniques, the hill-climbing strategy has higher computational complexity. Given a network with n nodes and m edges, at each step of the hill-climbing technique, it needs to consider all the possible network partitions. Moreover, it requires $\log n$ steps to reach a local maximum in the modularity density space. In addition, the computation of the modularity density of each network partition requires $O(m)$ basic operations. In the worst case, the hill-climbing strategy is necessary to consider n^2 network partitions. Therefore, the overall time cost of the hill-climbing technique at each generation is $O(mn^2 \log n)$. For a sparse network $n \log n \approx m$, the time complexity of the hill-climbing strategy is $O(m^2 n)$, which is far greater than that of GA, $O(N_p m)$, where N_p is the size of the population. Due to the high computational complexity $(O(m^2 n))$, Meme-Net can only handle small-scale networks. For the proposed algorithm MLCD, its low computational complexity $(O(m(\log n)^2)$ makes it possible to tackle larger scale networks.

In addition, Meme-Net and MLCD have different initialization process. In the initialization of Meme-Net, the cluster identifiers of αn nodes are assigned to all of their neighbors. The diversity and the quality of the generated initial population are controlled by the parameter α. For the MLCD, it no longer needs a controlling parameter to adjust the diversity and the quality of the generated initial population. The initialized solutions are generated by assigning the cluster identifier of each node with one of its neighbors'.

Finally, the experimental comparisons between MLCD and Meme-Net demonstrate that MLCD has superior performance than Meme-Net in terms of the quality of the detected community structure, the stability and the time consumption.

2.3.7 Experimental Results

MLCD is tested on the extension of GN and LFR benchmarks networks and 12 real-world networks. The comparisons between MLCD and its three variants, M-two-phase, M-LPAm and GA, are made to illustrate the effectiveness of each level learning strategy. The algorithms GA and M-two-phase are the variants of MLCD by removing the multi-level learning strategies and the partition-level learning strategy, respectively. The algorithm M-LPAm is the variant of MLCD by removing both the partition-level learning and the community-level learning strategies. Moreover, the comparisons between M-two-phase (M-LPAm) and two-phase (LPAm) are given to show the effectiveness of the hybrid technique. The algorithm two-phase (LPAm) is the variant of M-two-phase (M-LPAm) by removing the genetic algorithm. In addition, the results obtained by the representative community detection algorithms FM [6], BGLL [3], Infomap [24], MOGA-Net [22], MODPSO [11] and Meme-Net [13] are also given for comparisons.

The parameter settings of MLCD and some compared algorithms are listed in Table 2.5. The key parameters, including the population size, the chosen population size, the crossover rate, the mutation rate and the maximum generation, are set to the same values. For each network, all algorithms are independently run 50 times.

As the two classes of benchmarks networks have known partitions, the evaluation criterion Normalize Mutual Information (NMI) (Eq. 2.3) is also adopted to estimate the similarity between the true partition and the detected one.

Table 2.5 Parameter settings of the compared algorithms

Parameter	Meaning	MLCD	MOGA-Net	MODPSO	Meme-Net
S_{pop}	The population size	300	300	300	300
S_{pool}	Size of the mating pool	15	15	15	15
P_c	Crossover rate	0.9	0.9	–	0.9
P_m	Mutation rate	0.15	0.15	0.15	0.15
G_{max}	The number of iterations	200	200	200	200

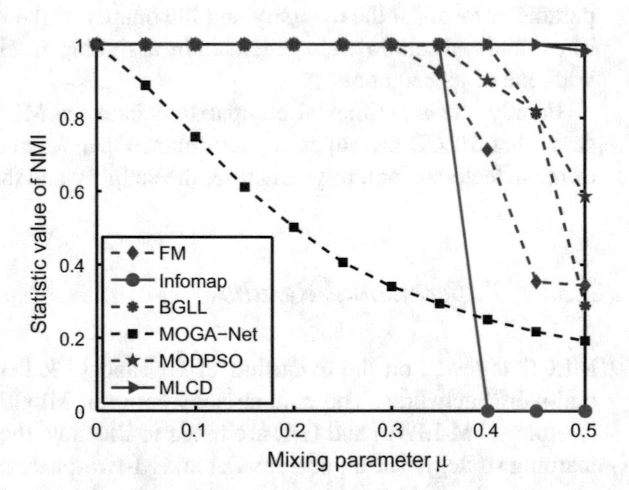

Fig. 2.14 Performance comparisons between FM, Infomap, BGLL, MOGA-net, MODPSO, and MLCD on the GN benchmark networks with different mixing parameters

2.3.7.1 Experiments on Artificial Generated Network

The extension of GN benchmark network model is introduced in Chap. 1. For GN benchmark networks, the maximum NMI values averaged over 50 independent runs obtained by MLCD and the compared algorithms are recorded in Figs. 2.14 and 2.15.

As is shown in Figs. 2.14 and 2.15, LPAm, MOGA-Net, FM, Infomap, and BGLL cannot reveal the true community structure of these networks when the mixing parameter μ is larger than 0.10, 0.10, 0.30, 0.35, and 0.40, respectively. The proposed algorithm MLCD can detect the true community division of networks when $\mu \leq 0.45$. When $\mu = 0.5$, the algorithm MLCD uncovers a community structure which is the closest to the true one. These indicate that compared with the classical community detection algorithms, MLCD has remarkable performance on detecting the community structure of the GN benchmark networks.

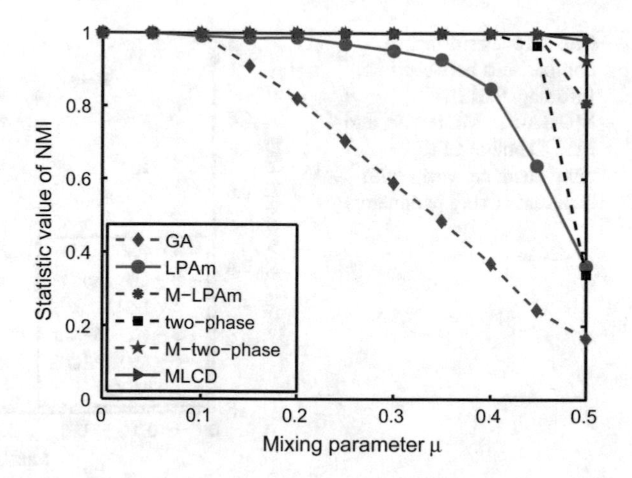

Fig. 2.15 Performance comparisons between MLCD, GA, LPAm, M-LPAm two-phase, and M-two-phase on the GN benchmark networks with different mixing parameters

The results in Fig. 2.15 also show that GA cannot detect the true community structure of networks when $\mu \geq 0.10$. MLCD, M-two-phase and M-LPAm are able to uncover the true community division of these networks when $\mu \leq 0.45$. When $\mu = 0.5$, communities detected by the algorithms which have higher level learning strategies are closer to the true one.

As is shown in Fig. 2.15, GA and LPAm cannot discover the true community structure of networks even if these networks have clear community structures ($0.1 \leq \mu \leq 0.4$). The algorithm M-LPAm, which combines GA and LPAm, can effectively find their true community structures. With μ increased, the ratio of the intra-community links to the inter-community links decreases, and thus the community structure of networks becomes increasingly fuzzier. When $\mu = 0.45$, two-phase and LPAm cannot find the true community structure of the network, while M-two-phase and M-LPAm can still uncover its true community division. These comparisons demonstrate the effectiveness of the hybrid techniques.

For the LFR benchmark networks, the maximum NMI values averaged over 50 independent trials obtained by MLCD and the compared algorithms are recorded in in Figs. 2.16 and 2.17.

The results in Figs. 2.16 and 2.17 show that except $\mu = 0.6$, the partitions obtained by MLCD are the closest to the real ones in the compared algorithms. Therefore, compared with the classical community detection algorithms, the proposed algorithm MLCD has a competitive advantage in discovering the best community divisions of the LFR benchmark networks.

The results in Fig. 2.17 also show that LPAm and two-phase can discover the true community division of networks when $\mu \leq 0.4$. With μ increased, they become more and more difficult to discover the true community divisions. The hybrid algorithms M-LPAm and M-two-phase can effectively find the true partition of these networks when $\mu \leq 0.50$. Even for $\mu \geq 0.55$, the values of NMI obtained by M-LPAm and M-two-phase are larger than those by LPAm and two-phase, respectively. These

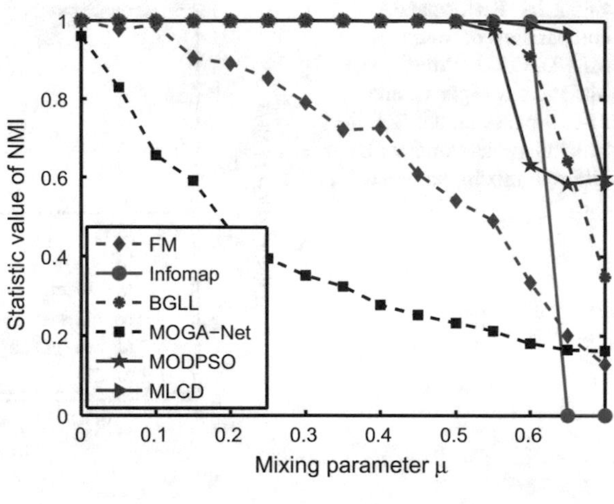

Fig. 2.16 Performance comparisons between FM, Infomap, BGLL, MOGA-net, MODPSO, and MLCD on the LFR benchmark networks with different mixing parameters

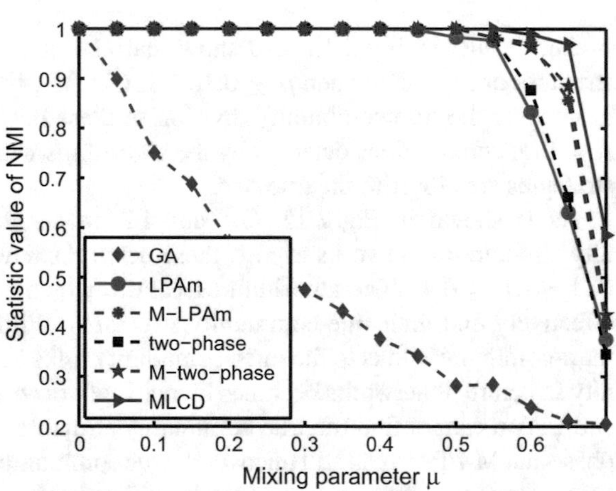

Fig. 2.17 Performance comparisons between MLCD, GA, LPAm, M-LPAm two-phase and M-two-phase on the LFR benchmark networks with different mixing parameters

indicate that the community divisions revealed by the hybrid techniques M-LPAm and M-two-phase are closer to the true ones than those by LPAm and two-phase.

Compared with the results in Fig. 2.17, it is found that MLCD, M-two-phase and M-LPAm can effectively uncover the true community division of these networks when $\mu \leq 0.50$. When $\mu > 0.50$, the algorithm which has higher level learning strategies can detect a community division with a higher value of *NMI*. This demonstrates that the higher level learning strategies can effectively escape from the local optima obtained by the lower level learning strategies.

2.3.7.2 Experiments on Real-World Networks

In this part, the proposed algorithm MLCD is tested on 12 real-world networks coming from different application fields. These networks include karate, dolphins, football, polbooks, jazz, elegans, e mail, power, Geom, ca_grqc, pgp, and Internet. These networks are considered as undirected and unweighted. The basic information of these real-world networks is shown in Table 1.1.

First, the comparisons between MLCD, M-two-phase, M-LPAm and GA are given to illustrate the effectiveness of each level learning strategy. The comparison results are shown in Table 2.6. The results in Table 2.6 show that for the karate, dolphins, football and polbooks networks, GA can hardly discover the optimal community divisions of these networks. The algorithm M-LPAm, which incorporates the node-level learning strategy into GA, has the ability to find the optimal community divisions of these networks. However, the algorithm M-LPAm is easy to fall into poor network partitions. After incorporating the community-level learning strategy into M-LPAm, the algorithm M-two-phase can always find the optimal network partitions for these networks. These demonstrate the effectiveness of the node-level and the community-level learning strategies. The results in Table 2.6 illustrate that for the elegans, e-mail, power, ca_jrqc, Geom, pgp and Internet networks, although the maximal and mean modularity values obtained by M-two-phase are larger than those by M-LPAm, M-two-phase are also easy to trap into poor network partitions. The algorithm MLCD, which incorporates the partition-level learning strategy into M-two-phase, can find the community divisions of these networks with higher modularity values than M-two-phase. This demonstrates the effectiveness of the partition-level learning strategy. The comparison results demonstrate that each level learning strategy can improve the effectiveness of the proposed algorithm.

Moreover, the comparisons between M-LPAm, LPAm, M-two-phase and two-phase are made to illustrate that the hybrid algorithms, M-LPAm and M-two-phase, can enhance both the accuracy and stability of LPAm and two-phase, respectively. The comparison results are recorded in Table 2.6. The results in Table 2.6 show that for most small-scale networks which have less than 5000 nodes, the Q_{max} and Q_{avg} values obtained by M-LPAm and M-two-phase are larger than those by LPAm and two-phase, respectively. Meanwhile, the Q_{std} values obtained by M-LPAm and M-two-phase are smaller than those by LPAm and two-phase, respectively. These indicate that the hybrid techniques, M-LPAm and M-two-phase, can enhance both the accuracy and stability of LPAm and two-phase, respectively, on the small-scale networks. The results in Table 2.6 also illustrate that for these networks which have more than 5000 nodes, the Q_{max} and Q_{avg} values obtained by M-LPAm are larger those by LPAm. However, the Q_{max} and Q_{avg} values obtained by M-two-phase are smaller than those by two-phase. This is possible because that with the network size increased, the searching spaces generated by GA are more and more difficult to cover all the preferable ones. Moreover, in the optimization of modularity, the number of local maxima is exponentially growing with the size of networks. Therefore, it is possible that the optimal solution around the search space generated by GA is worse than that by a random way. This can result in this situation in which the community divisions generated by M-two-phase are worse than those by two-phase when they work on the networks with more than 5000 nodes. Note that, after incorporating the partition-level learning strategy into M-two-phase, the community divisions of these networks uncovered by MLCD are better than those by M-two-phase and two-phase.

Table 2.6 The maximum, average, and standard deviation of modularity values (Q_{max}, Q_{avg} and Q_{std}) obtained by GA, LPAm, M-LPAm, two-phase, M-two-phase, MLCD, MOGA-Net, and MODPSO on the real-world networks. Values are recorded under fifty independent trials. "—" indicates that the corresponding algorithm cannot tackle it

P	Network	Criterion	GA	LPAm	M-LPAm	two-phase	M-two-phase	MLCD	MOGA-Net	MODPSO
1	Karate	Q_{max}	**0.4198**	0.4052	**0.4198**	**0.4198**	**0.4198**	**0.4198**	0.4159	**0.4198**
		Q_{avg}	0.4176	0.3564	**0.4198**	0.4165	**0.4198**	**0.4198**	0.3945	0.4182
		Q_{std}	0.0047	0.0285	**0**	0.0077	**0**	**0**	0.0089	0.0079
2	Dolphins	Q_{max}	0.5258	0.5071	**0.5285**	**0.5285**	**0.5285**	**0.5285**	0.5034	0.5265
		Q_{avg}	0.5138	0.4938	0.5281	0.5232	**0.5285**	**0.5285**	0.4584	0.5255
		Q_{std}	0.0070	0.0114	0.0013	0.0029	**0**	**0**	0.0163	0.0061
3	Polbooks	Q_{max}	0.5267	0.5145	**0.5272**	**0.5272**	**0.5272**	**0.5272**	0.4993	0.5264
		Q_{avg}	0.5213	0.4976	0.5271	0.5267	**0.5272**	**0.5272**	0.4618	0.5263
		Q_{std}	0.0037	0.0158	0.0001	0.0017	**0**	**0**	0.0129	0.0007
4	Football	Q_{max}	0.4577	0.6032	**0.6046**	**0.6046**	**0.6046**	**0.6046**	0.4325	**0.6046**
		Q_{avg}	0.4530	0.5777	0.6044	0.6043	**0.6046**	**0.6046**	0.3906	0.6038
		Q_{std}	0.0210	0.0199	0.0004	0.0009	**0**	**0**	0.0179	0.0011
5	Jazz	Q_{max}	0.3641	0.4448	**0.4451**	**0.4451**	**0.4451**	**0.4451**	0.2929	0.4421
		Q_{avg}	0.2873	0.4360	0.4450	0.4443	**0.4451**	**0.4451**	0.2952	0.4419
		Q_{std}	0.0312	0.0098	0.0012	0.0013	0.0001	**0**	0.0084	0.0001
6	Elegans	Q_{max}	0.2750	0.4121	0.4439	0.4498	0.4507	**0.4532**	0.2977	0.3963
		Q_{avg}	0.2139	0.3781	0.4362	0.4411	0.4473	**0.4527**	0.2805	0.3829
		Q_{std}	0.0105	0.0155	0.0042	0.0046	0.0019	**0.0004**	0.0084	0.0181
7	E-mail	Q_{max}	0.3137	0.5415	0.5724	0.5814	0.5820	**0.5828**	0.3007	0.5193
		Q_{avg}	0.2866	0.5349	0.5650	0.5756	0.5805	**0.5822**	0.2865	0.3493
		Q_{std}	0.0090	0.0024	0.0039	0.0036	0.0008	**0.0006**	0.0075	0.0937
8	Power	Q_{max}	0.6345	0.5397	0.7663	0.9382	0.9401	**0.9408**	0.6914	0.8543
		Q_{avg}	0.6208	0.5344	0.7551	0.9370	0.9377	**0.9401**	0.6864	0.8510
		Q_{std}	0.0045	0.0022	0.0051	0.0006	0.0004	**0.0002**	0.0022	0.0056
9	Ca_grqc	Q_{max}	0.4229	0.7146	0.7114	0.8655	0.8635	**0.8682**	0.6839	—
		Q_{avg}	0.4049	0.7074	0.7003	0.8644	0.8617	**0.8679**	0.6674	—
		Q_{std}	0.0080	0.0029	0.0055	0.0008	0.0010	**0.0002**	0.0063	—
10	Geom	Q_{max}	0.5736	0.6625	0.8046	0.8067	0.8052	**0.8110**	0.6269	0.7007
		Q_{avg}	0.5566	0.6494	0.7957	0.8044	0.8018	**0.8102**	0.6208	0.6954
		Q_{std}	0.0077	0.0066	0.0031	0.0016	0.0016	**0.0007**	0.0030	0.0077
11	Pgp	Q_{max}	0.5988	0.7178	0.8129	0.8848	0.8839	**0.8867**	0.6180	0.3324
		Q_{avg}	0.5863	0.7068	0.8063	0.8834	0.8815	**0.8865**	0.5866	0.3279
		Q_{std}	0.0099	0.0070	0.0030	0.0013	0.0011	**0.0002**	0.0021	0.0090
12	Internet	Q_{max}	0.3892	0.4684	0.5839	0.6759	0.6680	**0.6782**	—	—
		Q_{avg}	0.3846	0.4618	0.5658	0.6741	0.6653	**0.6770**	—	—
		Q_{std}	0.0049	0.0041	0.0075	0.0010	0.0022	**0.0006**	—	—

Table 2.7 The modularity values obtained by FM, BGLL and Infomap on the real-world networks

P	Networks	FM	BGLL	Infomap
1	Karate	0.3807	0.4188	0.4020
2	Dolphins	0.4955	0.5188	0.3978
3	Polbooks	0.5020	0.4986	0.5268
4	Football	0.5773	0.6046	0.6005
5	Jazz	0.4389	0.4431	0.4423
6	Elegans	0.4062	0.4322	0.4168
7	E-mail	0.5116	0.5412	0.5354
8	Power	0.9341	0.7756	0.8302
9	Ca_grqc	0.8122	0.8405	0.8009
10	Geom	0.7767	0.7775	0.7255
11	Pgp	0.8521	0.8604	0.8127
12	Internet	0.6360	0.6464	0.5755

Finally, the comparisons between MLCD, FM, LPAm, BGLL, Infomap, MOGA-Net and MODPSO on the real-world networks are given, and the comparison results are recorded in Tables 2.6 and 2.7. The results show that for all real networks, the modularity values obtained by MLCD are the largest in the compared algorithms. Moreover, except the algorithms of FM, BGLL and Infomap which can stably reveal a network partition, the Q_{std} values obtained by MLCD are smaller than those by LPAm, MOGA-Net and MODPSO. These indicate that compared with the classical community detection algorithms, MLCD has remarkable performance on solving the community detection problem of networks.

2.3.7.3 Experimental Comparisons Between MLCD and Meme-Net

The results of MLCD and Meme-Net on the extension of GN benchmark and 5 real-world networks are recorded in Fig. 2.18 and Tables 2.6 and 2.8. The comparison results show that for the GN benchmark networks, when $\mu \leq 0.35$, both Meme-Net and MLCD can find their true network partitions. When $0.35 < \mu < 0.5$, Meme-Net cannot detect their community structures while MLCD can find their true network partitions. For the real-world networks, the Q_{max} and Q_{avg} values obtained by Meme-Net are smaller than those by MLCD, which means that MLCD has better performance than Meme-Net.

Moreover, the results in Tables 2.6 and 2.8 also show that the Q_{std} values of these networks obtained by Meme-Net are larger than those by MLCD, which indicates that MLCD has superior performance than Meme-Net in terms of the stability. The box plots of the statistic values of the maximal modularity obtained by GA, Meme-Net and MLCD over 50 runs on the 12 real-world networks are shown in Fig. 2.19.

Table 2.8 The maximum, average and standard deviation of modularity obtained by Meme-Net on the real-world networks over 50 independent runs

P	Networks	Q_{max}	Q_{avg}	Q_{std}
1	Karate	0.4020	0.4020	0
2	Dolphins	0.5185	0.5096	0.0061
3	Polbooks	0.5232	0.5218	0.0031
4	Football	0.6044	0.6023	0.0015
5	Jazz	0.4376	0.4330	0.0011

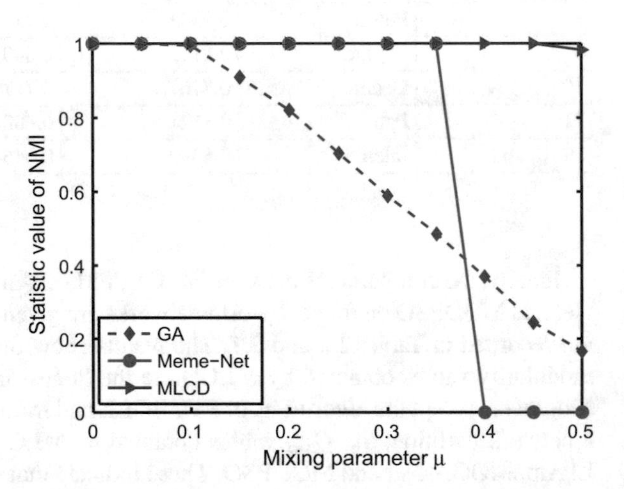

Fig. 2.18 Performance comparisons between GA, Meme-Net and MLCD on the GN benchmark networks with different mixing parameters

The results in Fig. 2.19 show that the variability of Q obtained by MLCD is the smallest, which further validates that MLCD is more stable than GA and Meme-Net.

Compared with Meme-Net, the proposed algorithm MLCD has a great advantage in terms of computational complexity. The average elapsed time of Meme-Net and MLCD on real-world networks over 50 times is recorded in Fig. 2.20. The results in Fig. 2.20 clearly show that for the elegans network with 453 nodes, Meme-Net cannot tackle it within a reasonable period of time. However, MLCD can detect communities on the Internet network with 22936 nodes in a reasonable time. This demonstrates that the proposed algorithm has a great advantage over Meme-Net in computing speed.

2.3.8 Conclusions

In this section, a fast multi-level learning-based memetic algorithm was proposed to optimize modularity for revealing the potential community structures of complex networks. The proposed algorithm is a hybrid global–local heuristic search methodology

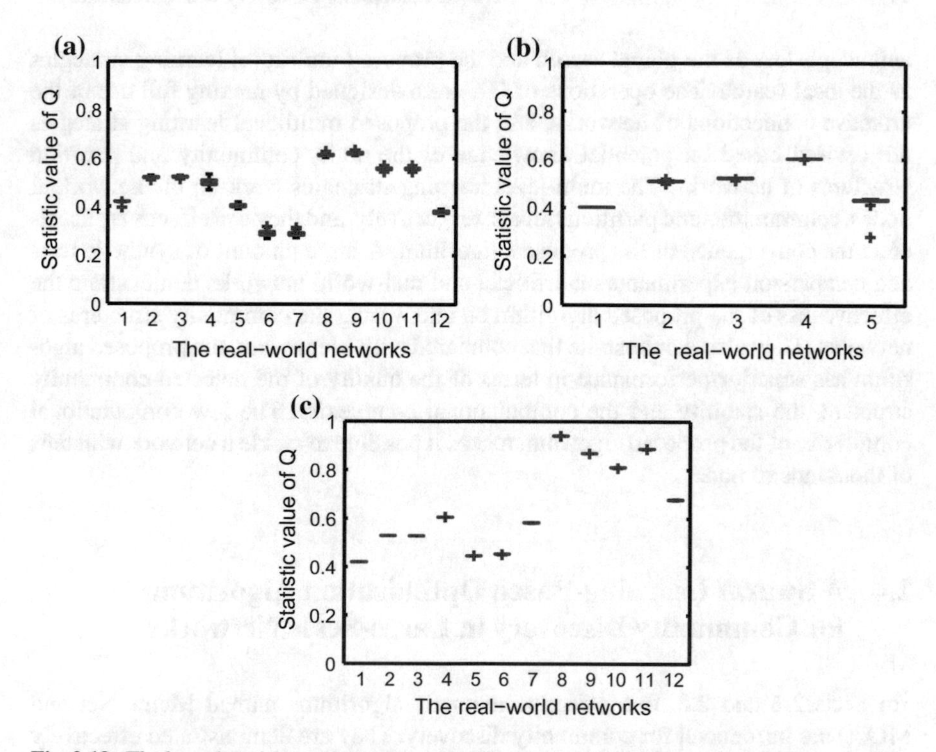

Fig. 2.19 The box plot of the statistic Q over the fifty trials on the twelve real-world networks. The box plots are adopted here to explain the distribution of the Q values obtained by GA, Meme-Net, and MLCD. On each box, the *red line* denotes the median and the symbol + represents outliers. The box plot for **a** GA, **b** Meme-Net, and **c** MLCD

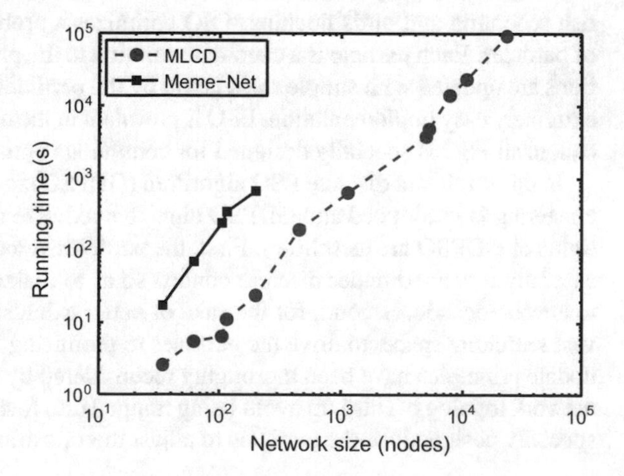

Fig. 2.20 A log-log plot about a comparison of running time between MLCD and Meme-Net in real-world networks with different sizes

and adopts GA as the global search and the proposed multilevel learning strategies as the local search. The operations of GA are redesigned by making full use of the effective connections of networks, and the proposed multilevel learning strategies are devised based on potential knowledge of the node, community and partition structures of networks. The multi-level learning strategies work on the network at nodes, communities and partitions levels, respectively, and they can effectively accelerate the convergence of the proposed algorithm. A large amount of synthetic tests and comparison experiments on artificial and real-world networks demonstrate the effectiveness of the proposed algorithm on uncovering the community structures of networks. They also demonstrate that compared with Meme-Net, the proposed algorithm has superior performance in terms of the quality of the detected community structure, the stability and the computational complexity. The low computational complexity of the proposed algorithm makes it possible to tackle a network with tens of thousands of nodes.

2.4 A Swarm Learning-Based Optimization Algorithm for Community Discovery in Large-Scale Networks

[2]In Sects. 2.2 and 2.3, two effective memetic algorithms named Meme-Net and MLCD are introduced for community discovery. They are demonstrated effectively to discover community structures. Here, an another computational intelligence algorithm, swarm intelligence optimization, is used for community discovery. Among swarm intelligence optimization, the outstanding paradigm is particle swarm optimization (PSO) [16]. PSO originated from the behavior of social animals, such as fish schooling and birds flocking. PSO optimizes a problem by employing a group of particles. Each particle is a candidate solution to the problem. The candidate solutions are updated with simple rules learnt by the particles. Due to its efficacy and its extremely easy implementation, PSO is prevalent in the optimization field. However, canonical PSO is specially designed for continuous optimization problems.

In this section, a discrete PSO algorithm (GDPSO) for large-scale social network clustering is introduced and GDPSO aims to maximize modularity. The main highlights of GDPSO are as follows. First, the particles velocity and position have been carefully redefined under discrete context so as to make them as easier as possible to encode/decode. Second, for the sake of better exhaustive global searching in the vast searching space, to drive the particles to promising regions, the particle-status-update principles have been thoroughly reconsidered by taking the advantage of the network topology. Third, to avoid being trapped into local optima, a greedy strategy specially designed for the particles to adjust their positions is newly suggested.

---

[2]Acknowledgement: Reprinted from Information Science, 316, Cai, Q., Gong, M., Ma, L., Ruan, S., Yuan, F., Jiao, L., Greedy discrete particle swarm optimization for large-scale social network clustering, 503-516, Copyright(2015), with permission from Elsevier.

2.4.1 *Greedy Particle Swarm Optimization for Network Community Discovery*

The whole framework of the proposed GDPSO algorithm for community discovery is given in Algorithm 8.

Algorithm 8 Framework of the proposed GDPSO algorithm.

Parameters: particle swarm size $popsize$, number of iterations $gmax$, inertia weight ω, learning factors c_1 and c_2;
Input: network adjacency matrix A;
Output: best fitness, community structure of the network;
1: **Step 1)** Initialize the population: initialize position vectors $pop[].X$; initialize velocity vectors $pop[].V = 0$; initialize the $Pbest$ vectors $Pbest[].X = pop[].X$;
2: **Step 2)** Evaluate particle fitness $pop[].fit$;
3: **Step 3)** Update the $Gbest$ particle: $Gbest = pop[best].X$;
4: **Step 4)** Set $t = 0$;
5: **Step 5)** Update particle statuses;
6: **Step 6)** Resorting particle positions: $resorting(pop[].X)$;
7: **Step 7)** Evaluate particles fitness $pop[].fit$;
8: **Step 8)** Update the $Pbest$ particles: if $pop[].fit > Pbest[].fit$, then $Pbest[].X = pop[].X$;
9: **Step 9)** Update the $Gbest$ particle: $Gbest = pop[best].X$;
10: **Step 10)** If $t < gmax$, then $t++$ and go to **Step 5)**; otherwise, stop the algorithm and output.

2.4.2 *Particle Representation and Initialization*

To make the proposed GDPSO algorithm feasible for the network clustering problem, the terms position and velocity under the discrete scenario are redefined.

Definition 1 *(Position).* The position vector represents a partition of a network. The position permutation of the particle i is defined as $\boldsymbol{X}_i = \{x_i^1, x_i^2, ..., x_i^n\}$, where $x_i^j \in [1, n]$ is an integer.

In the above definition, x_i^j is called a label identifier, which carries the cluster information. If $x_i^j = x_i^k$, then nodes j and k belong to the same cluster. A graphical illustration of the particle representation is shown in Fig. 2.21.

It can be observed from Fig. 2.21 that the above discrete position definition is straightforward and easy to decode and will reduce the computational complexity, especially in the presence of large-scale networks because the dimensions of the fitness function are the same as the number of nodes in the networks.

Definition 2 *(Velocity)* The velocity permutation of particle i *is defined as* $\boldsymbol{V}_i = \{v_i^1, v_i^2, ..., v_i^n\}$, where $v_i^j \in \{0, 1\}$. If $v_i^j = 1$, then the corresponding element x_i^j in the position vector will be changed; otherwise, x_i^j maintains its original state.

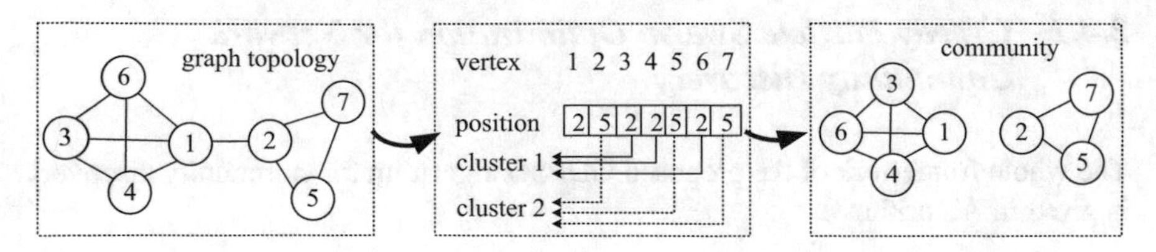

Fig. 2.21 A schematic illustration of the particle representation

In the canonical version of PSO, there is a threshold V_{max} that is used to inhibit particles from flying apart because there exists a situation whereby when the speed of a particle is substantial, it will fly out of the boundaries. To define the velocity permutation in the above style, we no longer need this parameter, which is hard to tune.

In the initialization step, the position vectors are initialized using a heuristic method introduced in previous work in [13]. The velocity vectors are initialized as all-zero vectors. The *pbest* vectors are initialized in the same manner as the position vectors, and the *gbest* vector is set as the best position vector in the original population.

2.4.3 Particle-Status-Updating Rules

In the proposed GDPSO algorithm, the particle position and velocity vectors have been redefined in a discrete form; thus, the mathematical operators in the canonical version of PSO no longer fit the discrete context. The mathematical operators have been redefined as follows:

$$V_i = \omega V_i \oplus (c_1 r_1 \times (Pbest_i \ominus X_i) + c_2 r_2 \times (Gbest_i \ominus X_i)) \tag{2.7}$$

$$X_i = X_i \otimes V_i \tag{2.8}$$

In GDPSO, the inertia weight ω and the learning factors c_1 and c_2 are set using typical values of 0.7298, 1.4961 and 1.4961. It can be seen that the above equations have the same format as in canonical PSO but that the key components are different. In the next step, these components will be illustrated in detail.

Definition 3 *(Position $\ominus$ Position).* Given two position permutations $P_1 = \{p_1^1, p_1^2, ..., p_1^n\}$ and $P_2 = \{p_2^1, p_2^2, ..., p_2^n\}$, position $\ominus$ Position is a velocity vector, i.e., $P_1 \ominus P_2 = V = \{v_1, v_2, ..., v_n\}$. The element v_i is defined as

$$\begin{cases} v_i = 0 \; if \; p_1^i = p_2^i \\ v_i = 1 \; if \; p_1^i \neq p_2^i \end{cases} \tag{2.9}$$

The inspiration of the above defined operator comes from two aspects. First, from the perspective of swarm intelligence, a particle will adjust its velocity by learning from its neighbors. The learning process is actually a comparison between the positions; in other words, two position vectors generate a velocity vector. Second, from the viewpoint of graph theory, two position vectors represent two types of network community structures. The defined $\ominus$ operation actually reflects the difference between two network structures.

Definition 4 *(Coefficient × Velocity).* The operator $\times$ is the same as the basic arithmetical multiplication operator. For instance, given a coefficient $c \cdot r = 1.3$ and given a velocity vector $V = \{1, 0, 1, 1, 0\}$, $c \cdot r \times V = \{1.3, 0, 1.3, 1.3, 0\}$.

Definition 5 *(Velocity ⊕ Velocity).* Velocity $\oplus$ Velocity equals a velocity as well. Given two velocity vectors $V_1 = \{v_1^1, v_1^2, ..., v_1^n\}$ and $V_2 = \{v_2^1, v_2^2, ..., v_2^n\}$, $V_1 + V_2 = V_3 = \{v_3^1, v_3^2, ..., v_3^n\}$. The element v_3^i is defined as

$$\begin{cases} v_3^i = 1 \; if \; v_1^i + v_2^i \geq 1 \\ v_3^i = 0 \; if \; v_1^i + v_2^i < 1 \end{cases} \tag{2.10}$$

The definition of the operator $\oplus$ is straight forward, and the operation is easy to perform. Moreover, it can always make sure that the velocity is binary coded, which is easier for the position to work with.

Definition 6 *(Position ⊗ Velocity).* The operator $\otimes$ is the key component. A particle will update its position according to a new velocity, i.e., Position $\otimes$ Velocity generates a new position. A good operator $\otimes$ should drive a particle to a promising region. Given a position $P_{old} = \{p_{old}^1, p_{old}^2, ..., p_{old}^n\}$ and a velocity $V = \{v_1, v_2, ..., v_n\}$, $P_{old} \otimes V = P_{new} = \{p_{new}^1, p_{new}^2, ..., p_{new}^n\}$. The element of P_{new} is defined as follows:

$$\begin{cases} p_{new}^i = p_{old}^i \; if \; v_i = 0 \\ p_{new}^i = \arg\max_j \Delta Q(p_{old}^i, j \mid j \in L_i) \; if \; v_i = 1 \end{cases} \tag{2.11}$$

where $L_i = \{l_1, l_2, ..., l_k\}$ is the set of label identifiers of vertex i's neighbors. The ΔQ is calculated using the following equation:

$$\Delta Q(p_{old}^i, j \mid j \in L_i) = fit(P_{old} \mid p_{old}^i \leftarrow j) - fit(P_{old}) \tag{2.12}$$

Equation 2.11 can be regarded as a greedy local search strategy because a particle updates its position by choosing the label identifier that can generate the largest fitness increment. A graphical Fig. 2.22b gives a simple illustration of how a particle updates its status.

In Fig. 2.22b, X_i represents the current particle's position vector, V_1 and V_2 are intermediate velocity vectors determined using Eq. 2.9, and V_i is the new velocity calculated using Eq. 2.10. Under the guidance of the velocity V_i, the current particle updates its current position X_i using Eq. 2.11, and thus, a new position X_j is obtained.

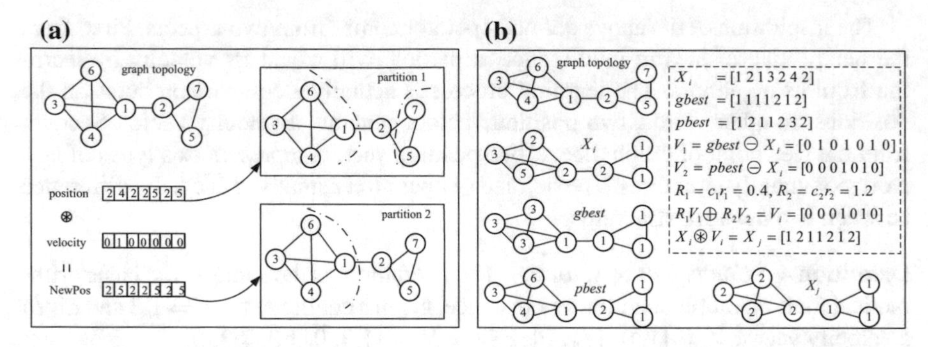

Fig. 2.22 A schematic illustration of **a** the ⊗ operator and **b** the Particle-status-updating rules

It can be observed from the above description that (1) the proposed GDPSO algorithm has a concise framework, (2) the newly defined particle representation is direct and easy to decode, and (3) the redefined updating rules are easy to realize. All these merits make this advanced algorithm capable of addressing large-scale networks.

2.4.4 Particle Position Reordering

In the proposed algorithm, to avoid unnecessary computing, a particle position reordering operator is designed. Given that $X = \{x_1, x_2, ..., x_n\}$ is a position vector of a particle, the particle position reordering operator acts on X and outputs a new vector X'. The operator reorders the elements in X with a starting value of 1. The pseudocode of the reordering operator is given in Algorithm 9.

Algorithm 9 Pseudocode of the reordering operator.

Input: an integer permutation $X = \{x_1, x_2, ..., x_n\}$;
Output: reordered permutation $X' = \{x'_1, x'_2, ..., x'_n\}$;

1: set $counter = 1, X' \leftarrow X$;
2: **for** $i = 1; i \le n; i++$ **do**
3:     **if** $x'_i \ne -1$ **then**
4:         **for** $j = i+1; j \le n; j++$ **do**
5:             **if** $x'_j = x'_i$ **then**
6:                 $x_j = counter, x'_j = -1$;
7:             **end if**
8:         **end for**
9:         $x'_i = -1, x_i = counter, counter++$;
10:     **end if**
11: **end for**
12: $X' \leftarrow X$;

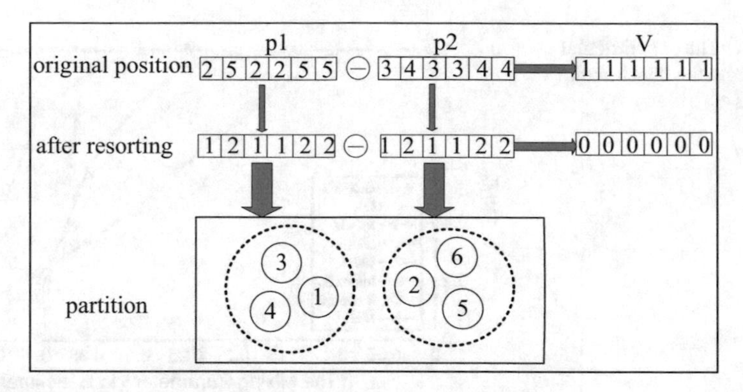

Fig. 2.23 A schematic illustration of the reordering operation

Figure 2.23 gives a graphical illustration of the reordering operator. As illustrated in Fig. 2.23, *p1* and *p2* are structure-equivalent (they correspond to the same partition). If we do not design the resorting operation, then according to Eq. 2.9, a non-zero velocity vector would be obtained, and according to Eq. 2.11, a new position would need to be calculated, which would require computing time. However, if we design the operation so that after resorting, *p1* and *p2* are the same, then according to Eq. 2.9, the obtained velocity will be a zero vector; thus, it would not need to calculate the new position, resulting in reduced computing time.

2.4.5 Experimental Results

Here, GDPSO is tested on both artificial generated networks and real-world networks and is compared with several state-of-the-art community detection algorithms: GA [21], MOGA [22], LPA [1], CNM [7] and Informap [24]. To enable a fair comparison, in the experiments, the objective function used in the GA is replaced by the modularity, and the modularity is used to choose the ultimate solution from the Pareto front for the MOGA. The population size and the maximum number of iterations for the GDPSO, GA, and MOGA are set to 100, the crossover and mutation probabilities for the GA and the MOGA are set to 0.9 and 0.1, respectively. The LPA and Informap are two iterative methods; the maximum number of iteration is set to 5 and 100, respectively. In addition, a comparison is performed with two canonical data clustering methods: the k-means method proposed in [15] and the normalized cut (Ncut) approach proposed in [26].

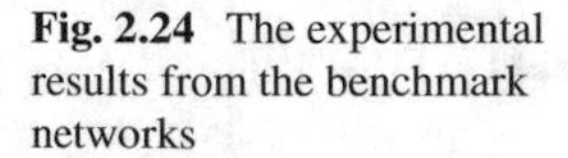

Fig. 2.24 The experimental results from the benchmark networks

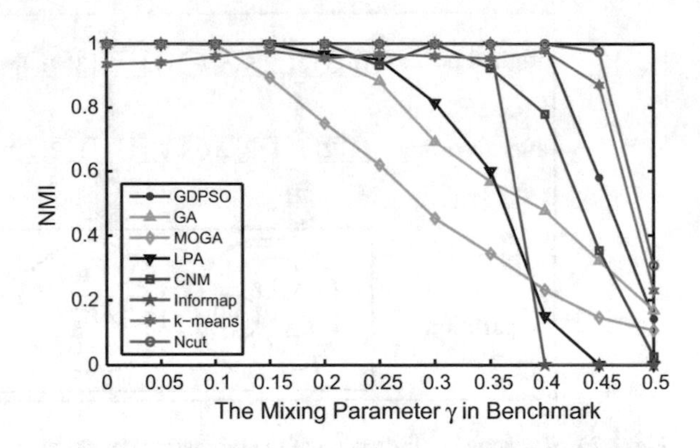

2.4.5.1 Results on Artificial Generated Network

All the algorithms are performed on the extension of GN benchmark networks. Figure 2.24 shows the statistical results averaged over 30 runs for different algorithms.

As shown in Fig. 2.24, when the mixing parameter is no larger than 0.15, all the methods, except the MOGA and k-means methods, can determine the true community structures of the networks ($NMI = 1$). As the mixing parameter increases, the GA, LPA and CNM fail to discover the correct partitions. For $\gamma = 0.4$, our proposed method and Ncut still obtain $NMI = 1$, whereas the other methods cannot. When γ is larger than 0.4, from the curve, it can be observed that the *NMI* values quickly decrease and that our method fails to detect the ground truths. One reason that can account for this is that when γ is larger than 0.4, the community structure of the network is rather vague, that is, there is no community structure. Moreover, optimizing the modularity has been proven to be NP hard. Therefore, it is very hard for an optimization method to uncover any communities under this scenario.

The experiments on the computer-generated benchmark networks prove that the proposed GDPSO framework for network clustering is feasible. The proposed algorithm is capable of uncovering community structures in social networks. Although from the curves, Ncut seems to perform the best; however, Ncut needs to specify the clusters in advance, which is very inconvenient.

2.4.5.2 Results on Real-World Networks

Here, the performance of the GDPSO is tested on four small-scale networks and on four large-scale networks: Karate, dolphin, football, SFI, e-mail, netscience, power grid, PGP. The parameters of each network are listed in Table 1.1. For each social network, each algorithm is run 30 times.

First the algorithms are tested on the four small-scale networks. A statistical analysis using the Wilcoxon rank sum test is performed. In the right-tailed rank sum test experiments, the significance level is set to 0.05 with the null hypothesis that

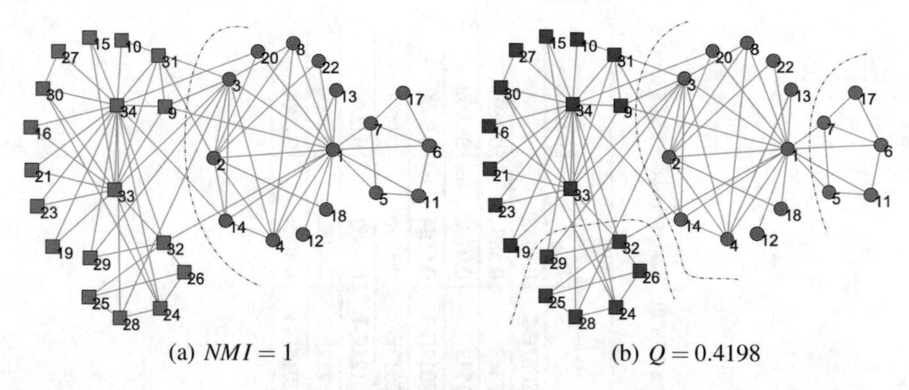

Fig. 2.25 Community structure of the Karate network. **a** Ground truth. **b** Detected structure

the two independent samples (the two modularity sets) come from distributions with equal means. Tables 2.9 and 2.10 show the experimental results and the hypothesis test p-values.

From the two tables, it is can noted that GDPSO performs remarkably well in terms of the modularity values and the computational time. The small p-values indicate that GDPSO is substantially better than the compared methods. GDPSO framework possesses excellent global exploration ability. On the four small-scale networks, GDPSO outperforms the other methods, except for the GA. The GA optimizes the same objective as does the GDPSO. From the experiments, it is shown that the GA and the GDPSO perform well, but the GDPSO converges faster. In the next step, the community structures discovered by the GDPSO on the four small-scale networks will be analyzed.

The Karate network is a network of relations between 34 members of a karate club. Figure 2.25 shows the real community structure and the detected structure of the network. It can be observed that GDPSO has discovered four clusters, which are the subdivisions of the real ones. This phenomenon also occurs in the dolphin network. Figure 2.26 clearly displays the discovered structure of the dolphin network. GDPSO divides one of the two clusters in the real structure into four smaller sections. It cannot determine if this division makes sense to the dolphins; from the perspective of optimization, GDPSO simply finds the best objective function value.

In football network, although GDPSO obtains the largest Q value, it could find that several vertices, such as the numbers 29, 43, 37, 81, 60, 91 and 63, are misclassified. This is mainly caused by the nuances in the scheduling of games.

Figure 2.27 exhibits the discovered communities in SFI network. From the figure, it could notice that the GDPSO splits the network into eight strong communities, with divisions running principally along disciplinary lines. The sub-communities that are subdivisions of the original three large-scale groups are centered around the interests of leading members.

Experiments on the four small-scale social networks indicate that the proposed GDPSO algorithm is effective. The algorithm exhibits an outstanding search abil-

Table 2.9 Experimental results on the Karate and dolphin networks. The parameter k used in k-means and Ncut is set to 4 for the Karate network and 5 for the dolphin network

Network	Karate					Dolphin				
Index	Q_{max}	Q_{avg}	T_{avg}	I_{avg}	p-value	Q_{max}	Q_{avg}	T_{avg}	I_{avg}	p-value
GDPSO	0.4198	0.4198	2.5800E-2	0.6881	×	0.5285	0.5284	5.8621E-2	0.8935	×
GA	0.4198	0.4198	1.9871E-1	0.6873	0.5000	0.5285	0.5280	2.3868	0.5892	0.1591
MOGA	0.4198	0.4160	1.3975	1	0.0000	0.5085	0.4098	3.1713	0.9442	0.0000
LPA	0.4151	0.3264	3.3133E-3	0.6623	0.0000	0.5258	0.4964	3.8011E-3	0.6194	0.0000
CNM	0.3800	0.3800	5.0700E-2	0.6920	0.0000	0.4950	0.4950	2.2517E-2	0.5730	0.0000
Informap	0.4020	0.4020	2.1331E-1	0.6995	0.0000	0.5247	0.5247	3.7121E-1	0.4662	0.0000
K-means	0.1429	0.0351	2.6121E-2	0.6059	0.0000	0.4796	0.3787	4.5717E-2	0.4282	0.0000
Ncut	0.4198	0.4198	4.2272E-3	0.6873	0.5000	0.5068	0.5047	8.2015E-3	0.5084	0.0000

Table 2.10 The experimental results on the football and SFI networks. The parameter k used in k-means and in Ncut is set to 12 for the football network and to 7 for the SFI network

Network	Football					SFI			
Index	Q_{max}	Q_{avg}	T_{avg}	I_{avg}	p-value	Q_{max}	Q_{avg}	T_{avg}	p-value
GDPSO	0.6046	0.6041	9.4210E-2	0.8889	×	0.7506	0.7449	4.5217E-2	×
GA	0.5929	0.5829	4.9276	0.8227	0.0000	0.7506	0.7505	6.8536	1.0000
MOGA	0.5280	0.5173	4.1731	0.7883	0.0000	0.7430	0.7323	1.9742	0.0000
LPA	0.6030	0.5848	8.2001E-3	0.8735	0.0000	0.7341	0.7095	5.3731E-3	0.0000
CNM	0.5770	0.5770	1.1292E-1	0.7620	0.0000	0.7335	0.7335	4.5661E-2	0.0000
Informap	0.6005	0.6005	1.2133	0.9242	0.0000	0.7334	0.7334	7.7563E-1	0.0000
k-means	0.5783	0.5130	5.4805E-2	0.8512	0.0000	0.4376	0.2780	3.4512E-2	0.0000
Ncut	0.6031	0.6007	8.8133E-3	0.9233	0.0000	0.7478	0.7470	9.5000E-3	0.9990

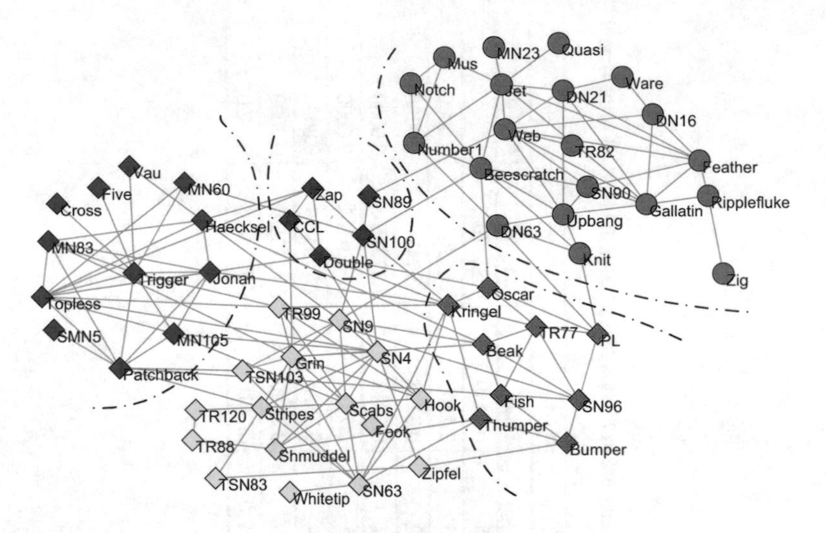

Fig. 2.26 Our detected community structure of the dolphin network ($Q = 0.5285$)

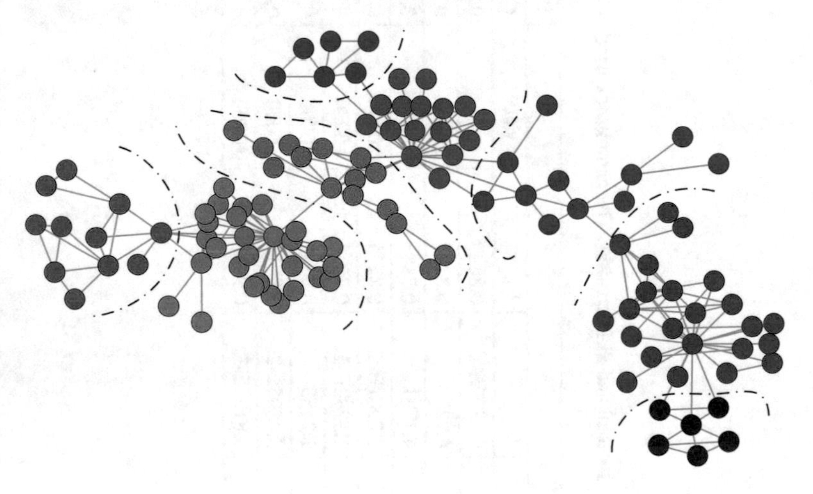

Fig. 2.27 Community structure of the SFI network ($Q = 0.7506$)

ity when addressing moderate-scale optimization problems. To further verify its optimization ability, next the algorithm is performed on four large-scale networks. Tables 2.11 and 2.12 list the statistical results.

The rank sum test results indicate that GDPSO is superior to the GA, MOGA, LPA, and k-means algorithm for the four large networks and that it outperforms Informap for the four big networks, except for the e-mail network. The CNM and Informap methods are two deterministic methods. For the four large networks, the CNM seems to perform the best from the angle of modularity, LPA is the fastest, and the k-means algorithm performs poorly. The MOGA, k-means and Ncut methods can hardly handle large-scale networks. For the PGP network, the MOGA cannot provide output after four hours, and the k-means and Ncut methods run out of computer memory. Through GDPSO, good modularity values could be found within a reasonable amount of time. From the tables, it can be observed that our method still falls into local optimum solutions; for example, for the netscience and power grid networks, the modularity values obtained by the GDPSO are smaller than those obtained by the CNM. One reason that can account for this is the designed greedy mechanism-based particle-position-update principle possibly causing prematurity. In the next subsection, the particle-position-update principle will further be discussed.

2.4.6 Additional Discussion on GDPSO

For the proposed GDPSO algorithm, the population size and the maximum number of iterations affect the performance of the algorithm. Tables 2.13 and 2.14 show the influence of these two parameters on the performance of the algorithm.

The two parameters *gmax* and *popsize* are normally set empirically. Small values of the two parameters may not result in convergence, whereas large values will require substantial amounts of computing time. From the above experiments, as shown in Tables 2.13 and in 2.14, it could observe that the impact of the two parameters on the performance of the algorithm is not salient. To obtain a tradeoff between the convergence and the computation time, $gmax = popsize = 100$ is set as the default configuration of GDPSO.

The experiments on small networks demonstrate the effectiveness of GDPSO algorithm; however, its performance on large networks remains unsatisfactory. There are two key reasons why the GDPSO obtains unsatisfactory results. On the one hand, for the large scale of networks clustering problem, the diversity preservation is insufficient. However, it needs to find a better strategy that can both preserve diversity and ensure fast convergence. On the other hand, the designed particle-position-update principle (Eq. 2.11) is based on a simple greedy mechanism, which may lead to prematurity.

Equation 2.11 is the key component of GDPSO. It updates the label identifier of a vertex with the neighbor identifier that generates the largest increase in the objective function value. From the perspective of graph theory, it is natural to update the vertex identifier with the two different methods shown in Fig. 2.28a.

The two different label-identifier-update principles shown in Fig. 2.28b, c make sense from the perspective of sociology. For example, the maxD principle is in accordance with the social phenomenon whereby people prefer to learn from or simply imitate the one who is the most attractive and talented amongst their friends, and the dominated principle complies with the social phenomenon that one would like to follow the state that is kept by the majority of his or her friends. Table 2.15 shows a comparison of the three update principles.

From the table, it can observe that the maxD and the dominated principles are computationally faster than that of ΔQ, and the maxD principle is the fastest. How-

Table 2.11 The experimental results for the e-mail and netscience networks. The parameter k used in k-means and Ncut is set to 4 for the e-mail network and to 100 for the netscience network

Network	E-mail				Netscience			
Index	Q_{max}	Q_{avg}	T_{avg}	p-value	Q_{max}	Q_{avg}	T_{avg}	p-value
GDPSO	0.5487	0.4783	2.4717E+1	×	0.9540	0.9512	3.5212E+1	×
GA	0.3647	0.3500	3.1015E+2	0.0000	0.9086	0.9003	1.9967E+2	0.0000
MOGA	0.3283	0.3037	5.3716E+2	0.0000	0.8916	0.8810	7.0681E+1	0.0000
LPA	0.2055	0.0070	6.5033E-2	0.0000	0.9266	0.9202	3.9117E-2	0.0000
CNM	0.4985	0.4985	1.5255E+1	1.0000	0.9555	0.9555	2.0414E+1	1.0000
Informap	0.5355	0.5355	2.3200	1.0000	0.9252	0.9252	1.5577	0.0000
K-means	0.3681	0.3600	5.2580	0.0000	0.6510	0.6359	5.6187E+1	0.0000
Ncut	0.4841	0.4749	4.7000E-2	0.2384	0.9293	0.9268	3.9302E-1	0.0000

Table 2.12 The experimental results for the power grid and PGP networks. The parameter k used in k-means and Ncut is set to 200 for the power grid network and to 300 for the PGP network

Network	Power grid				PGP			
Index	Q_{max}	Q_{avg}	T_{avg}	p-value	Q_{max}	Q_{avg}	T_{avg}	p-value
GDPSO	0.8382	0.8368	4.7818E+2	×	0.8050	0.8013	6.3609E+2	×
GA	0.7161	0.7124	3.4485E+3	0.0000	0.6576	0.6473	3.7798E+4	0.0000
MOGA	0.7035	0.6949	4.9177E+3	0.0000	—	—	—	—
LPA	0.7602	0.7476	1.7373E-1	0.0000	0.7949	0.7845	7.0391E-1	0.0000
CNM	0.9229	0.9229	4.9121E+2	1.0000	0.8481	0.8481	7.8562E+3	1.0000
Informap	0.8140	0.8140	6.3715	0.0000	0.7777	0.7777	1.2313E+1	0.0000
K-means	Out of Memory			×	Out of Memory			×
Ncut	0.8875	0.8866	4.4442	1.0000	Out of Memory			×

Table 2.13 The influence of the population size on the performance of the algorithm. The modularity values over 30 independent runs are recorded in this table

($Gmax = 100$)	$Popsize = 20$		$Popsize = 60$		$Popsize = 100$		$Popsize = 140$		Popsize $= 180$	
	Q_{max}	Q_{avg}	Q_{max}	Q_{avg}	Q_{max}	Q_{avg}	Q_{max}	Q_{avg}	Q_{max}	Q_{avg}
Karate	0.4198	0.4085	0.4198	0.4174	0.4198	0.4198	0.4198	0.4198	0.4198	0.4198
Dolphin	0.5269	0.5265	0.5277	0.5265	0.5285	0.5284	0.5285	0.5284	0.5285	0.5285
Football	0.6046	0.6027	0.6046	0.6035	0.6046	0.6041	0.6046	0.6041	0.6046	0.6043
SFI	0.7484	0.7408	0.7484	0.7439	0.7506	0.7449	0.7506	0.7456	0.7506	0.7453
E-mail	0.5361	0.4162	0.5384	0.4632	0.5487	0.4783	0.5476	0.4819	0.5437	0.4864
Netscience	0.9323	0.9275	0.9535	0.9492	0.9540	0.9512	0.9537	0.9507	0.9547	0.9511
Power grid	0.7685	0.7627	0.8377	0.8311	0.8382	0.8368	0.8383	0.8357	0.8401	0.8372
PGP	0.8028	0.7974	0.8047	0.8002	0.8050	0.8013	0.8047	0.8011	0.8053	0.8017

Table 2.14 The influence of the maximum iteration number *gmax* on the performance of the algorithm. The modularity values over 30 independent runs are recorded in this table

($Popsize = 100$)	$Gmax = 50$		$Gmax = 100$		$Gmax = 150$		$Gmax = 200$		$Gmax = 250$	
	Q_{max}	Q_{avg}	Q_{max}	Q_{avg}	Q_{max}	Q_{avg}	Q_{max}	Q_{avg}	Q_{max}	Q_{avg}
Karate	0.4198	0.4180	0.4198	0.4198	0.4198	0.4198	0.4198	0.4198	0.4198	0.4198
Dolphin	0.5277	0.5267	0.5285	0.5284	0.5285	0.5285	0.5285	0.5285	0.5285	0.5285
Football	0.6046	0.6040	0.6046	0.6041	0.6046	0.6037	0.6046	0.6039	0.6046	0.6043
SFI	0.7487	0.7447	0.7506	0.7449	0.7506	0.7454	0.7506	0.7449	0.7506	0.7441
E-mail	0.5157	0.4731	0.5487	0.4783	0.5237	0.4687	0.5431	0.4864	0.5454	0.4997
Netscience	0.9531	0.9482	0.9540	0.9512	0.9537	0.9507	0.9547	0.9511	0.9507	0.9491
Power grid	0.8365	0.8331	0.8382	0.8368	0.8383	0.8357	0.8401	0.8372	0.8404	0.8380
PGP	0.8041	0.8012	0.8050	0.8013	0.8050	0.7991	0.8053	0.8031	0.8053	0.8008

Table 2.15 Comparison of different particle-position-update principles

Algorithm index	GDPSO (ΔQ)			GDPSO (maxD)				GDPSO (dominated)			
	Q_{max}	Q_{avg}	T_{avg}	Q_{max}	Q_{avg}	T_{avg}	p-value	Q_{max}	Q_{avg}	T_{avg}	p -value
Karate	0.4198	0.4198	2.5800E-2	0.4156	0.4080	1.7724E-3	0.0000	0.4156	0.4089	3.3976E-3	0.0000
Dolphin	0.5285	0.5284	5.8621E-2	0.5268	0.5265	4.4154E-3	0.0000	0.5268	0.5265	2.8781E-2	0.0000
Football	0.6046	0.6041	9.4210E-2	0.6046	0.6038	1.0939E-2	0.3349	0.6046	0.6038	3.6844E-2	0.5000
SFI	0.7506	0.7449	4.5217E-2	0.7484	0.7433	8.7936E-3	0.0153	0.7497	0.7458	2.3563E-1	0.8812
E-mail	0.5487	0.4783	2.4717E+1	0.4941	0.3108	4.3074E-1	0.0000	0.3670	0.2453	1.1583	0.0000
Netscience	0.9540	0.9512	3.5212E+1	0.9522	0.9478	2.2385	0.0000	0.9334	0.9296	1.2988E+1	0.0000
Power grid	0.8382	0.8368	4.7818E+2	0.8382	0.8368	4.5868E+1	0.5000	0.8145	0.8116	2.0348E+2	0.0000
PGP	0.8050	0.8013	6.3609E+2	0.8021	0.7977	1.7526E+1	0.0000	0.8145	0.8093	1.0025E+3	1.0000

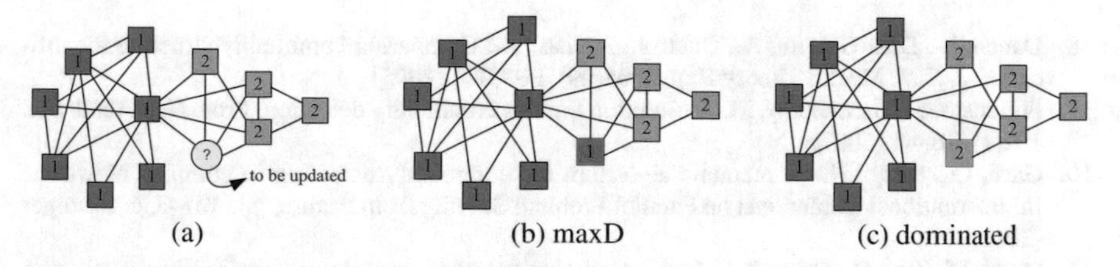

Fig. 2.28 **a** Different vertex-label-identifier-update principles. **b** Choose the identifier of the neighboring vertex that has the largest degree. **c** Choose the dominated identifier from the neighbor vertices

ever, from the modularity index perspective, the maxD and the dominated principles all fall into local optima solutions. The rank sum test results indicate that the ΔQ mechanism works better than do the other two particle-position-update principles.

2.4.7 Conclusions

In this section, a discrete PSO method designed to discover community structures in social networks was introduced. In GDPSO, first the particle position and velocity are redefined in a discrete form and subsequently redesigned the particle-update rules based on the network topology; consequently, a discrete PSO framework was established. When applying GDPSO to solve the network clustering problem, because the scale of a real-world social network is especially large most of the time, to alleviate prematurity, a greedy local-search-based mechanism was specially designed for the particle-position-update rule. Experiments on both synthetic and real-world networks demonstrated that the proposed algorithm for network clustering is effective and promising.

References

1. Bagrow, J.P., Bollt, E.M.: Local method for detecting communities. Phys. Rev. E **72**(4), 046,108 (2005)
2. Barber, M.J., Clark, J.W.: Detecting network communities by propagating labels under constraints. Phys. Rev. E **80**(2), 026,129 (2009)
3. Blondel, V.D., Guillaume, J.L., Lambiotte, R., Lefebvre, E.: Fast unfolding of communities in large networks. J. Stat. Mech.: Theory Exp. **2008**(10), P10,008 (2008)
4. Blum, C., Puchinger, J., Raidl, G.R., Roli, A.: Hybrid metaheuristics in combinatorial optimization: a survey. Appl. Soft Comp. **11**(6), 4135–4151 (2011)
5. Cai, Q., Gong, M., Ma, L., Ruan, S., Yuan, F., Jiao, L.: Greedy discrete particle swarm optimization for large-scale social network clustering. Inf. Sci. **316**, 503–516 (2015)
6. Clauset, A., Newman, M.E., Moore, C.: Finding community structure in very large networks. Phys. Rev. E **70**(6), 066,111 (2004)
7. Cormen, T.H., Leiserson, C.E., Rivest, R.L.: Clifford stein. Introduction to algorithms (2001)

8. Danon, L., Diaz-Guilera, A., Duch, J., Arenas, A.: Comparing community structure identification. J. Stat. Mech.: Theory Exp. **2005**(09), P09,008 (2005)
9. Fortunato, S., Barthelemy, M.: Resolution limit in community detection. Proc. Nat. Acad. Sci. **104**(1), 36–41 (2007)
10. Gach, O., Hao, J.K.: A memetic algorithm for community detection in complex networks. In: International Conference on Parallel Problem Solving from Nature, pp. 327–336. Springer (2012)
11. Gong, M., Cai, Q., Chen, X., Ma, L.: Complex network clustering by multiobjective discrete particle swarm optimization based on decomposition. IEEE Trans. Evol. Comput. **18**(1), 82–97 (2014)
12. Gong, M., Cai, Q., Li, Y., Ma, J.: An improved memetic algorithm for community detection in complex networks. In: 2012 IEEE Congress on Evolutionary Computation (CEC), pp. 1–8. IEEE (2012)
13. Gong, M., Fu, B., Jiao, L., Du, H.: Memetic algorithm for community detection in networks. Phys. Rev. E **84**(5), 056,101 (2011)
14. Gong, M., Ma, L., Zhang, Q., Jiao, L.: Community detection in networks by using multiobjective evolutionary algorithm with decomposition. Physica A: Stat. Mech. App. **391**(15), 4050–4060 (2012)
15. Jain, B.J., Obermayer, K.: Elkans k-means algorithm for graphs. In: Mexican International Conference on Artificial Intelligence, pp. 22–32 (2010)
16. Kennedy, J.: Particle swarm optimization. In: Encyclopedia of Machine Learning, pp. 760–766. Springer (2011)
17. Li, Z., Zhang, S., Wang, R.S., Zhang, X.S., Chen, L.: Quantitative function for community detection. Phys. Rev. E **77**(3), 036,109 (March 2008)
18. Ma, L., Gong, M., Liu, J., Cai, Q., Jiao, L.: Multi-level learning based memetic algorithm for community detection. Appl. Soft Comput. **19**, 121–133 (2014)
19. Newman, M.E., Girvan, M.: Finding and evaluating community structure in networks. Phys. Rev. E **69**(2), 026,113 (2004)
20. Ong, Y.S., Lim, M.H., Chen, X.: Memetic computationłpast, present & future [research frontier]. IEEE Comput. Intell. Mag. **5**(2), 24–31 (2010)
21. Pizzuti, C.: Ga-net: a genetic algorithm for community detection in social networks. In: International Conference on Parallel Problem Solving from Nature, pp. 1081–1090 (2008)
22. Pizzuti, C.: A multiobjective genetic algorithm to find communities in complex networks. IEEE Trans. Evol. Comput. **16**(3), 418–430 (2012)
23. Rosvall, M., Axelsson, D., Bergstrom, C.T.: The map equation. Eur. Phys. J. Spec. Top. **178**(1), 13–23 (2009)
24. Rosvall, M., Bergstrom, C.T.: Maps of random walks on complex networks reveal community structure. Proc. Nat. Acad. Sci. **105**(4), 1118–1123 (2008)
25. Shang, R., Bai, J., Jiao, L., Jin, C.: Community detection based on modularity and an improved genetic algorithm. Phys. A: Stat. Mech. Appl. **392**(5), 1215–1231 (2013)
26. Shi, J., Malik, J.: Normalized cuts and image segmentation. IEEE Trans. Pattern Anal. Mach. Intell. **22**(8), 888–905 (2000)

Chapter 3
Network Community Discovery with Evolutionary Multi-objective Optimization

Abstract As described in the previous chapters, the community discovery problems can be formulated as single-objective optimization problems. But it is difficult for single-objective optimization algorithms to reveal community structures at multiple resolution levels. The multi-resolution communities can effectively reflect the hierarchical structures of complex networks. In this chapter, we model the multi-resolution community detection problems as multi-objective optimization problems. And thereafter, we use four different evolutionary multi-objective algorithm for solving the multi-resolution community detection based multi-objective optimization problems. Among the four algorithms, three algorithms adopt the framework of MOEA/D, MODPSO, and NNIA to detect multi-resolution communities in undirected and static networks, and an algorithm uses the framework of MOEA/D to detect multi-resolution communities in dynamic networks.

3.1 Review on the State of the Art

As shown in Chap. 2, most community discovery problems can be modeled as single-objective optimization problems through optimizing the objective functions, such as modularity or modularity density. In the absence of information on the community size of a network, a method should be able to discover community structures at different resolution scales, to make sure that it will eventually identify the right communities. In addition, one single fixed community partition returned by the single-objective algorithms may not be suitable for the networks with multiple potential structures (e.g., hierarchical and overlapping structures).

Multi-resolution community discovery problem can naturally be formulated as a multi-objective optimization problem. It always can find some network partitions corresponding to a trade-off between different objectives. There are also some multi-objective community discovery algorithms [15–17, 26, 29, 30]. In [26], Pizzuti proposed a multi-objective genetic algorithm to uncover community structure in complex networks. The algorithm adopts the framework of NSGA-II and optimizes two objective functions able to identify densely connected groups of nodes having sparse interconnections. One of the objectives is Community Score, the other

M. Gong et al., *Computational Intelligence for Network Structure Analytics*,
DOI 10.1007/978-981-10-4558-5_3

objective is Community Fitness. In [30], Shi et al. also designed an MOEA for community detection and proposed two strategies for selecting solution from the final trade-off solutions. The method uses an effective EMO algorithm PESA-II and optimizes two conflicting complementary objectives, which are decomposed from modularity. Besides, two model selection methods were proposed, which were used to select solutions from the trade-off Pareto solutions. In [15], a novel multi-objective community detection algorithm based on a multi-objective evolutionary algorithm with decomposition was proposed. This algorithm aims to simultaneously optimize the two new contradictory objectives, negative ratio association [1] and ratio cut [32]. In [16], a multi-objective evolutionary algorithm for revealing multi-resolution community structures was proposed. This algorithm also simultaneously optimizes negative ratio association and ratio cut. A discrete framework of the particle swarm optimization algorithm was proposed in [17].

There has been a large body of works on analyzing communities in static social networks, but only a few studies examined the dynamics of communities in evolving social networks. Folino et al. proposed a dynamic multi-objective genetic algorithm (DYN-MOGA) to discover communities in dynamic networks by employing genetic algorithm [8]. Gong et al. proposed a dynamic community discovery algorithm based on NNIA [15].

This chapter is organized into five sections. Section 3.2 introduces a basic multi-objective evolutionary algorithm for community discovery. Section 3.3 introduces a multi-objective algorithm to identify multi-resolution network structures. An efficient multi-objective discrete particle swarm optimization for multi-resolution community discovery is introduced in Sect. 3.4. Section 3.5 introduces a multi-objective algorithm for community discovery in dynamic networks.

3.2 A Decomposition Based Multi-objective Evolutionary Algorithm for Multi-resolution Community Discovery

[1]This section introduces a community discovery algorithm based on a multi-objective evolutionary algorithm with decomposition, termed as MOEA/D-Net. MOEA/D-Net aims to simultaneously optimize the two new contradictory objectives, negative ratio association [1] and ratio cut [32].

To optimize these two objective functions, the multi-objective evolutionary algorithm based on decomposition (MOEA/D) proposed by Zhang and Li [34] is used. MOEA/D decomposes a multi-objective optimization problem into a number of scalar optimization subproblems and optimizes them simultaneously by evolving a population of solutions. At each generation, the population is composed of the best

[1]Acknowledgement: Reprinted from Physica A: Statistical Mechanics and its Applications, 391(15), Gong, M., Ma, L., Zhang, Q., Jiao, L., Community detection in networks by using multi-objective evolutionary algorithm with decomposition, 4050–4060, Copyright(2012), with permission from Elsevier.

solution found so far (i.e., since the start of the run of the algorithm) for each subproblem. The neighborhood relations among these subproblems are defined based on the distances between their aggregation coefficient vectors. The optimal solutions to two neighboring subproblems should be very similar. Each subproblem (i.e., scalar aggregation function) is optimized in MOEA/D by using information from its neighboring subproblems. There are several methods for constructing aggregation functions. The most popular ones among them include the weighted sum approach and Tchebycheff approach.

The advantages of MOEA/D-Net are as follows. MOEA/D-Net aims to simultaneously optimize the two new contradictory objectives, negative ratio association and ratio cut. The negative ratio association and the ratio cut have the potential to balance each other's tendency to increase or decrease the number of communities, and that both of the two objectives are related to the density of subgraphs to overcome the resolution limit. MOEA/D-Net selectively explores the search space without the need to know in advance the exact number of communities, and returns just not a single partitioning of the network, but a set of solutions. Each of these solutions corresponds to a different trade-off between the two objectives and thus to diverse partitioning of the network consisting of a various number of clusters.

3.2.1 Multi-objective Evolutionary Algorithm for Community Discovery

First, a summary description of MOEA/D-Net algorithm for community discovery is given. MOEA/D-Net aims to optimize negative ratio association [1] and ratio cut [32]. Then, multi-resolution community discovery is formulated as a two-objective optimization problem. MOEA/D-Net first decomposes multi-resolution community discovery into a number of scalar objective optimization subproblems. In MOEA/D-Net, Tchebycheff approach is used for constructing aggregation functions. The scalar optimization problem is in the form

$$\begin{aligned} \min \quad & g^{te}(x|\lambda, z^*) = \max_{1 \le i \le m} \lambda_i |f_i(x) - z_i^*| \\ Subject\ \ to \quad & x \in \Omega \end{aligned} \tag{3.1}$$

where $z^* = (z_1^*, z_2^*, \cdots, z_m^*)$ is the reference point $z_i^* = \{\min f_i(x) | x \in \Omega\}$, for each $i = 1, 2, \cdots, m$. For each nondominated point x^*, there exist a weight vector λ such that x^* is the optimal solution of (3.1) and each optimal solution of (3.1) is a nondominated solution. Therefore, one is able to obtain different nondominated solutions by altering the weight vector.

The general framework of MOEA/D-Net is as follows.

Algorithm 10 Framework of the proposed MOEA/D-Net.

Input: the number of decomposed sub-problems: N_p, crossover probability: p_c, mutation probability: p_m, the size of neighborhood: N_q, the update size: N_u and the number of generations: g_{max}.

Output: Pareto front solutions. Each solution corresponds to a partition of a network.

Step 1) Initialization: Generate an initial population, $x = \{x_1, x_2, ..., x_{N_p}\}$; set $z_i^* = 10^6$ where $i = 1, 2$; set $t \longleftarrow 1$.

Step 2) for $g^{te}(x|\lambda_i, z^*), i = 1, 2, ..., N_p$, **do**

Step 2.1) Genetic operators: Randomly select two subproblems x_j and x_k from the neighbors of x_i, and then generate two new solutions x'_j and x'_k from x_j and x_k by using genetic operators.

Step 2.2) Select update solution: select the solution with the minimum value of g_i^{te} as x_u.

Step 2.3) Update of reference point z^*: for each $j = 1, 2$, if $z_j^* > f_j(x_u)$, $z_j^* = f_j(x_u)$.

Step 2.4) Update of neighboring solutions: Select N_u subproblems from the neighboring subproblems of x_i, where $g_j^{te}(x_j|\lambda_j, z^*) > g_j^{te}(x_u|\lambda_j, z^*)$, and then update their solutions as x_u.

Step 3) Stopping criteria: If $t < g_{max}$, then $t++$ and go to **Step 2**, otherwise, stop the algorithm.

3.2.2 *Problem Formation*

An undirected graph can be given $G = (V, E)$ with $|V| = n$ vertexes and $|E| = m$ edges. The adjacency matrix is A. If V_1 and V_2 are two disjoint subsets of V, $L(V_1, V_2) = \sum_{i \in V_1, j \in V_2} A_{ij}$ and $L(V_1, \overline{V_1}) = \sum_{i \in V_1, j \in \overline{V_2}} A_{ij}$. Given a partition $S = (V_1, V_2, \cdots, V_k)$ of the graph, where V_i is the vertex set of subgraph G_i for $i = 1, 2, \cdots, k$, the modularity density is defined as

$$D = \sum_{i=1}^{k} \frac{L(V_i, V_i) - L(V_i, \overline{V_i})}{|V_i|} \tag{3.2}$$

In this equation, each summand means the ratio between the difference of the internal and external degrees of the subgraph G_i and the size of the subgraph. The first term of D is equivalent to the ratio association [1] and the second term is equivalent to the ratio cut [32]. The larger the value D, the more accurate a partition is. To maximize the modularity density D, it needs to maximize the first term and minimize the second term. Generally, maximizing the ratio association often divides a network into small communities with high densely interconnected [1], while minimizing the ratio cut often divides a network into large communities with sparsely connected with the rest. Therefore, these two complementary terms reflect two fundamental aspects of a good partition, and the modularity density is an intrinsic trade-off between these two objectives.

In MOEA/D-Net, to formulate the problem as a minimum optimization problem, the first term is revised. The two criteria NRA and RC can be described as two contradictory multi-objective functions. The corresponding multi-objective optimization problem can be described as

$$\begin{cases} \min \quad f_1 = NRA = -\sum_{i=1}^{k} \frac{L(V_i, V_i)}{|V_i|} \\ \min \quad f_2 = RC = \sum_{i=1}^{k} \frac{L(V_i, \overline{V_i})}{|V_i|}. \end{cases} \tag{3.3}$$

3.2.3 Representation and Initialization

MOEA/D-Net adopts the locus-based adjacency representation [19]. In this graph-based representation, each individual g of the population consists of N genes $g_1, g_2, \cdots, g_N$ and each g_i can take allele values j in the range $1, 2, \cdots, N$. Genes and alleles represent nodes of the graph $G = (V, E)$ modeling a network. Thus, a value of j assigned to the ith gene is then interpreted as a link between the node i and j. This means that in the resulting clustering solution, they will be in the same cluster. The decoding of this representation requires the identification of all connected components. All nodes belonging to the same connected component are then assigned to one cluster. This decoding step can be performed in linear time as observed in [19]. A main advantage of this representation is that there is no need to fix the number of clusters in advance, as it is automatically determined in the decoding step. Figure 3.1 illustrates the locus-based adjacency scheme for a network of 7 nodes.

In order to avoid uninteresting divisions containing unconnected nodes, the initialization process takes into account the effective connections of the network. For each individual, the allele value j assigned to the ith gene is randomly selected from the neighbors of node i. This initialization process improves the convergence of the algorithm because the space of the possible solutions is restricted.

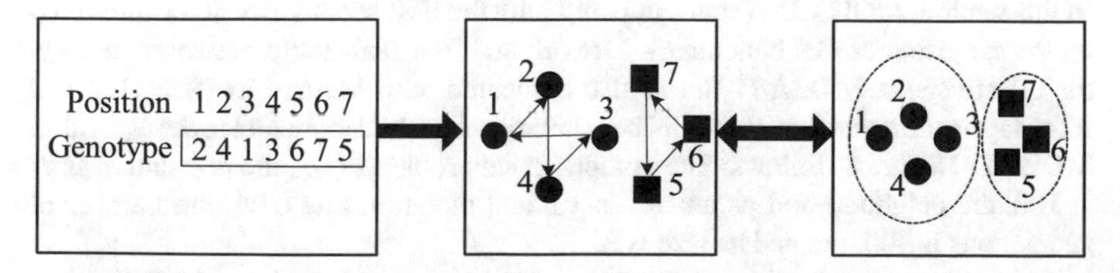

Fig. 3.1 Illustration of the locus-based adjacency scheme. *Left* One possible genotype. *Middle* Translate the genotype into the graph structure (the graph is shown as directed only to aid in understanding how it originates from the genotype). *Right* The final clusters (every connected component is interpreted as an individual cluster)

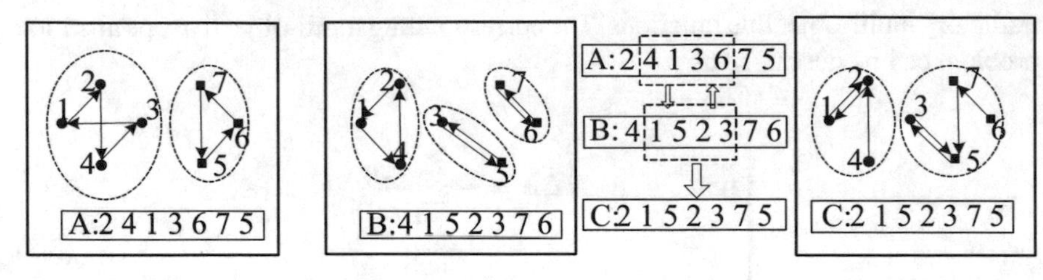

Fig. 3.2 A and B are two parent genotypes and their corresponding graph structures. A random two-point crossover of the genotypes yields the child C, which has inherited much of its structure from its parents, but differs from both of them

3.2.4 Genetic Operators

Crossover. The two-point crossover in favor of uniform crossover is adopted because the two-point crossover can better maintain the effective connections of the nodes in the network. Given two parents A and B, first randomly select two points i and j (i.e., $1 \le i \le j \le N$), and then everything between the two points is swapped between the parents (i.e., $A_k \leftrightarrow B_k, \forall k \in \{k|i \le k \le j\}$). An example of the operation of two-point crossover on the encoding employed is shown in Fig. 3.2.

Mutation. In this process, randomly pick a chromosome C to be mutated. Then, employ one-point neighbor-based mutation on this chromosome: a gene i is selected randomly on the chromosome, then the possible values of its allele are restricted to the neighbors of gene i (i.e., $C_i \leftarrow j, j \in \{j|a_{ij} = 1\}$). The neighbor-based mutation guarantees that, in a mutated child, each node is linked only with one of its neighbors. This can avoid the useless exploration of the search space because of the same above observation in the process of initialization.

3.2.5 Experimental Results

In this section, MOEA/D-Net is compared with the Fast Modularity algorithm (FM) on the extension of GN benchmark network and four real-world networks to show the effectiveness. MOEA/D-Net is also compared with Meme-Net (Sect. 2.2) and InfoMap on the extension of the GN benchmark network. Parameters in the algorithm MOEA/D-Net are as follows: The number of subproblems (i.e., the population size) is 100, the neighborhood parameter is 10, and mutation rate 0.06, the number of generations is 400, the update size is 2.

In order to evaluate the quality of the partitioning obtained, in the following, Normalized Mutual Information (NMI) in Eq. 2.3 and modularity (Q) in Eq. 1.6 are used.

3.2.5.1 Experimental Results on Artificial Generated Network

In this section, MOEA/D-Net is tested on the computer-generated networks, which has a known community structure, to illustrate the proposed algorithm can recognize and discover its community structure. Then, a comparison between the results obtained by MOEA/D-Net and that obtained by Meme-Net is made, which shows the multi-objective algorithm based on these two components of modularity density has better performances than the single-objective algorithms based on the modularity density optimization. MOEA/D-Net is also compared with FM and InfoMap algorithms to verify the effectiveness of MOEA/D-Net.

The network used here is the extension of GN benchmark network described in Chap. 1. For each network, NMI over 30 independent runs is computed. Figure 3.3 shows the average NMI obtained by MOEA/D-Net, Meme-Net, FM and InfoMap algorithms. As is shown in Fig. 3.3, when the value of mixing parameter is small ($\mu <= 0.35$) which means the fuzziness of the community in the network is low, MOEA/D-Net and InfoMap algorithm find the true partition correctly (NMI equals 1). When the mixing parameter increases, those algorithms are more difficult to detect the true partition, but the detected partition by MOEA/D-Net is the most close to the true one (NMI is 0.9919, 0.7911, and 0.4005 when $\mu = 0.40$, $\mu = 0.45$ and $\mu = 0.50$, respectively). Therefore, MOEA/D-Net obtains higher NMI value than Meme-Net, FM and InfoMap obtained, corresponding to the proposed method MOEA/D-Net has better performances than Meme-Net, FM and InfoMap on the extension of GN Benchmark network.

This experiment shows MOEA/D-Net is a high valid method for revealing community structure on the extension of GN Benchmark.

3.2.5.2 Experimental Results on Real-World Networks

In this section, MOEA/D-Net is tested on four real-world networks: karate, dolphins, football, and the polbooks.

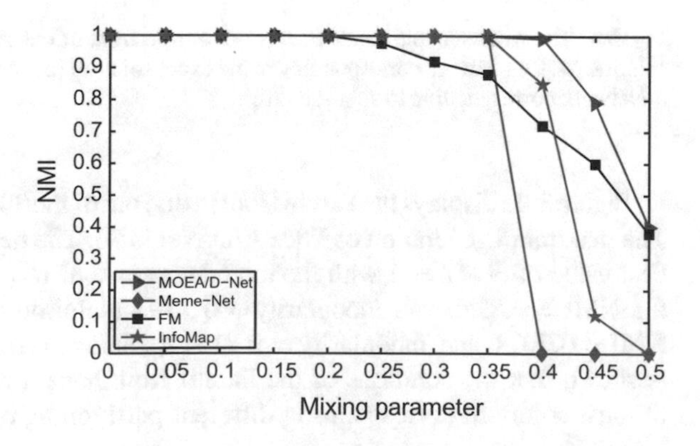

Fig. 3.3 *NMI* obtained by MOEA/D-Net, Meme-Net, FM and InfoMap on the extension of the classical GN Benchmark

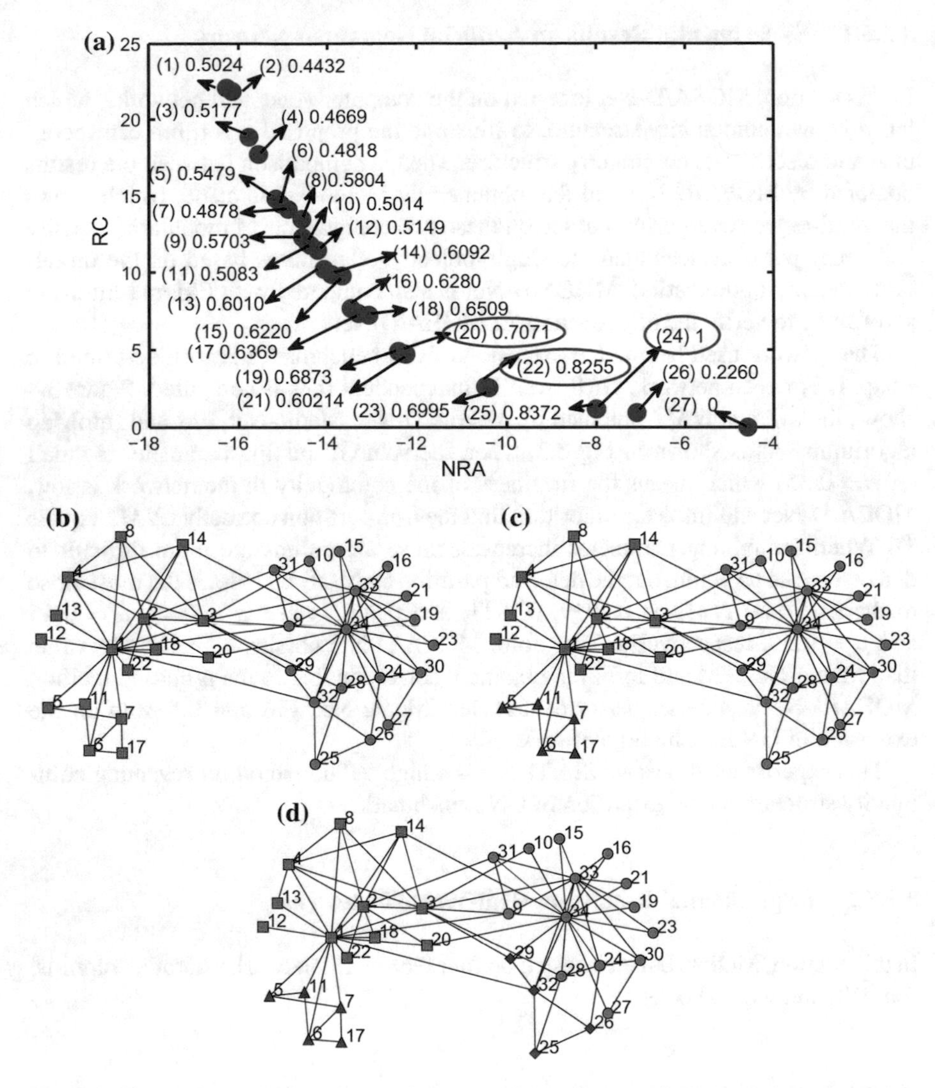

Fig. 3.4 The results on the karate network. **a** Pareto front of one run. **b** Network corresponding to solution (24). **c** Network corresponding to the exact solution (node number (22) on the Pareto front. **d** Network corresponding to solution (20)

Figure 3.4a displays the Pareto front in one out of the 30 runs on the karate network. The maximum generation of MOEA/D-Net is 50. The network corresponding to the best value of $NMI = 1$ with the modularity = 0.3715 (solution (24)), the one with the NMI = 0.8255 and modularity = 0.3391 (solution (22)) and the one with the NMI = 0.7071 and modularity = 0.4151 (solution (20)) are shown in Fig. 3.4. It is shown that the solutions of the Pareto front have a hierarchical structure. Each of these solutions corresponds to different partitioning of the network consisting of various clusters. The true partitioning, which is displayed in Fig. 3.4b, consists of two modules obtained by the split of the two main groups. It is shown in Fig. 3.4c that the left subgraph is divided into two smaller ones and in Fig. 3.4d that both the subgraphs are divided into two smaller ones, respectively. Figure 3.4 shows that MOEA/D-Net can produce a set of solutions which represent different divisions to the karate network at different hierarchical levels. The number of subdivisions is automatically determined by the nondominated individuals resulting from MOEA/D-Net.

In the following, for each network, MOEA/D-Net is run 30 times, and the average value of best NMI (I_{avg}) and its corresponding value of modularity ($Mod(Q)$) over 30 runs are recorded in Table 3.1. The average values of best modularity (Q_{avg}) and its corresponding to the value of NMI ($I(Q)$) are also recorded in Table 3.2. The maximum generation was 50. At each run, the solutions, which have the maximum value of NMI and modularity, are selected. The average results of 30 times obtained by MOEA/D-Net and FM on those four real-world networks are shown in Table 3.1. The statistic values of best NMI and the statistic values of best modularity Q over the 30 runs on the four real-world networks in terms of box plots are also shown in Fig. 3.5, which can illustrate the stability of MOEA/D-Net. As Fig. 3.5 shows, on each of the four networks, the variability of NMI and Q values obtained over the 30 runs is relatively small.

Table 3.1 reports the average of the best NMI (I_{avg}), the average modularity value ($I(Q)$) corresponding to the solutions having the best NMI, the NMI value of the solution found by FM (FMI_{avg}) and the modularity value of the solution found by FM($FM_{M}od$).

As is shown in Table 3.1, the average of the best NMI of the first two real-world networks obtained by MOEA/D-Net is 1. This means the true partitions to both real-world networks can be obtained at each run of MOEA/D-Net. The modularity

Table 3.1 The results of 30 runs of best NMI obtained by our method and fast modularity algorithm for the real-world datasets

Network	I_{avg}	$Mod(Q)$	FMI_{avg}	$FM(Q)$
Zachary's karate club	1	0.371	0.693	0.380
Bottlenose Dolphins	1	0.373	0.573	0.495
American college football	0.925	0.599	0.762	0.577
Books about US politics	0.596	0.481	0.530	0.502

Table 3.2 The results of 30 runs of best modularity obtained by MOEA/D-Net

Network	$I(Q)$	Q_{avg}
Zachary's karate club	0.687	0.420
Bottlenose Dolphins	0.623	0.520
American college football	0.891	0.604
Books about US politics	0.574	0.527

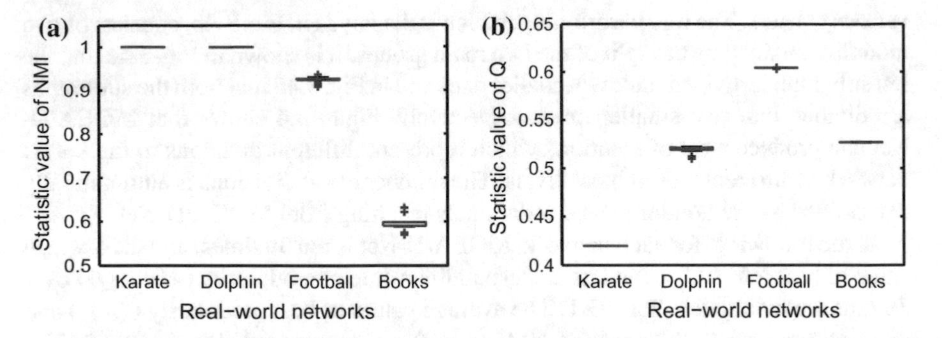

Fig. 3.5 The box plot of the statistic value of NMI and Q Over the 30 runs on the four real-world networks. **a** the statistic value of NMI. **b** the statistic value of Q. Here, box plots are used to illustrate the distribution of the NMI obtained by MOEA/D-Net. On each box, the red line is the median, the edges of the box are one-fourth and three-fourths, the whisker extends to the most extreme data points the algorithm considers to be not outliers, and the outliers are plotted individual. Symbol + denotes outliers

value of the true partitioning to the karate network is 0.371 and the modularity value of the true partitioning to the dolphins network is 0.373. The true partitioning of the dolphins network is shown in Fig. 3.6. However, on karate club and dolphins networks, the fast modularity algorithm found a solution with a NMI value of 0.693 and 0.573, respectively.

The football network and polbooks network are more difficult, especially the last one. MOEA/D-Net and FM cannot find the true partitioning to them. However, it is clearly known from Table 3.1 that highest average the best value of NMI is obtained by employing MOEA/D-Net. The fast modularity algorithm found a solution with a NMI value of 0.762, while MOEA/D-Net found a solution with average best NMI value of 0.925 for 30 runs.

The last real-world network is the most complex one. Its structure is not very clear as the first two networks, so it is difficult to detect the community within it. However, it is also clearly known from Table 3.3 that highest average the best value of NMI is obtained by employing MOEA/D-Net. The fast modularity algorithm found a solution with a NMI value of 0.502, while MOEA/D-Net finds a solution with average best NMI value of 0.596 for 30 runs.

Tables 3.1 and 3.2 clearly show the average best value of modularity by MOEA/D-Net are larger than obtained by fast modularity algorithm on four real-world networks. It also shows that except for the karate club networks, the value of NMI, corresponding to the best value of modularity, is larger than obtained by fast modularity algorithm. Therefore, MOEA/D-Net has better performances than FM on the four real-world networks. For the karate network, the reason the value of NMI, corresponding to the best value of modularity, obtained by MOEA/D-Net is smaller than the obtained by fast modularity algorithm is there exists the resolution limit in optimizing modularity to reveal community structure of networks. As is shown in Fig. 3.4, the karate club network has several kinds of hierarchical structure. There-

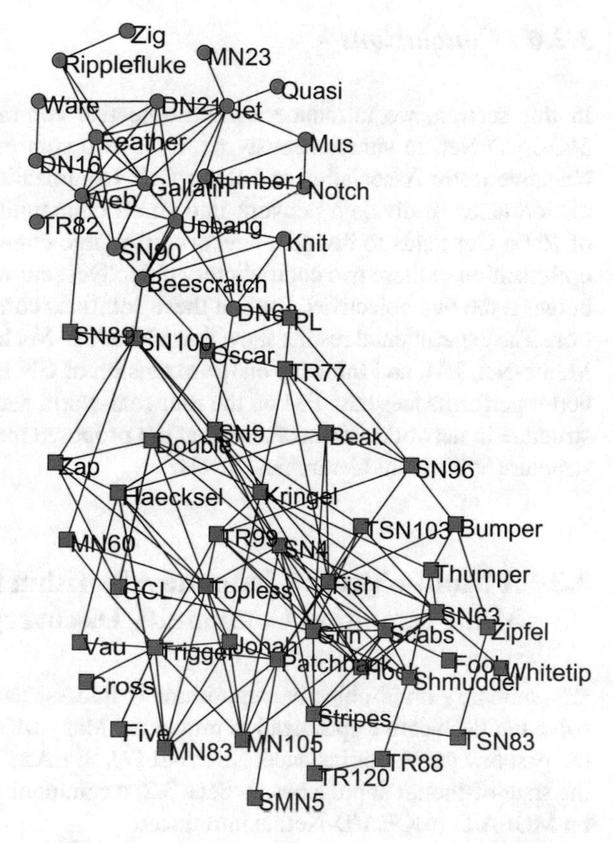

Fig. 3.6 The true partitioning of the Bottlenose Dolphins network

Table 3.3 Networks parameters in MICD

Parameter	Meaning	Value
G_{max}	The number of iterations	200
n_D	Maximum size of dominate population	200
n_A	Maximum size of active population	40
n_C	Number of cloning population	200
p_m	Probability of hypermutation	0.01

fore, although the best value of Q obtained by MOEA/D-Net is larger than that obtained by fast modularity algorithm, there also exists a value of NMI obtained by MOEA/D-Net that is smaller than the value obtained by fast modularity algorithm when the modularity is chosen as the criterion to reveal community structure in networks.

It is clearly known from this experiment that MOEA/D-Net is a more effective algorithm compared with fast modularity algorithm.

3.2.6 Conclusions

In this section, we introduce a multi-objective community discovery algorithm, MOEA/D-Net, to simultaneously optimize two contradictory objective functions, Negative Ratio Association and Ratio Cut. Optimization of Negative Ratio Association tends to divide a network into small communities, while the optimization of Ratio Cut tends to divide a network into large communities. The simultaneous optimization of these two contradictory objectives returns a set of trade-off solutions between the two objectives. Each of these solutions corresponds to a network partition. The experimental results show that MOEA/D-Net has better performances than Meme-Net, FM, and InfoMap on the extension of GN Benchmark network and has better performances than FM on the four real-world networks to reveal community structure in networks. It also shows that the proposed method can reveal community structure at different hierarchical levels.

3.3 A Multi-objective Immune Algorithm for Multi-resolution Community Discovery

[2]Evolutionary multi-objective algorithms (MOEAs) have proved to be effective to solve multi-objective optimization problems. Many MOEAs have been proposed in the past few years. For instance, NSGA-II [7], SPEA2 [35], MOEA/D [34], etc., are the state-of-the-art approaches. In Sect. 3.2, a community discovery algorithm based on MOEA/D (MOEA/D-Net) is introduced.

In this section, a multi-resolution communities discovery algorithm (MICD) based on multi-objective immune algorithm (NNIA) [14]. The advantages of MICD are as follows. This method simultaneously optimizes two objective functions. One of the objective functions is called Modified Ratio Association (MRA) [1]. The optimization of MRA tends to divide a network into small communities. The other objective function is called Ratio Cut (RC) [32]. The optimization of RC tends to divide a network into large communities. These two objective functions are conflicting to each other. Thus, there is a trade-off between the two objectives, which represents the network partitions at different scales. In order to simultaneously optimize the two objective functions, the framework of multi-objective immune algorithm NNIA with some modifications in individual representation and search operators is used. MICD is able to find a set of trade-off solutions between the two objectives. Each of these solutions corresponds to a network partition at one resolution level.

---

[2]Acknowledgement: Reprinted from Applied Soft Computing, 13(4), Gong, M., Chen, X., Ma, L., Zhang, Q., Jiao, L., Identification of multi-resolution network structures with multi-objective immune algorithm, 1705–1717, Copyright(2013), with permission from Elsevier.

3.3.1 Multi-objective Immune Optimization for Multi-resolution Communities Identification

In this section, we will describe the multi-objective immune algorithm for multi-resolution community detection, MICD. Following the framework of NNIA, there are three populations in MICD. One is called dominant population, which is the set of nondominated individuals. The second population is active population, which is composed of the individuals selected from the nondominated individuals for they are less-crowded. These individuals are selected to do proportional cloning, recombination and hypermutation. The population storing clones is called clone population. The dominant population, active population, and clone population at the generation k are represented by variable matrices $\mathbf{D}_k$, $\mathbf{A}_k$ and $\mathbf{C}_k$, respectively. The main procedure of MICD is as follows. In MICD, the representation and initialization are the same as those in MOEA/D-Net, so that we do not describe them again in this section.

Algorithm 11 Framework of the proposed MICD.

Input: The adjacency matrix of the network: A, The maximum number of generations: $gens$, the maximum size of dominant population: n_D, the maximum size of active population: n_A, and the size of clone population: n_C.

Output: Pareto front solutions. Each solution corresponds to a partition of a network.

Step 1): Initialization: Generate an initial population $\mathbf{B}_0$. with size n_D, set $k = 0, \mathbf{D}_0 = \varphi, \mathbf{A}_0 = \varphi$, and $\mathbf{C}_0 = \varphi$.

Step 2): Update Dominant Population: Identify nondominated individuals in $\mathbf{B}_k$. Copy all the nondominated individuals to form the temporary dominant population (denoted by $\mathbf{DT}_{k+1}$). If the size of $\mathbf{DT}_{k+1}$ is not greater than n_D, let $\mathbf{D}_{k+1} = \mathbf{DT}_{k+1}$. Otherwise, calculate the crowding-distance values of all individuals in $\mathbf{DT}_{k+1}$, sort them in descending order of crowding-distance, and choose the first n_D individuals to form $\mathbf{D}_{k+1}$.

Step 3): Termination: If $k \geq gens$ is satisfied, go to Step8; Otherwise, $k = k + 1$.

Step 4): Nondominated Neighbor-Based Selection: If the size of $\mathbf{D}_k$ is not greater than n_A, let $\mathbf{A}_k = \mathbf{D}_k$. Otherwise, calculate the crowding-distance values of all individuals in $\mathbf{D}_k$, sort them in descending order of crowding-distance, and choose the first n_A individuals to form $\mathbf{A}_k$.

Step 5): Proportional Cloning: Get the clone population $\mathbf{C}_k$ by applying proportional cloning to $\mathbf{A}_k$.

Step 6): Recombination and Hypermutation: Perform recombination and hypermutation operators on $\mathbf{C}_k$ to generate $\mathbf{C}_k'$, which represents the resulting population.

Step 7): Get the antibody population $\mathbf{B}_k$ by combining $\mathbf{C}_k'$ and $\mathbf{D}_k$; go to Step2. Step8: Apply a new metric for selecting solution at each resolution scale. Step9: Get the partition results by decoding the individuals in $\mathbf{D}_{k+1}$, stop.

3.3.2 Problem Formation

In Sect. 3.2, MOEA/D-Net aims to optimize negative ratio association and ratio cut. In MICD, ratio association is transformed as modified ratio association (MRA), which is

$$MRA = 2(n-k) - \sum_{i=1}^{k} \frac{L(V_i, V_i)}{|V_i|} \tag{3.4}$$

Then, MICD aims to simultaneously optimize the following two objective functions.

$$\begin{cases} \min \quad MRA = 2(n-k) - \sum_{i=1}^{k} \frac{L(V_i,V_i)}{|V_i|} \\ \min \quad RC = \sum_{i=1}^{k} \frac{L(V_i,\overline{V_i})}{|V_i|}. \end{cases} \tag{3.5}$$

3.3.3 Proportional Cloning

The proportional cloning operation $\mathbf{P}^C$ on the active population $\mathbf{A} = \{\mathbf{a}_1, \mathbf{a}_2, ..., \mathbf{a}_{|\mathbf{A}|}\}$ can be defined as

$$\begin{aligned} &\mathbf{P}^C(\mathbf{a}_1 + \mathbf{a}_2 + ... + \mathbf{a}_{|A|}) \\ &= \mathbf{P}^C(\mathbf{a}_1) + \mathbf{P}^C(\mathbf{a}_1) + \cdots + \mathbf{P}^C(\mathbf{a}_{|A|}) \\ &= \{\mathbf{a}_1^1 + \mathbf{a}_1^2 + \cdots + \mathbf{a}_1^{q_1}\} + \{\mathbf{a}_2^1 + \mathbf{a}_2^2 + \cdots + \mathbf{a}_2^{q_2}\} \\ &+ \cdots + \{\mathbf{a}_{|A|}^1 + \mathbf{a}_{|A|}^2 + \cdots + \mathbf{a}_{|A|}^{q_{|A|}}\} \end{aligned} \tag{3.6}$$

where $\mathbf{P}^C(\mathbf{a}_i) = \{\mathbf{a}_i^1 + \mathbf{a}_i^2 + \cdots + \mathrm{a}_i^{q_i}\}, \mathbf{a}_i^j = \mathbf{a}_i, i = 1, 2, \cdots, |A|, j = 1, 2, \cdots, q_i$. q_i is self-adaptive parameter. In Eq. 3.13, the symbol + is only used to separate the antibodies. q_i denotes the number of cloning on $\mathbf{a}_i$. The crowding-distance of a dominate antibody $\mathbf{d} \in \mathbf{D}$ is defined as

$$\rho(\mathbf{d}, \mathbf{D}) = \sum_{k=1}^{s} \frac{\rho_k(\mathbf{d}, \mathbf{D})}{f_k^{\max} - f_k^{\min}} \tag{3.7}$$

where $f_k^{\max}$ and $f_k^{\min}$ are the maximum and minimum value of the kth objective and

$$\rho_k(\mathbf{d}, \mathbf{D}) = \begin{cases} \infty & if \ \ f_k(\mathbf{d}) = mor M \\ min\{f_k(\mathbf{d}') - f_k(\mathbf{d}'')\} & others \end{cases} \tag{3.8}$$

$m = \min\{f_k(\mathbf{d}')|\mathbf{d}' \in \mathbf{D}\}$, $M = \max\{f_k(\mathbf{d}')|\mathbf{d}' \in \mathbf{D}\}$, $\mathbf{d}', \mathbf{d}'' \in \mathbf{D} : f_k(\mathbf{d}'') < f_k(\mathbf{d}) < f_k(\mathbf{d}')$. Based on the crowding-distance $\rho(\mathbf{d}, \mathbf{D})$, the individual with greater crowding-distance value has a larger cloning scale. The values of q_i are calculated as

$$q_i = \left\lceil n_C \times \frac{\rho(\mathbf{a}_i, \mathbf{A})}{\sum_{j=1}^{|\mathbf{A}|} \rho(\mathbf{a}_j, \mathbf{A})} \right\rceil \tag{3.9}$$

where $\rho(\mathbf{a}_i, \mathbf{A})$ denotes the crowding-distance value of the active antibodies $\mathbf{a}_i$, n_C is an expectant value of the size of the clone population. The subsequent update of the dominated population and nondominated neighbor-based selection make sure the size of dominate population and the active population are not greater than n_D and n_A, respectively.

3.3.3.1 Recombination and Hypermutation

One of the heuristic search operators in NNIA is called recombination operation. Let **A** indicate the clone population, and $\mathbf{C} = (\mathbf{c}_1, \mathbf{c}_2, \ldots, \mathbf{c}_{|\mathbf{C}|})$ represent the resulting population after applying proportional cloning to **A**. The recombination of $\mathbf{P}^R$ on the clone population **C** can be defined as

$$\begin{aligned} &\mathbf{P}^R(\mathbf{c}_1 + \mathbf{c}_2 \cdots + \mathbf{c}_{|\mathbf{C}|}) \\ &= \mathbf{P}^R(\mathbf{c}_1) + \mathbf{P}^R(\mathbf{c}_2) + \cdots + \mathbf{P}^R(\mathbf{c}_{|\mathbf{C}|}) = crossover(\mathbf{c}_1, \mathbf{A}) \\ &+ crossover(\mathbf{c}_2, \mathbf{A}) + \cdots + crossover(\mathbf{c}_{|\mathbf{C}|}, \mathbf{A}) \end{aligned} \tag{3.10}$$

where $crossover(\mathbf{c}_i, \mathbf{A})$ represents selecting one individual from two offspring, which are generated by a crossover operator on clone $\mathbf{c}_i$ and an active antibody selected randomly from A. The crossover operator used here is two-point crossover, because this crossover operator can better maintain the effective connective links of the vertices in network. For the two selected individuals from A and C, randomly generate two points x_1, x_2 (assume $x_1 < x_2$) and a random number r within (0 1). If the value of r is smaller than 0.5, swap all the alleles between $[x_1, x_2]$, otherwise, swap all the alleles before point x_1 and after point x_2. From the genetic representation used here, each individual of the population is valid. Thus, using this operator, the generated child individuals generated are also valid. For hypermutation, we call it the neighbor-based hypermutation, and the possible values of an allele are restricted to the neighbors of gene i. For each allele in one individual, we generate a random number τ in the range [0, 1] with uniform distribution. If τ is smaller than pm (the probability of Hypermutation), then the allele is randomly changed to one of its neighbors. It can be described as $\mathbf{c}_i(j) = \{\forall k | k \in Neighbor(j)\}$. This neighbor-based hypermutation guarantees that each node in a mutated child is randomly linked with one of its neighbors, which can avoid the useless exploration of the search space.

3.3.3.2 Select Solution at Each Scale

By the optimization of MRA and RC, it can acquire a set of Pareto-optimal solutions, which represent the network partitions at different resolution scales. However, for a certain resolution scale (that is to say, the number of community is a fixed value), there may be more than one solution, and it cannot decide which one is a relatively better one only by these two objectives. So there exists a solution selection problem at each resolution scale. In MICD, modularity (Eq. 1.6) is used to find a Pareto front solution at each resolution scale.

3.3.4 Analysis of Computational Complexity

Assume that the maximum size of the dominated population is n_D, the maximum size of active population is n_A, clone population is n_C, the number of vertex is n, and the number of edge is m. The time complexity of one generation of MICD can be calculated as follows: The time complexity for identifying nondominated individuals in the population is $O((n_C + n_D)^2)$; the worst time complexity for updating the dominate population is $O((n_C + n_D)\log(n_C + n_D))$; the worst time complexity for nondominated neighbor-based selection is $O(n_D \log(n_D))$; the time complexity for cloning is $O(n_C)$; the time complexity for recombination and mutation is $O(n_C)$, the decoding of the genetic representation of the population and the calculation of objective functions has the time complexity $O(n_C(m + n))$. According to the operational rules of the symbol O, the worst time complexity of one generation for MICD can be simplified as $\max\{O((n_C + n_D)^2),\ O(n_C(m + n))\}$. As the size of network increases, $O(n_C(m + n))$ will be much larger than $O((n_C + n_D)^2)$. For large-scale networks, the complexity of one generation can be depicted as $O(n_C(m + n))$. So the whole time complexity of MICD of g generations is $O(g(m + n)n_C)$.

3.3.5 Experimental Results

The values of the parameters used in MICD are summarized in Table 3.3. The NMI (Eq. 2.3) is also used to evaluate the performance of MICD.

3.3.5.1 Experimental Results on Artificial Generated Networks

In order to test the ability of MICD, the extension of the GN benchmark network is also used (in Chap. 1).

In this experiment, MOGA-Net [26], MOCD [30], Meme-Net (in Sect. 2.2) and CNM [4] are adopted as comparison algorithms. MOGA-Net and MOCD are two community detection algorithms using different MOEAs to optimize different objectives, and CNM is a popular greedy algorithm optimizing modularity. For these three GA-based algorithms, the parameters are set similarly to MICD: the population size in MOGA-Net and Meme-Net is 200; the internal population size and the external population size are both 200 in MOCD; the crossover and mutation rate are both 0.80 and 0.01 in the three methods; the numbers of generation are both 200.

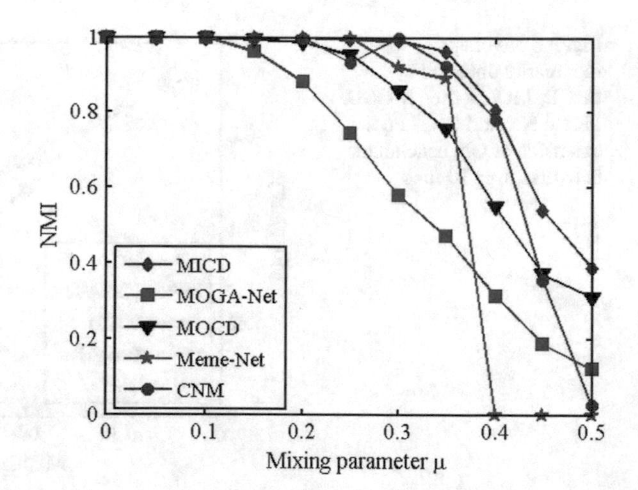

Fig. 3.7 Average NMI obtained by MICD, MOGA-Net, MOCD, Meme-Net and CNM on the extension of GN benchmark networks over 10 runs

Figure 3.7 shows the average NMI obtained by MICD, MOGA-Net, MOCD, Meme-Net, and CNM. As is shown, when $\mu <= 0.30$, only MICD and CNM can almost obtain the true partitions (NMI equals 1). With the mixing parameter increasing, the network becomes fuzzier, it is more difficult for these algorithms to detect the true partitions, but the proposed MICD performs best. As is shown in Fig. 3.7, we can see that MICD is more accurate than the three GA-based algorithms and the popular CNM. The most important feature of MICD is that it can generate a lot of solutions, and these solutions may contain partitions at different resolution scales. While, for these benchmark networks, only the largest Q for each network is selected. The best values of Q for each network of each algorithm are recorded in Fig. 3.8. As the mixing parameter μ increases from 0 to 0.5, the values of modularity obtained for each algorithm are lower and lower. However, the proposed method MICD and CNM are the better than the other algorithms. Although the results of CNM are similar to that obtained by MICD, the latter are more close to the true partitions (in Fig. 3.7).

To verify the ability of MICD on detecting community structure of a network at different resolution levels, a hierarchical scale-free network with 25 vertices, RB 25, is used, which was proposed by Ravasz and Barabasi [27]. The RB 25 network can be generated in the following way: first, the starting point is a small cluster of five densely linked vertices; second, generate four replicas of this hypothetical cluster and connect the four external vertices of the replicated clusters to the central vertex of the old cluster, obtaining a large module of 25 vertices. Some hierarchical structures of RB 25 network are shown in Fig. 3.10a, and the true partition is 5 clusters.

In [3], the authors built a toy model network with a simple topology. However, it is difficult for community detection algorithms to detect this model network, because it includes communities of different sizes, some of them are sparse and others dense. The network model is small enough to have a clear vision of the clusters, and the ideal partition is 5 communities. Some representative structures of the toy model network can be seen in Fig. 3.10b. The third network is the FB network proposed

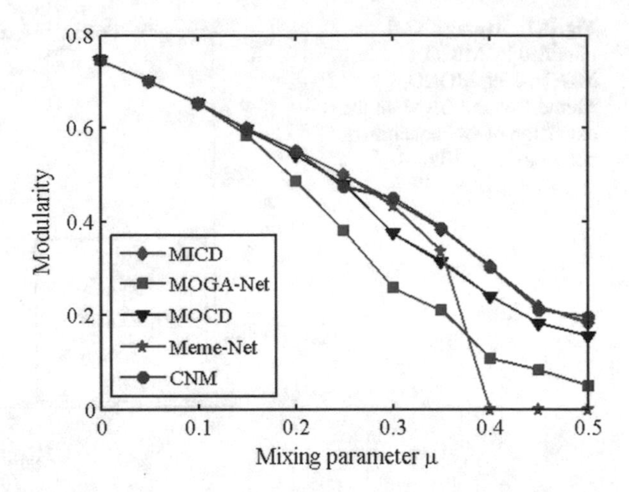

Fig. 3.8 Average modularity obtained by MICD, MOGA-Net, MOCD, Meme-Net and CNM on the extension of GN benchmark networks over 10 runs

by Fortunato and Barthlemy [10]. It consists of two cliques of 20 vertices linked with two small cliques of 5 vertices. In order to compute NMI, the best partition is recognized as four groups, with each clique as one group. It has been demonstrated that modularity cannot separate the two small cliques due to the resolution limit of modularity.

The fourth artificial network used here is the H 13-4 network [2]. The network corresponds to a homogeneous in degree network with two predefined hierarchical structures, the number of vertices being 256, the number of links of each vertex with the most internal community being 13 (formed by 16 vertices), the number of links with the most internal community being 4 (four groups of 64 vertices), and the number of links with any other vertex at random in the network being 1. There are two representative hierarchical structures of the network, one is the hierarchical structure consisting of 4 groups of 64 vertices (in this study, this partition is recognized as the true partition), and the other consisting of 16 groups of 16 vertices. The H 13-4 network is used here to show MICD can reveal community structures at multiple scales. The trade-off fronts after selection are shown in Fig. 3.9. The proposed method can acquire a set of solutions at different resolution scales. The arrow symbol in Fig. 3.9 indicates the representative community structures corresponding to the number of communities obtained by MICD. Table 3.4 summarizes the number of solutions obtained for each Pareto front solution, the max value of Q, and the NMI value computed by considering each desired division as the true one. Table 3.4 shows that MICD can generate community partitions at different resolution scales for each network. For these four networks, the values of max NMI obtained are all 1, which demonstrates that MICD can get the true partition of the networks.

In order to illustrate the reliability and the correctness of the selected solutions, just some representative scales corresponding to relevant trade-off fronts solutions obtained by MICD are displayed. In Fig. 3.10a, the solution with 2 communities corresponding to dashed line I divides RB 25 network into 2 communities; the solution

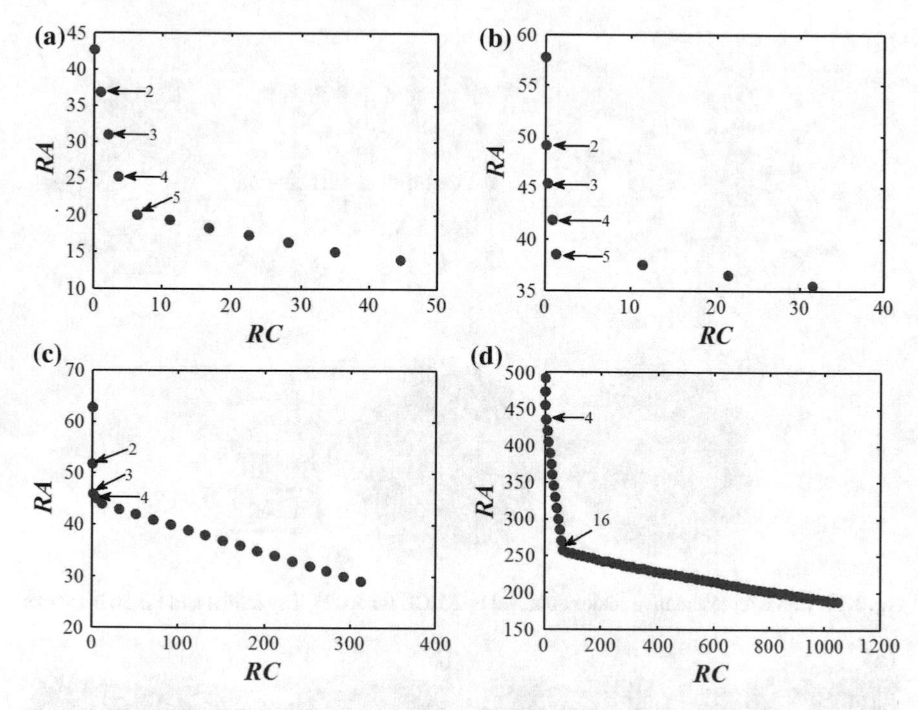

Fig. 3.9 The trade-off fronts after selection at each resolution scale for different networks. **a**, **b**, **c** and **d** are the trade-off fronts corresponding to community structures of resolution scales for RB 25 network, toy model network, FB network, and H 13-4 network, respectively

Table 3.4 The Pareto solutions of the four artificial networks

Network	Vertices	Edges	NS	Max Q	Max NMI
RB25	25	66	11	0.5510	1
Toy	32	67	8	0.4590	1
FB	50	404	20	0.5426	1
H 13-4	256	2308	71	0.7165	1

with 3 communities corresponding to dashed line I and II divide the network into 3 communities; the solution with 4 communities corresponding to dashed line I, II, and III divide the network into 4 communities; and the solution with 5 communities corresponding to dashed line I, II, III, and IV divide the network into 5 communities, which is also the ideal partition. The analysis is the same for Fig. 3.10b, c, and MICD also can get the ideal partitions with toy model network 5 communities and FB network 4 communities. In Fig. 3.11, only the community structures at two representative sales of H 13-4 network are plotted. The network is divided into 4 communities in Fig. 3.11a, with each community containing 64 vertices, and in Fig. 3.11b, the network is divided into 16 communities with each community containing 16 vertices.

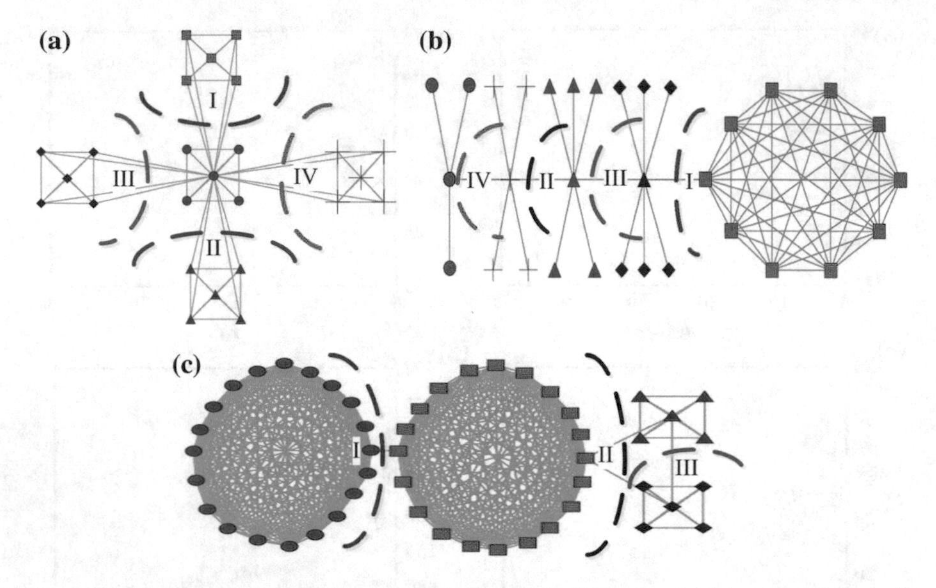

Fig. 3.10 Some representative scales obtained by MICD for RB25, Toy model and FB 50 networks

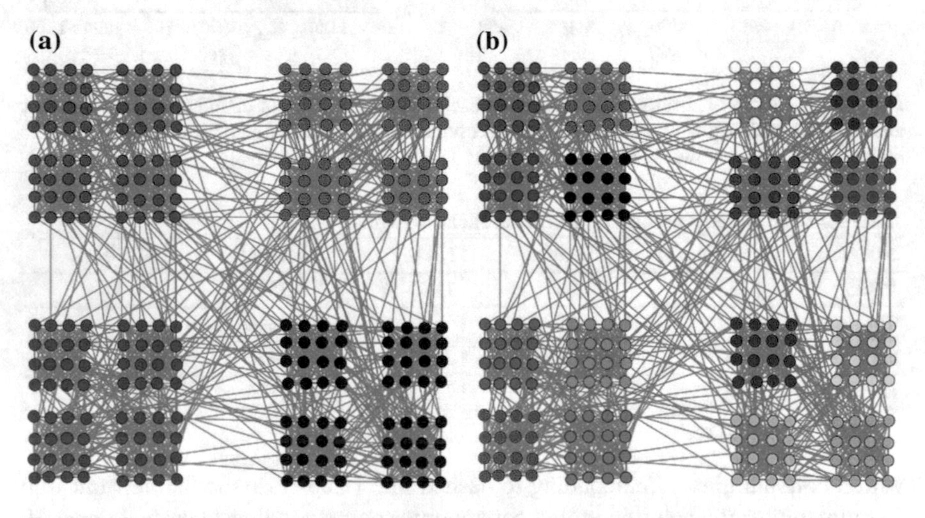

Fig. 3.11 Community structures at two representative scales of H 13-4 network

From the experiments above, MICD has shown the ability of revealing multi-resolution community structures in artificial hierarchical networks. In addition, the selected representative community structures account for the correctness of MICD for these artificial networks.

Table 3.5 The results obtained by MICD for real-world networks

Network	Nodes	Edges	NS	Avg NMI	Avg Q	NC
Journal	40	189	16.6	1	0.4783	4
Dolphin	62	159	24.4	1	0.5187	4.9
Football	115	613	39.7	0.93	0.6042	10
SFI	118	200	40.7	*	0.7490	7.2
Science	1589	2742	167.6	*	0.9556	282.5

3.3.5.2 Experimental Results on Real-World Networks

In this part, MICD is tested on several real-world networks: karate, dolphins, Journal, football, SFI, netscience. For each network, MICD is run 10 times. Table 3.5 shows the average results of the proposed method on each network over 10 runs. It summarizes the number of solutions at different resolution scales (denoted by NS), average max NMI, average max Q, and the number of communities corresponding to max Q (denoted by NC). As observed, for the first three networks, the proposed method can obtain true partitions (NMI equals 1) and some community structures at different resolution scales at the same time. Because SFI and Science networks do not have the true partitions, so we cannot compute the value of NMI ($*$ represents the corresponding network does not have the true partition). For each network, the average max Q and the community number (corresponding to max Q) are recorded. For example, the average max Q of Science network is 0.9556 and the average community number is 282.5.

Figure 3.12a displays the Pareto front of Journal network in one of ten runs. As observed, MICD generated 17 solutions (corresponding to a variety of resolution levels). The solution 2 in Fig. 3.12a splits the network into 2 communities as shown in Fig. 3.12b, with physical and chemical journals as a community, and biological and ecological journals as another community; solution 3 divides the network into 3 communities, with ecological and biological journals separated, but physical and chemical journals remain together in the same community (Fig. 3.12c); in Fig. 3.12d, the network is divided into 4 communities correctly by solution 4, with each field journals as one community, and it is ideal partition of this network (corresponding NMI equals 1). Besides, Fig. 3.13 describes the results for SFI network. This network does not have the true partition, so Q is used to determine the division of network partitions. As can be seen from Fig. 3.13b–d, SFI network is divided into 4, 6, and 8 groups, respectively. In fact, the partition corresponding to Fig. 3.13d is the best partition (max Q) among all the partitions. In all, the proposed method can obtain community structures at different resolution scales. It shows the ability and effectiveness of revealing multi-resolution community structure in both artificial networks and real-world networks.

In this section, the results of the MICD are compared with a multi-resolution community detection method Meme-Net and two other multi-objective community detection methods MOGA-Net and MOCD.

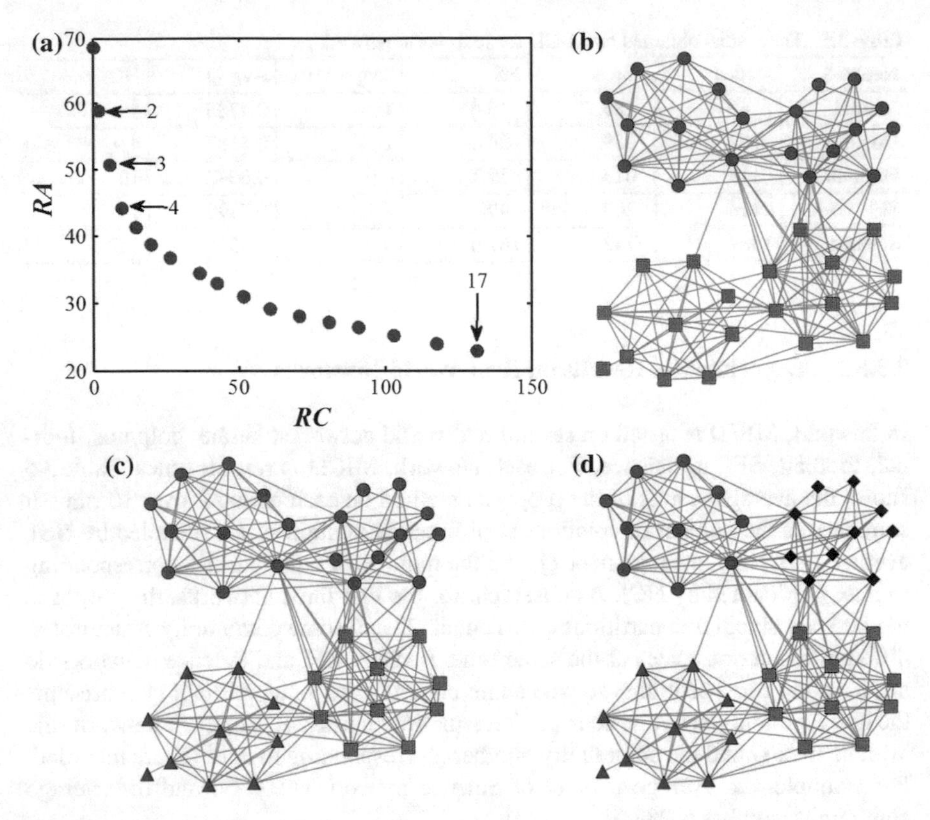

Fig. 3.12 The results for journal network. **a** Pareto front at each resolution scale of one run. **b–d** represent some representative community structures at some resolution scales

A comparison between MICD and Meme-Net, for two real-world networks, is given in Table 3.6.

Because the networks listed in Table 3.6 have true partitions, so NMI metric is used to evaluate the community structures obtained. The average value and standard deviation of NMI (I_{avg}^{MICD} and I_{std}^{MICD}) over the 10 runs are computed. As seen from Table 3.6, the values of (I_{avg}^{MICD}) obtained by MICD are better than that obtained by Meme-Net. The results on the two real-world networks show that the proposed approach can reveal community structures at different resolution scales in just a single run and at the same time ensure the accuracy of classification at each scale.

In order to further verify the efficiency of MICD, MICD is compared with the other two multi-objective approaches: MOGA-Net and MOCD. Both of the above methods could generate some Pareto front solutions. Four real-world networks are tested in this experiment, and these networks are widely used as benchmarks in community detection. In Table 3.7, the average number of solutions at different resolution scales (NS), the average best NMI and the average max Q obtained by MOGA-Net, MOCD and MICD are reported. For each network, the algorithms are executed 10 times.

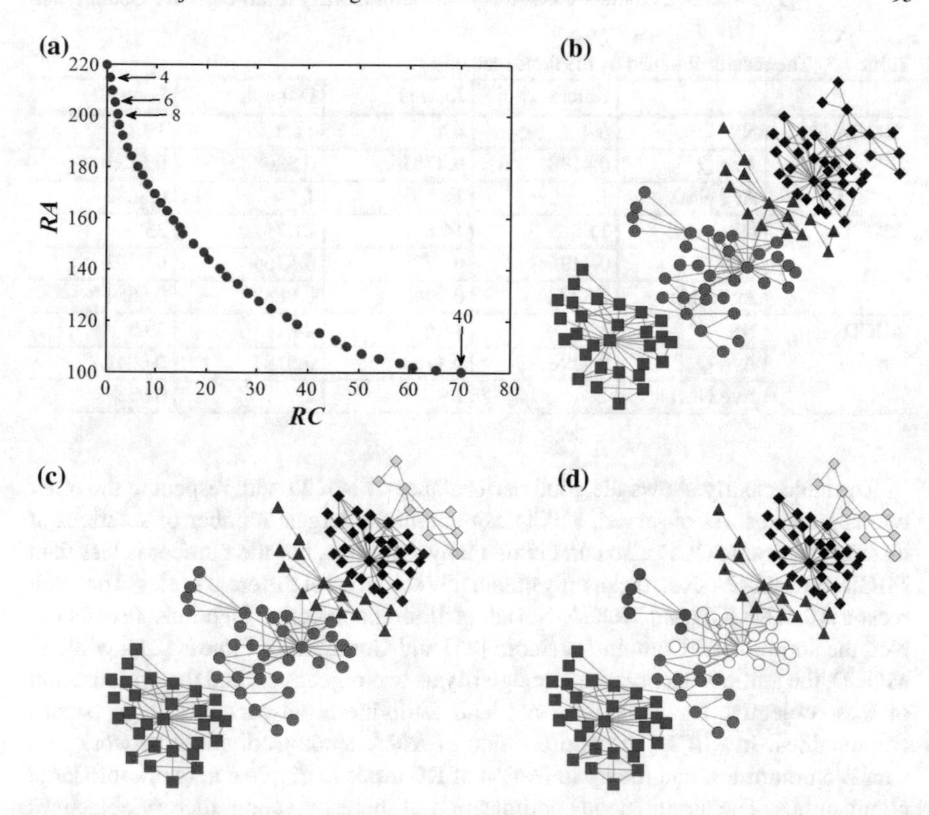

Fig. 3.13 The results for SFI network. **a** Pareto front at each resolution scale of one run. **b–d** represent some representative community structures at some resolution scales

Table 3.6 The results of Meme-Net and MICD on two real-world networks at different resolution scales

Network	$N_{cluster}$	I_{avg}^{Meme}	I_{avg}^{MICD}	I_{std}^{Meme}	I_{std}^{MICD}
Karate	2	1	1	0	0
	3	0.740	0.826	0	0
	4	0.690	0.707	0.007	0
	5	0.651	0.651	0.038	0
Dolphins	2	1	1	0	0
	3	0.787	0.923	0.073	0
	5	0.569	0.721	0.035	0.048
	9	0.467	0.489	0.048	0.019
	11	0.400	0.443	0.034	0.015
	14	0.346	0.390	0.017	0.011

Table 3.7 The results obtained by the three methods

		Karate	Journal	Dolphin	Football
MOGA-Net	NS	6.4	4.4	11.2	10.1
	Avg Q	0.4198	0.4783	0.5208	0.5835
	Avg NMI	1	1	1	0.8682
MOCD	NS	11.6	14.8	21.7	33
	Avg Q	0.4196	0.4773	0.5264	0.5792
	Avg NMI	0.8372	0.9940	0.8888	0.8684
MICD	NS	12.5	16.6	24.4	39.7
	Avg Q	0.4198	0.4783	0.5187	0.6042
	Avg NMI	1	1	1	0.933

The table clearly shows the good performance of MICD with respect to the other two approaches. As observed, MICD can obtain the largest number of solutions at different scales. MOCD also can obtain many solutions, but the number is less than MICD. For MOGA-Net, it can only obtain few solutions at different scales. The main reason is that MOCD and MOGA-Net adopt different objective functions. In MOGA-Net, the author used Community Score [25] and Community Fitness [21], while in MOCD, the authors decomposed modularity as two objectives, and the optimization of these objective functions does not tend to divide a network into large (small) communities. In MICD, the optimization of MRA tends to divide a network into small communities, and the optimization of RC tends to divide a network into large communities. The simultaneous optimization of these two contradictory objectives returns a set of trade-off solutions between the two objectives. So that is the reason why MICD has the ability of revealing community structure than the other two methods. The average best NMI over the four networks are computed 10 runs. From Table 3.7, the average best NMI obtained by MICD and MOGA-Net are both 1 for Karate, Journal and Dolphin networks, while MOCD cannot get the true partitions for these networks. As for Football network, all the methods cannot get the true partition, but MICD performs the best (with NMI equals 0.93). Besides, the max values of Q obtained by the three methods are also recorded for each network. As can be seen, MICD perform best for Karate, Journal and Football networks, while MOCD has the best result on Dolphin network (Q equals 0.5264). By comparison, the results show that MICD has more powerful ability of revealing community structure at different resolution levels and at the same time ensures the accuracy of community structures for different networks.

3.3.6 Conclusions

In this section, a multi-resolution community discovery algorithm was introduced. First, multi-resolution community discovery is formulated as a multi-objective optimization problem and then a new approach MICD for revealing multi-resolution community structures is designed. MICD optimizes two objective functions, one of which is Modified Ratio Association, and the other is Ratio Cut. The optimization of MRA usually divides a network into sets of some small communities, and the optimization of RC often divides a network into sets of some large communities. The simultaneous optimization of these two contradictory objectives returns a set of trade-off solutions between the two objectives. Each of these solutions corresponds to a partition of a network at one scale. Experiments on artificial networks and real-world networks showed that MICD can generate some topology structures (different numbers of communities) in only a single run. Some examples of the representative structures obtained by MICD are analyzed, and the reliability of MICD is verified.

3.4 An Efficient Multi-objective Discrete Particle Swarm Optimization for Multi-resolution Community Discovery

In Sects. 3.2 and 3.3, two multi-resolution communities discovery algorithms are introduced and they are based on different evolutionary multi-objective algorithms (MOEA/D and NNIA). In this section, a multi-resolution community discovery algorithm based on multi-objective discrete particle swarm optimization (MODPSO) will be introduced.

The advantages of MODPSO are as follows. Based on network topology, the particles velocity and position vectors are redefined in discrete form. All the arithmetical operators between velocity and position vectors are also redefined. A problem-specific particle swarm initialization approach and a turbulence operator are introduced. In MODPSO algorithm, decomposition mechanism is adopted. The objective functions used in MODPSO have been extended to signed version. MODPSO minimizes two objectives termed as Kernel K-Means and Ratio Cut. But their neglect of the signed features of networks confounds their applications. In order to handle signed networks, the two objectives have been extended to signed version.

3.4.1 Multi-objective Discrete Particle Swarm Optimization for Multi-resolution Community Discovery

In MODPSO, the adopted decomposition method is the widely used Tchebycheff approach (Eq. 3.1). The whole framework of MODPSO algorithm is given in Algorithm 12.

Algorithm 12 Framework of the proposed MODPSO.

Parameters: max generation: $maxgen$, number of decomposed sub-problems: ns, swarm size: pop, mutation probability pm, inertia weight: ω, the learning factors: c_1, c_2.
Input: The adjacency matrix A of a network. **Output**: Pareto front solutions. Each solution corresponds to a partition of a network.

Step 1) Initialization
Step 1.1) Position initialization: $P = \{x_1, x_2, ..., x_{pop}\}^T$.
Step 1.2) Velocity initialization: $V = \{v_1, v_2, ..., v_{pop}\}^T$.
Step 1.3) Generate a well-distributed weighted vectors: $W = \{w_1, w_2, ..., w_{pop}\}^T$.
Step 1.4) Personal best position initialization: $Pbest = \{pbest_1, pbest_2, ..., pbest_{pop}\}^T$, $pbest_i = x_i$.
Step 1.5) Initialize reference point z^*.
Step 1.6) Initialize neighborhood N based on Euclidean distance, i.e., $N = \{n_1, n_2, ..., n_{pop}\}^T$.
Step 2) set $t = 0$. // the number of flight cycles
Step 3) Cycling
for $i = 1, 2, ..., pop$, **do**
Step 3.1) Randomly select one particle from the neighbors as the $gbest$ particle, i.e., $gbest \leftarrow random(x_i.neighbor)$.
Step 3.2) Calculate new velocity v_i^{t+1} for the ith particle according to Eq. 3.12.
Step 3.3) Calculate new position x_i^{t+1} for the ith particle according to Eq. 3.15.
Step 3.4) If $t < maxgen \cdot pm$, Turbulence operation on x_i^{t+1}, see Algorithm 14 for more information.
Step 3.5) Evaluation of x_i^{t+1}.
Step 3.6) Update neighborhood solutions: for the jth ($j = 1, 2, ..., N$) particle in the neighborhood of the ith particle, if $g^{te}(x_i^{t+1}|w_j, z^*) \leq g^{te}(x_i^t|w_j, z^*)$, then $x_i^t = x_i^{t+1}$, $F(x_i^t) = F(x_i^{t+1})$.
Step 3.7) Update reference point z^*.
Step 3.8) Update personal best solution $pbest_i$.
Step 4) Stopping criteria: If $t < maxgen$, then $t++$ and go to **Step 3**, otherwise, stop the algorithm and output.

3.4.2 Problem Formation

The objective functions in MODPSO are also functions (Eq. 3.5) used in Sect. 3.3. Equation 3.5 could just be used for unsigned networks.

In order to handle signed networks, objective functions Eq. 3.5 are revised in signed version. Consequently, the signed network clustering problem is reformulated as the following optimization problem:

$$\min \begin{cases} SRA = -\sum_{i=1}^{k} \frac{L^+(V_i,V_i)-L^-(V_i,V_i)}{|V_i|} \\ SRC = \sum_{i=1}^{k} \frac{L^+(V_i,\overline{V_i})-L^-(V_i,\overline{V_i})}{|V_i|} \end{cases} \tag{3.11}$$

where $L^+(V_i, V_j) = \sum_{i\in V_i, j\in V_j} A_{ij}$, $(A_{ij} > 0)$ and $L^-(V_i, V_j) = \sum_{i\in V_i, j\in V_j} |A_{ij}|$, $(A_{ij} < 0)$.

To minimize *SRA* and *SRC*, the positive links within a community are dense while the negative links between communities are also dense, which is in accordance with the feature of signed community.

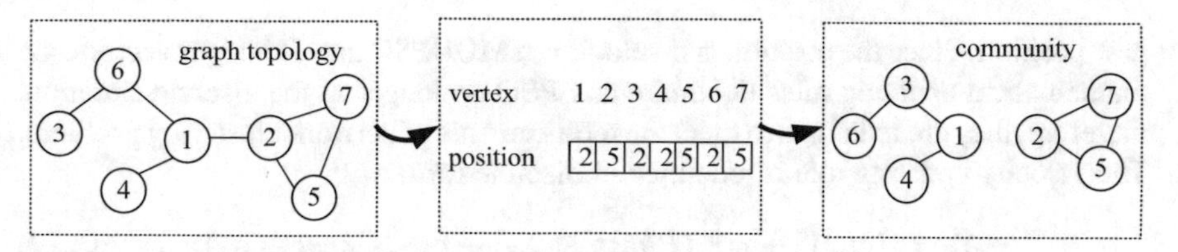

Fig. 3.14 A generic illustration of particle representation

3.4.3 Definition of Discrete Position and Velocity

In order to solve complex network discovery problem, the term position and velocity are redefined in discrete form. The definitions are as follows:

- **Definition of position**: In PSO, the position vector represents a solution to the optimized problem. For the network discovery problem, the position permutation of a particle i is defined as $X_i = \{x_1, x_2, ..., x_n\}$. Each dimension of position is a random integer between 1 and n, i.e., $x_i \in [1, n]$, where n is equal to the total vertices number of the network. If $x_i = x_j$, then we take it that node i and j belong to the same cluster.

Figure 3.14 gives an illustration of how the discrete position of a particle is coded and decoded.

The motive behind the definition of the position vector is that, it is straight forward and easy to decode which will lower down the computational complexity.

- **Definition of velocity**: Velocity works on the position sequence and it is rather crucial. A good velocity gives the particle a guidance and determines whether the particle can reach its destination and by how fast it could. The discrete velocity of particle i is defined as $V_i = \{v_1, v_2, ..., v_n\}$. V_i is binary-coded, and if $v_i = 1$, the corresponding element x_i in the position vector will be changed, otherwise, x_i keeps its original state.

The first motivation of the velocity definition in the above style is to prevent particles from flying away. Because in general, it is necessary to set a threshold V_{max} to inhibit particles from flying out of the boundaries. But since the velocity is binary-coded, V_{max} parameter is not needed. The second motivation lies in the very definition of position. The defined position vector is integer coded, how to define a proper velocity to work on the position is nontrivial. The defined velocity actually reflects the differences between two position vectors.

3.4.4 Discrete Particle Status Updating

In the MODPSO, a velocity provides a particle with the moving direction and tendency. After updating the velocity, one particle makes use of the new velocity to build new position. Since the position and velocity in MODPSO are all integer vectors, the mathematical updating rules in continuous PSO no longer fit the discrete situation, therefore, they are redefined to meet the requirements of network clustering problem. The velocity updating rule is redefined in discrete form as:

$$V_i = sig(\omega V_i + c_1 r_1 (Pbest_i \oplus X_i) + c_2 r_2 (Gbest \oplus X_i)) \tag{3.12}$$

where ω is the inertia weight, c_1 and c_2 are the cognitive and social components, respectively. r_1 and r_2 are two random numbers with range [0,1].

In Eq. 3.12, $\oplus$ is defined as a XOR operator and the function $Y = sig(X)$, where $Y = (y_1, y_2, ..., y_n)$, $X = (x_1, x_2, ..., x_n)$, is defined as follows:

$$\begin{cases} y_i = 1 \; if \; rand(0,1) < sigmoid(x_i) \\ y_i = 0 \; if \; rand(0,1) \geq sigmoid(x_i) \end{cases} \tag{3.13}$$

where the *sigmoid* function is defined as:

$$sigmoid(x) = \frac{1}{1 + e^{-x}} \tag{3.14}$$

In MODPSO, to promote exploration and exploitation, the inertia weight ω is randomly generated between [0,1], and the cognitive and social components c_1 and c_2 are set to the typical value of 1.494.

Based on the newly defined discrete velocity updating rule, the position updating rule is defined as the following discrete form:

$$x_i^t = x_i^t \otimes v_i^t \tag{3.15}$$

In Eq. 3.15 the operator $\otimes$ is the key procedure in the particle status updating process. It directly affects the performance of the algorithm since the right operand of $\otimes$ determines the direction to which the particle flies. A good operator $\otimes$ should help to guide the particle to a better place that is much closer to the food rather than a bad position that is far away from it.

Given a position $X_1 = \{x_{11}, x_{12}, ..., x_{1n}\}$ and a velocity $V = \{v_1, v_2, ..., v_n\}$. position $\otimes$ velocity generates a new position which corresponds to a new solution to the optimization problem, i.e., $X_1 \otimes V = X_2$, $X_2 = \{x_{21}, x_{22}, ..., x_{2n}\}$. The element of X_2 is defined as follows:

$$\begin{cases} x_{2i} = x_{1i} \quad if \; v_i = 0 \\ x_{2i} = Nbest_i \; if \; v_i = 1 \end{cases} \tag{3.16}$$

where $Nbest_i$ is an integer. Suppose vertex i has a neighbor set $N = \{n_1, n_2, ..., n_k\}$, then $Nbest_i$ is calculated by:

$$Nbest_i = \arg\max_r \sum_{j \in N} \varphi(x_{1j}, r) \tag{3.17}$$

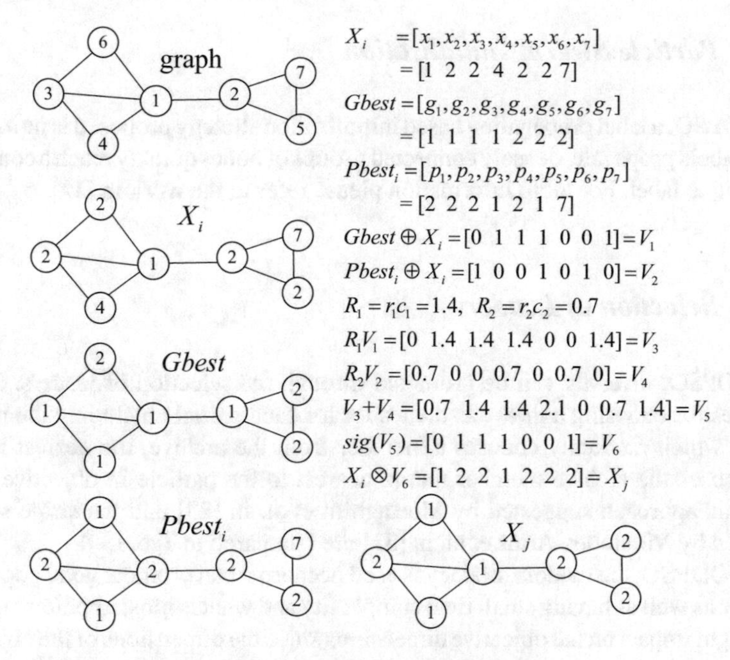

Fig. 3.15 A schematic example of how a particle updates its status

where $\varphi(i, j) = 1$, if $i = j$, otherwise, 0. The function $\arg\max_r f(x)$ returns the value of r that maximizes $f(x)$. So, in Eq. 3.17, $Nbest_i$ actually equals to the label identifier possessed by the majority of the neighbors of node i. To calculate $Nbest_i$ in this way makes sense, because in reality it is more possible for one member to join the community that is formed by the majority of its friends.

A schematic example of the detailed operations about the discrete particle status updating rules can be found in Fig. 3.15.

In Fig. 3.15, X_i and $Pbest_i$ are the current position and the personal best position of particle i, respectively. $Gbest$ is the global best solution of the swarm. V_1, V_2, V_3, V_4 and V_5 are the intermediate variables. V_6 and X_j are figured out by Eq. 3.12 and Eq. 3.15, respectively.

From what is illustrated above, the MODPSO framework has the following features:

- The definitions of discrete position and velocity are straight forward and very simple.
- The newly defined arithmetic operators are very easy to realize which greatly lower down the computational complexity.
- The MODPSO framework does not need to know the clusters of a network in advance, it can automatically determine it by itself.

The MODPSO framework seems to be very suitable for solving network discovery problem.

3.4.5 Particle Swarm Initialization

In MODPSO, a label propagation based initialization strategy proposed is introduced. As the labels propagate, densely connected groups of nodes quickly reach a consensus on a unique label. For more information please refer to the work in [12].

3.4.6 Selection of Leaders

In a MOPSO, diversity can be promoted through the selection of leaders. Several strategies for choosing a *gbest* as the leader for each particle including the random method which randomly chooses a member from the archive, the nearest method that chooses the archive member that is nearest to the particle in objective space, the sigma approach suggested by Mostaghim et al. in [23] and the stripe strategy advanced by Villalobos-Arias et al. in [31] are compared in Table 3.8.

In MODPSO, the random strategy is used because it can promote good population diversity as well as having small time complexity and what is most important is that it has a slight impact on the objective dimensions since the dimensions of the objectives are rather high. For one particle, there are *ns* corresponding neighbors (defined based on the Euclidean distances between the aggregation weight coefficient vectors), then one particle is randomly chosen from the neighbors as the leader to guide the flight.

In Step 3.8, the *pbest* is updated by using the concept of Pareto dominance, i.e., if the newly generated solution dominates the *pbest* solution, then update it with the new solution, if *pbest* dominates the newly produced solution, *pbest* remains its original state, if they are mutually nondominated, then aggregation method is used to determine whether to update *pbest* or not. The process is given in Algorithm 13.

Algorithm 13 *pbest* updating process.

1: **if** $x_i^{t+1} \prec pbest_i$ **then**
2: $\quad pbest_i = x_i^{t+1}$
3: **else**
4: $\quad$ **if** $x_i^{t+1} \succ pbest_i$ **then**
5: $\quad\quad$ do nothing;
6: $\quad$ **else**
7: $\quad\quad$ **if** $w_{i1} f(x_i^{t+1}) + w_{i2} f(x_i^{t+1}) < w_{i1} f(pbest_i) + w_{i2} f(pbest_i)$ **then**
8: $\quad\quad\quad pbest_i = x_i^{t+1}$;
9: $\quad\quad$ **end if**
10: $\quad$ **end if**
11: **end if**

Table 3.8 Comparison between different leader selection mechanisms. D is the objective dimension, A is the archive size, and N is the particle swarm size

Strategy	Time complexity	Diversity	Impact on dimension
Random	$O(N)$	Good	Slight
Nearest	$O(DAN)$	Normal	Middle
Sigma	$O((D(D-1)/2)AN)$	Bad	Great
Stripe	$O(N)$	Good	Two dimensions

3.4.7 *Turbulence Operator*

To preserve diversity and help a MOPSO to escape from local optima, many existing MOPSOs adopt the turbulence operator. For the network community discovery problem, in MODPSO, the adopted turbulence operator is called *NBM* (neighbor-based mutation). The procedure can be depicted as follows: first generate a pseudorandom number between 0 and 1, for each gene in every chromosome, if the random number is smaller than the mutation probability *pm*, the *NBM* process is applied to the gene, namely, assign its label identifier to all of its neighbors. The pseudocode is given in Algorithm 14.

Algorithm 14 Pseudo code of turbulence operation on one particle.

```
1: for i = 0; i < vertex; i++ do
2:     if rand(0, 1) < pm then
3:         for j = 0; j < node[i].neighborsize; j++ do
4:             x[node[i].neighbor[j]] = x[i];
5:         end for
6:     end if
7: end for
```

3.4.8 *Complexity Analysis*

1. *Space Complexity*: In MODPSO, two main memorizers are needed. The first one is the clustering data memorizer which needs a complexity of $O(n^2)$, n is the number of vertices of the network. The second memorizer is for the particles, say there are N particles, then the complexity is $O(Nn)$. Thus, the total space complexity of MODPSO is $O(n^2)$.
2. *Computational Complexity*: The main time complexity lies in Step 3 of MODPSO since Step 1 can be accomplished in linear time. Here, n and m are used to denote the vertex and edge numbers of the network, respectively. Step 3.1 and 3.7 need $O(1)$ basic operation, Step 3.2, 3.6 and 3.8 need $O(n)$ basic operations, and

Step 3.3 requires $O(D^2)$ basic operations where D is the averaged degree of a network. Step 3.4 needs $O(n)$ basic operations, Step 3.5 needs $O(m+n)$ basic operations. So, the worst case time complexity is $4O(n) + O(D^2) + O(m+n) + 2O(1)$. According to the operational rules of the symbol O, the worst case time complexity for MODPSO can be simplified as $O(pop \cdot maxgen \cdot (m+n))$, where pop and $maxgen$ are the population size and iteration number, respectively.

3.4.9 Experimental Results

Five EA based algorithms named as GA-net, Meme-net, MOGA-net, MOCD, MOEA/D-net and two MOPSO-based algorithms referred to as MOPSO-r1 and MOPSO-r2 are chosen to compare with the proposed approach. MODPSO is also compared with several other well known avenues named as GN, CNM and Informap. The experimental parameters of the algorithms are listed in Table 3.9.

The MOPSO algorithm proposed in [5] uses the concept of Pareto dominance to allow the heuristic to handle problems. It employs an external repository of particles that is used by other particles to guide their own flight. However, this algorithm is originally designed for continuous MOPs. In order to make it possible for comparison, based on the original framework of MOPSO in [5], the definitions of particles (position and velocity) and their mutual operations are made the same as that in the algorithm. Thus, a revised MOPSO (denoted as MOPSO-r1) for network clustering is introduced as a comparison algorithm. In [24], Palermo et al. put forward a DPSO algorithm for multi-objective design space exploration by making use of aggregation technique to extend the formulation of the PSO to the multi-objective domain.

Table 3.9 Parameters of the algorithms. pop represents the population size, $maxgen$ denotes the max iterations of the algorithm, pc and pm are the crossover and mutation possibility, respectively, and ns is the neighborhood size

Algorithm	pop	$maxgen$	pc	pm	ns	References
MODPSO (A1)	100	100	—	0.1	40	
MOPSO-r1 (A2)	100	100	—	0.1	—	[5]
MOPSO-r2 (A3)	100	100	—	0.1	—	[24]
GA-net (A4)	100	100	0.9	0.1	—	[25]
Meme-net (A5)	100	100	0.9	0.1	—	[13]
MOGA-net (A6)	100	100	0.9	0.1	—	[26]
MOCD (A7)	100	100	0.9	0.1	—	[30]
MOEA/D-net (A8)	100	100	0.9	0.1	40	[15]
GN (A9)	—	—	—	—	—	[11]
CNM (A10)	—	—	—	—	—	[4]
Informap (A11)	—	100	—	—	—	[28]

A revised version of this algorithm for network clustering is denoted as MOPSO-r2. In MOPSO-r1 and MOPSO-r2, the turbulence operators are the same as ours.

GA-net was proposed by Pizzuti, and it provided a feasible scheme to combine EA with network clustering problem. The author defined an evaluation index called community score to check the performance of the algorithm. The reason why it is chosen as comparison algorithm is to show EA-based algorithm is likely to become involved in local optima.

Meme-net is introduced in Sect. 2.2.

MOCD, MOGA-net, and MOEA/D-net are the three MOEA-based network community discovery algorithms. In order to make a comparison between the MOEA-based and the MOPSO-based algorithms, these three algorithms are chosen as the comparison approaches. In MODPSO, the decomposition strategy is utilized, so does MOEA/D-net, the main difference between the two approaches is the optimization strategy, MOEA/D-net uses genetic algorithm to optimize subproblems.

GN is a divisive hierarchical clustering algorithm proposed by Girvan and Newman. In this algorithm, the authors have first brought forward a well-known and widely traced network partitioning evaluation index, i.e., the modularity. The performance of GN is remarkably well though it is computationally complicated.

CNM was proposed by Clauset et al. [4]. This method is essentially a fast implementation of the GN approach.

Informap was proposed by Rosvall and Bergstrom in [28]. This algorithm is based on the information theory, it uses the probability flow of random walks on a network as a proxy for information flows in the real system and decomposes the network into modules by compressing a description of the probability flow.

For the case when the ground truth of a network is known, NMI (Eq. 2.3) is used. For the case when the ground truth of a network is unknown, the modularity Q and the signed modularity SQ are used as the evaluation indexes, and these two indexes are also used to select the ultimate clustering solution from the Pareto front.

3.4.9.1 Experiments on Artificial Generated Networks

The extension of GN benchmark network is also used to test the performance of MODPSO.

Figure 3.16 summarizes the maximum *NMI* values averaged over 30 runs for different algorithms when the mixing parameter γ increases from 0.0 to 0.5 with interval 0.05.

As is shown in Fig. 3.16, when the mixing parameter is no bigger than 0.35, almost all the methods, except GA-net, MOGA-net, MOPSO-r2 and MOCD, can figure out the true partitions (*NMI* equals 1). As the mixing parameter increases, the community structure in the network is becoming fuzzy gradually, it becomes harder and harder to figure out the true structures. When $\gamma = 0.4$, MODPSO gets $NMI = 1$ while MOPSO-r1 gets $NMI = 0.9952$. But when $\gamma = 0.45$, MODPSO still gets $NMI = 1$, MODPSO is still able to correctly classify the fuzzy network. From the curves, the proposed MODPSO algorithm is more powerful than the rest,

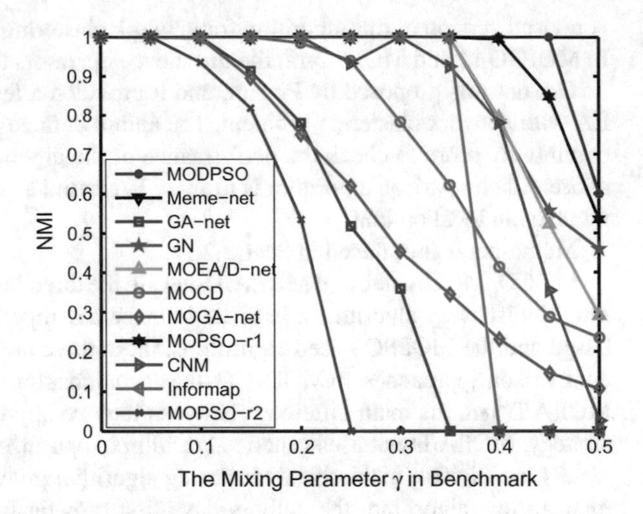

Fig. 3.16 Max *NMI* values averaged over 30 runs for different algorithms on GN extended benchmark data sets

though MOPSO-r1 approach is just a little shy to MODPSO, it is much slower since it keeps on updating the external archive which is time consuming.

In MODPSO, two diversity promotion mechanisms are introduced, i.e., label propagation based initialization (LPI) and turbulence operator. In order to show their impacts on the performance of the whole algorithm, some experiments are done here. In terms of the particle swarm initialization, here the introduced label propagation is introduced with random initialization and heuristic initialization introduced in [13]. Then, MODPSO is run with the arbitrary combinations between the initialization methods and turbulence operator. The statistical results are shown in Fig. 3.17.

It is clearly showed that the introduced two mechanisms can greatly improve the performance of the algorithm, especially the turbulence operator.

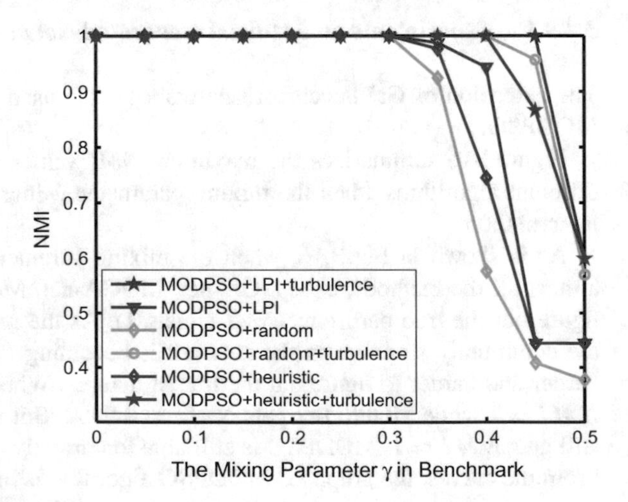

Fig. 3.17 Max *NMI* values averaged over 30 runs for MODPSO with arbitrary combinations between the initialization methods and turbulence operator

In MODPSO, the parameter ns is used to determine the neighborhood size. ns should be neither too small nor too large. To test its influence on the performance of MODPSO, the algorithm is run 30 times with different ns, and the averaged results are shown in Fig. 3.18. From the results, it is shown that $ns = 40$ seems perfect for MODPSO. Thus, in later experiments, ns is fixed as 40.

Apart from the parameter ns, the population size pop and the iteration number $maxgen$ parameters are also crucial. To test their influence on the performances of the algorithm, one of them is fixed as 100, and then the other one changes from 60 to 160 with interval 20. Figure 3.19 shows the statistical NMI values obtained by different pop and $maxgen$ numbers.

It is observed from the curves that when $maxgen$ is fixed as 100, $pop = 100$ seems perfect and when pop is fixed as 100, $maxgen = 140$ performs the best and $maxgen = 80, 100, 120, 160$ perform similarly well. Considering both the computational complexity and the algorithm performance, in MODPSO, pop and $maxgen$ are all set as 100.

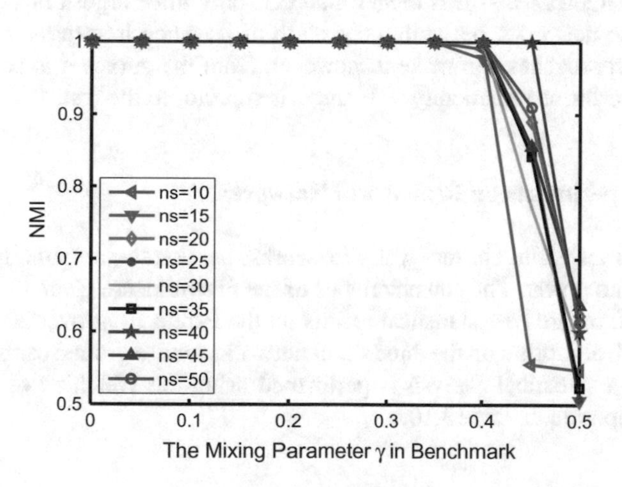

Fig. 3.18 Max *NMI* values averaged over 30 runs for MODPSO with different neighborhood size

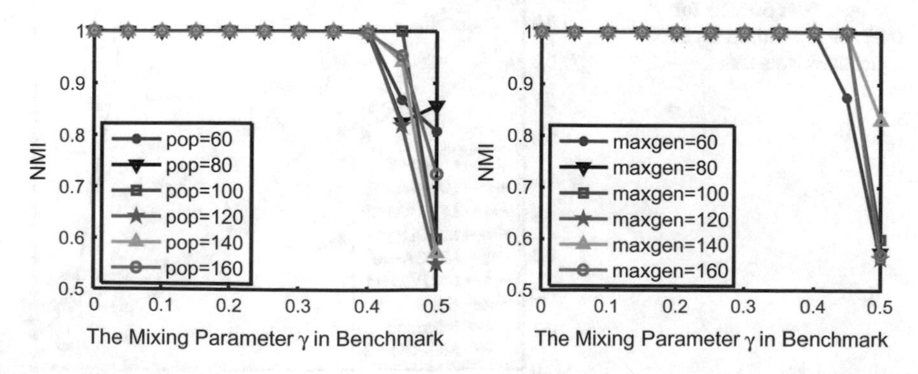

Fig. 3.19 Max *NMI* values averaged over 30 runs for MODPSO with different population size and iteration numbers

The above benchmark networks do not reflect some important features in graph representations of real systems, like the fat-tailed distributions of node degree and community size, because on those benchmark networks, all vertices have approximately the same degree, moreover, all communities have exactly the same size by construction. Therefore, new classes of benchmark graphs called LFR have been proposed by Lancichinetti et al. in [22], in which the distributions of node degree and community size are both power laws with tunable exponents. They assume that the distributions of degree and community size are power laws, with exponents τ_1 and τ_2, respectively. Each vertex shares a fraction $1 - \mu$ of its edges with the other vertices of its community and a fraction μ with the vertices of the other communities; $0 \leq \mu \leq 1$ is the mixing parameter. In experiments, 17 networks are generated with the mixing parameter increasing from 0 to 0.8 with an interval of 0.05. Each network contains 1000 nodes and the cluster size ranges from 10 to 50. $\tau_1 = 2$ and $\tau_2 = 1$, the averaged degree for each node is 20 and the max node degree is 50. Every algorithm is run 30 times and the statistical results are exhibited in Fig. 3.20.

From the figure, when μ is bigger than 0.1, only three algorithms can correctly classify these networks, but with the growth of μ, when it surpasses 0.6, none of these three approaches can make it, however, from the curves it is concluded that MODPSO performs remarkably well and it is superior to the rest.

3.4.9.2 Experiments on Real-World Networks

MODPSO is tested on six real-world networks, i.e., karate, dolphin, football, SFI, netscience and power. The characteristics of the networks are given in Table 1.1.

Table 3.10 records the statistical results on the Pareto front over 30 independent trials for each algorithm on the three small networks whose ground truths are known. In addition, a statistical analysis is performed using the Welch's t-test and the p-values are reported in Table 3.10.

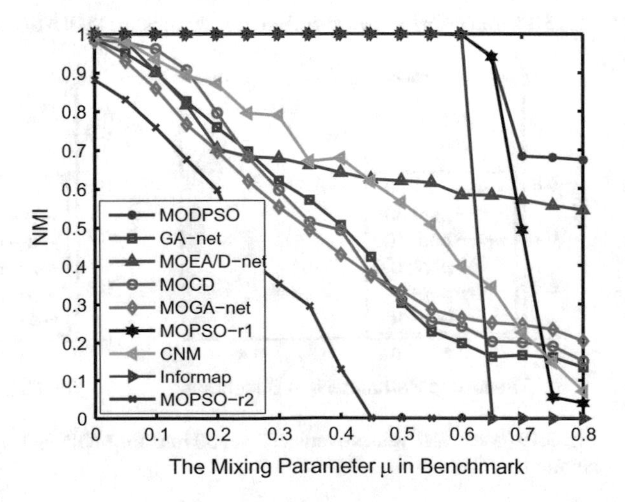

Fig. 3.20 Max *NMI* values averaged over 30 runs for different algorithms on LFR benchmark data sets

Table 3.10 Experimental results on the three networks with known ground truths. "×" means that there does not exist the value. The "*p*-value" is obtained over two modularity sets obtained by MODPSO and the comparison algorithm

Network	Index	A1	A2	A3	A4	A5	A6	A7	A8	A9	A10	A11
Karate	NMI_{max}	**1**	0.8372	**1**	0.6369	**1**	**1**	0.8372	**1**	0.8630	0.6920	0.6995
	NMI_{avg}	**1**	0.8372	0.8374	0.6369	0.8598	**1**	0.8372	**1**	0.8630	0.6920	0.6995
	Q_{max}	**0.4198**	0.4151	**0.4198**	0.4059	0.4020	**0.4198**	0.4188	**0.4198**	0.2330	0.3800	0.4020
	Q_{avg}	**0.4198**	0.4112	0.4189	0.4059	0.3857	0.4160	0.4188	**0.4198**	0.2330	0.3800	0.4020
	p-value	×	0.0000	0.0042	0.0000	0.0000	0.0000	0.0000	0.5000	0.0000	0.0000	0.0000
Dolphin	NMI_{max}	**1**	0.8379	**1**	0.4304	**1**	**1**	**1**	**1**	0.5540	0.5730	0.5622
	NMI_{avg}	**1**	0.8379	0.9566	0.4149	0.7853	0.9442	0.9901	**1**	0.5540	0.5730	0.5662
	Q_{max}	**0.5268**	0.5021	0.5258	0.5014	0.5155	0.5258	0.5259	0.5210	0.4060	0.4950	0.5247
	Q_{avg}	**0.5248**	0.4981	0.5237	0.4946	0.4838	0.5215	0.5210	0.5189	0.4060	0.4950	0.5247
	p-value	×	0.0000	0.0762	0.0000	0.0000	0.0000	1.0242e-4	0.0000	0.0000	0.0000	0.3491
Football	NMI_{max}	**0.9289**	0.9269	0.9064	0.9194	0.8616	0.8046	0.8928	0.9269	0.9210	0.7620	0.9242
	NMI_{avg}	**0.9278**	0.9269	0.8964	0.9001	0.7739	0.7883	0.8568	0.9264	0.9210	0.7620	0.9242
	Q_{max}	**0.6046**	**0.6046**	**0.6046**	0.5940	0.5888	0.5280	0.5958	0.6044	0.5350	0.5770	0.6005
	Q_{avg}	0.6035	**0.6044**	0.6012	0.5830	0.5512	0.5173	0.5785	0.6032	0.5350	0.5770	0.6005
	p-value	×	1	0.0491	0.0000	0.0000	0.0000	0.0000	0.2188	0.0000	0.0000	0.0000

In hypothesis testing experiments, the right-tailed t-test is performed with the null hypothesis that the two independent samples in the vectors X and Y (the two modularity sets) come from distributions with equal means and the significance level α is set as 0.05. $\alpha \geq p$ indicates that the null hypothesis can be rejected at the 5% significance level. The small p-values ($p \leq 0.05$) listed in Table 3.10 show that MODPSO outperforms almost all the comparison algorithms on the three small networks with known ground truths.

Table 3.10 shows that MODPSO can successfully detect the clustering ground truth of the network (correspond to $NMI = 1$). It also figures out the partition with highest Q value. These two clustering situations are displayed in Fig. 3.21a, b. It is obvious that Fig. 3.21b is a subdivision of Fig. 3.21a. On this network, MOEA/D-net and MODPSO perform the best from the angle of Q, but from the perspective of statistics, $p = 0.5$ indicates that the statistical difference between their performances is not significant.

On dolphin network, MODPSO and MOEA/D-net correctly figure out the true partition of the network, and from the perspective of modularity, MODPSO is superior to MOEA/D-net but inferior to that of MOCD. The detected results by MODPSO are displayed in Fig. 3.22.

As is shown in Fig. 3.22b that MODPSO separates the bottom part of the real network into three smaller parts but it misplaces vertex SN89. In the literatures, SN89 can be viewed as a fuzzy node, i.e., it can be either classified to the first cluster or to the second one. GA-net has found 9 clusters, but it misplaces node 8, 20, and 29. Meme-net discovers 5 clusters, but node 8, 20 have been misplaced. GN reveals 5 communities by dividing the bottom part into four smaller parts, but it is a pity that node 40 is wrongly classified. From the p-values listed in the table it is concluded that on this network, there are no significant statistical differences between the performances of MODPSO, MOPSO-r2 and Informap.

Due to the self-complicated structure of the football network, none of the algorithms can find the true partition. Meme-net and GA-net all wrongly separate several nodes. GN gets 12 communities which is rather close to the real ones though nodes 50, 60, 64, 98 have been forced to form a new group, and in [11] the author pointed out that this is because of the nuances in the scheduling of games. Figure 3.23 shows the clustering results of MODPSO on this complex network. For the partition with

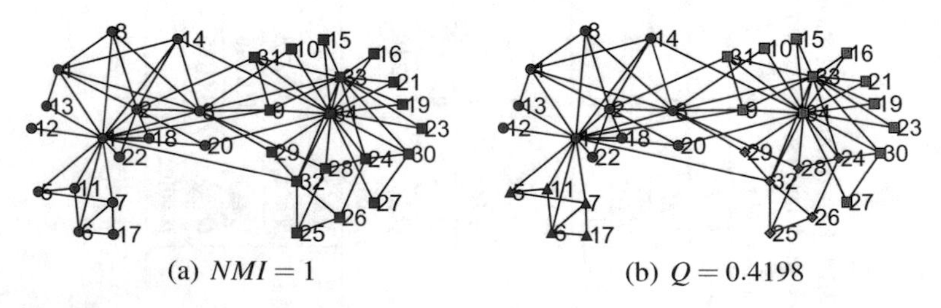

(a) $NMI = 1$ (b) $Q = 0.4198$

Fig. 3.21 The clustering results on karate club network by MODPSO. **a** is the real structure detected by MODPSO. **b** is the structure with highest Q value

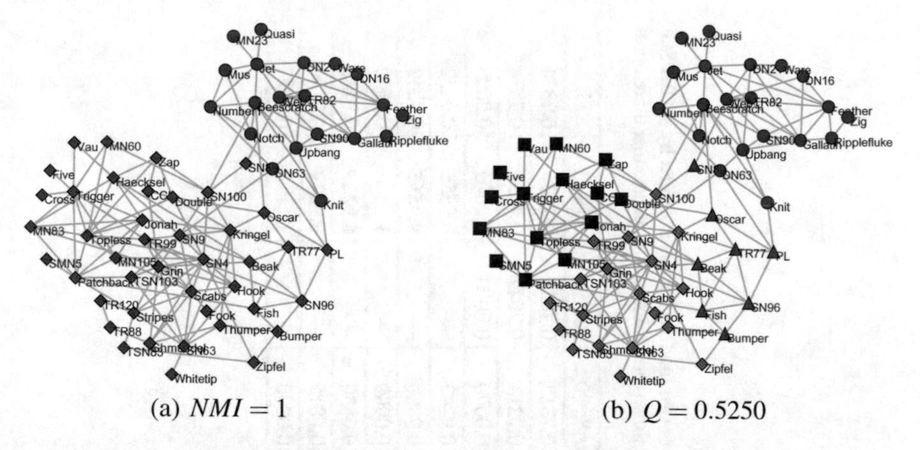

(a) $NMI = 1$ (b) $Q = 0.5250$

Fig. 3.22 Clustering results on dolphin network by MODPSO, **a** is the real structure detected by MODPSO, **b** is the structure with highest Q detected by MODPSO

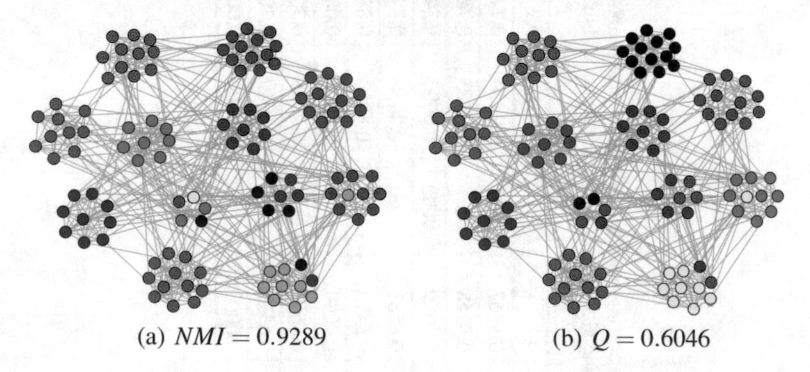

(a) $NMI = 0.9289$ (b) $Q = 0.6046$

Fig. 3.23 Results on football network by MODPSO, **a** is the structure with highest NMI value, **b** is the structure with highest Q value

highest modularity, MODPSO yields 10 clusters. By observation it is shown that it misplaces several vertices such as number 12, 25, 51, 36, 42, 58, 59, and 63, however, from the angles of *NMI* and modularity, MODPSO is still very effective and promising. From the p-values, it is noted that MOPSO-r1 is superior to MODPSO and MOEA/D-net performs similarly good as MODPSO while MODPSO outperforms the rest methods.

Table 3.11 records the statistical modularity values of the algorithms on the three complex networks with unknown ground truths. Statistical analysis using the Welch's t-test is also performed. MODPSO performs remarkably well and it is very fast especially when the size of the network is big. On the two large-scale networks, Meme-net and GN cannot give output within 40 minutes which indicates they possess high computational complexity.

From these very small p-values ($p \leq 0.05$) recorded in the table, one concludes that the difference is statistically significant. From the p-values, we note that

Table 3.11 Experimental results on the three networks with unknown ground truths. The p-value is obtained over two modularity sets obtained by MODPSO and the comparison algorithm. "—" denotes that the corresponding algorithm cannot give outputs within a given time (forty minutes). × means that there does not exist the value

Network	Index	A1	A2	A3	A4	A5	A6	A7	A8	A9	A10	A11
SFI	Q_{max}	**0.7484**	0.5877	0.5733	0.5867	0.7097	0.7430	**0.7493**	0.7312	0.7027	0.7335	0.7334
	Q_{Avg}	**0.7481**	0.5823	0.5528	0.5748	0.6818	0.7323	0.7474	0.7211	0.7027	0.7335	0.7334
	p-value	×	0.0000	0.0000	0.0000	0.0000	0.0000	**0.5460**	0.0000	0.0000	0.0000	0.0000
Netscience	Q_{max}	**0.9503**	0.9021	0.8381	0.8581	—	0.8916	0.8923	0.9143	—	**0.9555**	0.9252
	Q_{Avg}	**0.9493**	0.9016	0.7878	0.8473	—	0.8810	0.8886	0.9060	—	**0.9555**	0.9252
	p-value	×	0.0000	0.0000	0.0000	—	0.0000	0.0000	0.0000	—	**1**	0.0000
Power grid	Q_{max}	**0.8299**	0.6067	0.6362	0.6660	—	0.7035	0.7065	0.6880	—	**0.9229**	0.8140
	Q_{avg}	**0.8225**	0.6065	0.5987	0.6571	—	0.6949	0.7003	0.6815	—	**0.9229**	0.8140
	p-value	×	0.0000	0.0000	0.0000	—	0.0000	0.0000	0.0000	—	**1**	0.0217

MODPSO visibly outperforms almost all the comparison algorithms. On the SFI network, the p-value over the two modularity sets obtained by MODPSO and MOCD is 0.5460 which means that the corresponding algorithm outperforms MODPSO. On the netscience network and the power grid network, our method outperforms all the rest algorithms except that of CNM. One explanation is that CNM starts from a set of isolated nodes, and then the links of the original graph are iteratively added such to produce the largest possible increase of the modularity.

The biggest component of the SFI graph consists of 118 vertices and experiments are performed on this part.

In Fig. 3.24, the results are illustrated from the application of MODPSO to the largest component of the SFI network and compare it with that of GN. Vertices are drawn as different shapes and colors according to the primary divisions detected.

From the figure, it is found that the MODPSO splits the network into 8 strong communities, with the divisions running principally along disciplinary lines while GN gets 7. In Fig. 3.24a, community at the top (black diamond) represents a group of scientists using agent-based models to study problems in economics and traffic flow. The second community (green triangle) represents a group of scientists working on mathematical models in ecology. The third community (circle in yellow, pink, red, and blue) which is also the biggest one represents a group of scientists working primarily in statistical physics. MODPSO and GN subdivide this group into four small ones. The last community at the bottom of the figure is a group working primarily on the structure of RNA. MODPSO subdivides it into two small ones just like the author mentioned in [11] that it can be divided further into smaller subcommunities, centered once again around the interests of leading members. GA-net and Meme-net detect 26

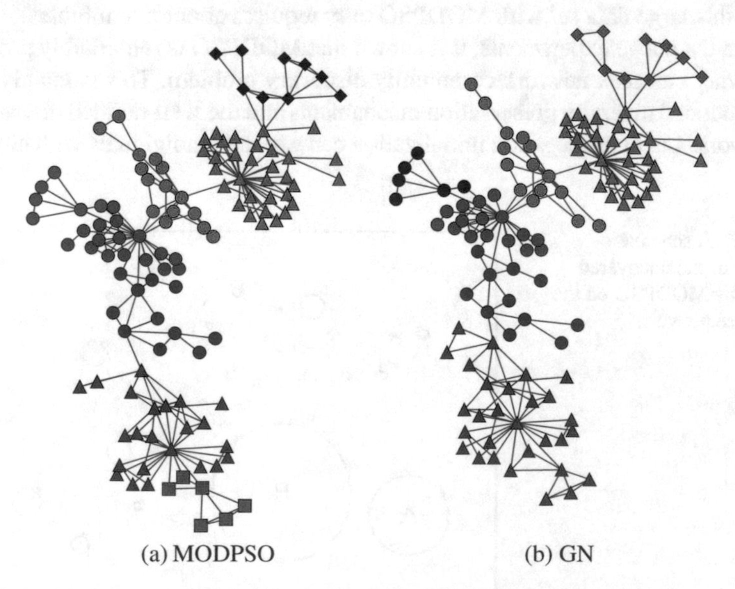

Fig. 3.24 Experiments on SFI network, **a** is the result of MODPSO, **b** is the one of GN. Different colors represent different communities

and 15 communities, respectively, but through the analysis, it is regarded that these results remain too much to be expected.Thus, MODPSO performs remarkably well.

The netscience network is weighted, in experiments, it is handled as an unweighted one. When running the GN and Meme-net algorithms, they cannot give outputs within 40 min which indicates that they have high computational complexity. The reason that leads to this phenomenon is that GN runs into calculating the edge betweenness which is time consuming while Meme-net is involved in greedy local search step which is very slow. However, MODPSO can get a high modularity within a very short time.

On this network, MODPSO obtains a partition of 286 clusters with a big modularity value of 0.9503 which indicates that MODPSO has discovered communities with strong structures. The clustering results can be represented by Fig. 3.25.

In Fig. 3.25, each circle denotes one cluster, the size of the circle indicates the size of the corresponding cluster. Circle A and B are the two largest clusters uncovered by our algorithm. Figures 3.26 and 3.27 exhibit the revealed structures of A and B.

On power network, all the algorithms except GN and Meme-net can yield high modularity values and MODPSO divides this big network into 255 strong small groups with a corresponding modularity value of 0.8299 though it is a little shy to that of CNM. CNM produces higher modularity value, but it still faces the modularity limitation problem while MODPSO just can avoid this. Besides, in [9] the author pointed out that a higher modularity does not necessarily mean a better partition of a network. What is more, it is believed that to obtain just one partition for a large-scale network is not sufficient. It is better to provide more clustering results for users to choose with different preferences and MODPSO can produce a set of different solutions. In addition, MODPSO is very fast, on the author's personal computer, testing this large data set with MODPSO only requires about five minutes.

From the above experiments, it is shown that MODPSO is remarkably promising for solving complex network community discovery problem. This is mainly due to the introduced diversity preservation mechanisms and the well-defined discrete PSO framework. Label propagation initialization can generate individuals with high clus-

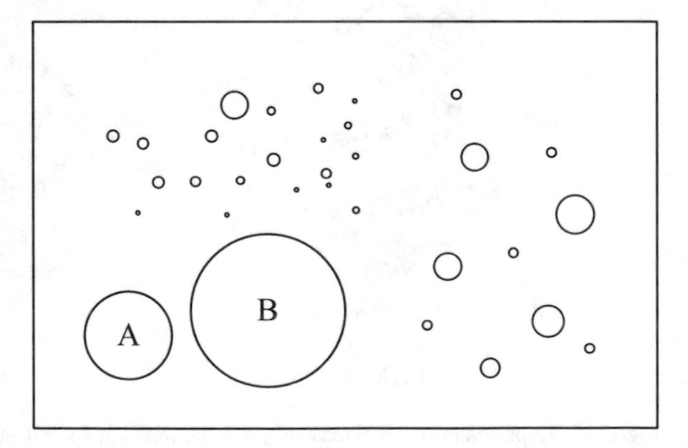

Fig. 3.25 A schematic example of the uncovered clusters by MODPSO on the netscience network

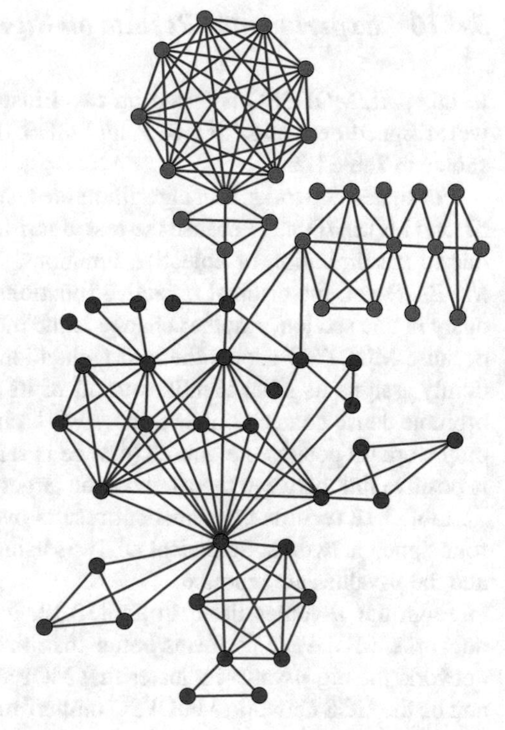

Fig. 3.26 Clustering result of our algorithm on the big part A of the netscience network

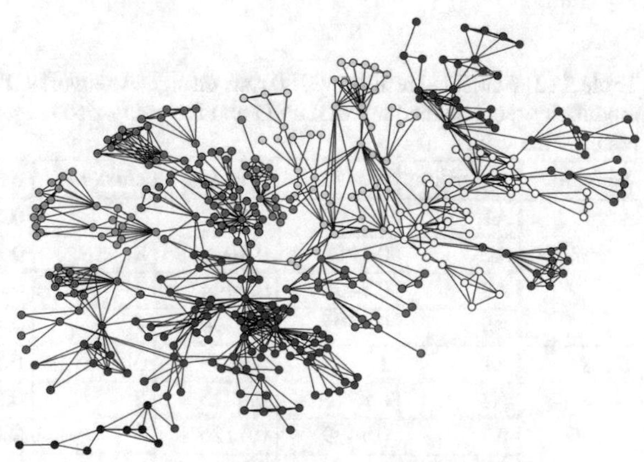

Fig. 3.27 Clustering result of our algorithm on the big part B of the netscience network

tering precision, turbulence operator helps to search optima solutions. The adopted decomposition strategy decomposes the MOP into a set of distinct scalar aggregation problems. Every particle solves the corresponding problem by applying priorities to each objective according to its weighting vector. This assists the optimization process to find potential solutions that are evenly distributed along the Pareto front and to mitigate against premature convergence.

3.4.10 Experimental Results on Signed Networks

In this part, MODPSO is tested on two illustrative signed networks and two real-world signed networks: SC, SPP, and GGS. The properties of these networks are shown in Table 1.2.

For signed networks, four algorithms are tested, MODPSO, MOPSO-r1, MOPSO-r2, and MOEA/D-net, because the rest algorithms cannot deal with signed networks due to the limitation of objective functions. To run MOPSO-r1, MOPSO-r2, and MOEA/D-net, the original objective functions are changed to those defined previously in this section. Another change is the mutation possibility which is set as 0.9, because MODPSO adopts the *NBM* mutation operation in which every node randomly assigns its label identifier to all of its neighbors, what is more, in order to promote dense connections within network clusters, the *NBM* process is limited to only work on positive neighbors (a node is said to have a positive neighbor if there is positive link between them), thus, the larger mutation possibility, the better.

Table 3.12 records the statistical results over 30 trials for the algorithms on the four signed networks. Statistical analysis using the Welch's t-test is also performed and the p-values are reported.

From the p-values listed in Table 3.12, it is noted that on the two illustrative networks, MODPSO performs better than the comparison algorithms, on the SPP network, the big p-value indicates that MOPSO-r1 performs better than MODPSO, and on the GGS network, MODPSO outperforms MOPSO-r1 and MOEA/D-net and

Table 3.12 Statistical results over 30 runs on signed networks. The "p-value" is obtained over two modularity sets obtained by MODPSO and the comparison algorithm. "×" means that there does not exist the value

Network	Algorithm	NMI_{max}	NMI_{avg}	*clusters*	SQ_{max}	SQ_{avg}	p-value
SC 1	A1	**1**	**0.9742**	3	**0.5213**	**0.5112**	×
	A2	0.7057	0.7022	8	0.3890	0.3868	0.0000
	A3	0.7495	0.7444	7	0.4305	0.4270	0.0000
	A8	0.7399	0.7399	7	0.4123	0.4115	0.0000
SC 2	A1	**1**	**0.9959**	3	**0.5643**	**0.5634**	×
	A2	0.8213	0.8213	5	0.5214	0.5214	0.0000
	A3	0.7439	0.7126	8	0.4691	0.4532	0.0000
	A8	0.8184	0.8184	5	0.5214	0.5214	0.0000
SPP	A1	**1**	0.9949	2	**0.4547**	0.4532	×
	A2	**1**	**1**	2	**0.4547**	**0.4547**	0.8393
	A3	**1**	0.9592	2	**0.4547**	0.4424	0.0054
	A8	0.8471	0.8471	3	0.4086	0.4086	0.0000
GGS	A1	**1**	**1**	3	**0.4310**	**0.4310**	×
	A2	0.9106	0.9106	4	0.3870	0.3870	0.0000
	A3	**1**	0.9940	3	**0.4310**	0.4281	0.0777
	A8	0.9106	0.9106	4	0.3870	0.3870	0.0000

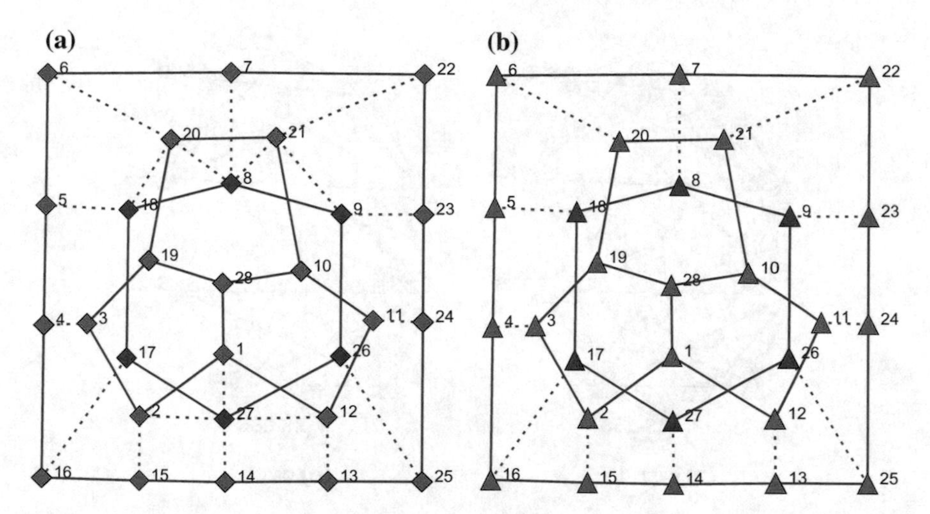

Fig. 3.28 Topology structures recognized by MODPSO on the two SC networks, **a** for SC1, **b** for SC2

the statistical differences between the performances of MODPSO, and MOPSO-r2 are not significant.

On the SC networks, only MODPSO can detect the ground truths of the networks. Figure 3.28a, b show the results of MODPSO on the two illustrative networks, where solid lines denote positive links and dashed lines denote negative links. Different colors represent different clusters. It has been found that the detected structures also have largest *SQ* values.

On the Pareto front yielded by MODPSO on the SPP network, there are three points representing three different solutions. One solution corresponds to the true partition with highest *SQ* value shown in Fig. 3.29b. Another solution separates the network as a single cluster and the third solution divides the network into three parts shown in Fig. 3.29b.

From Fig. 3.29b, node SNS has been isolated as a single group. This kind of segmentation is meaningful because node SNS is different from the rest nodes in its original group, it has both positive and negative links with the rest members.

The topology community structure of the GGS network recognized by MODPSO is shown in Fig. 3.30 where links represent political arrangements with positive (solid line) and negative (dash line) ties, MODPSO on this network is rather stable, this is mainly due to the intrinsic strong structure of the network since there only exists positive links within group and negative links between groups.

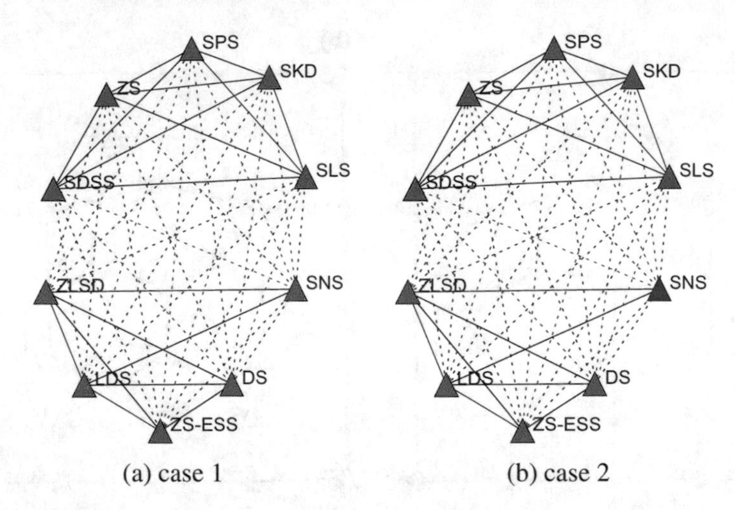

Fig. 3.29 Results of MODPSO on the SPP network. Case 1 and 2 are the two clustering topology structures that correspond to the two solutions on the Pareto front

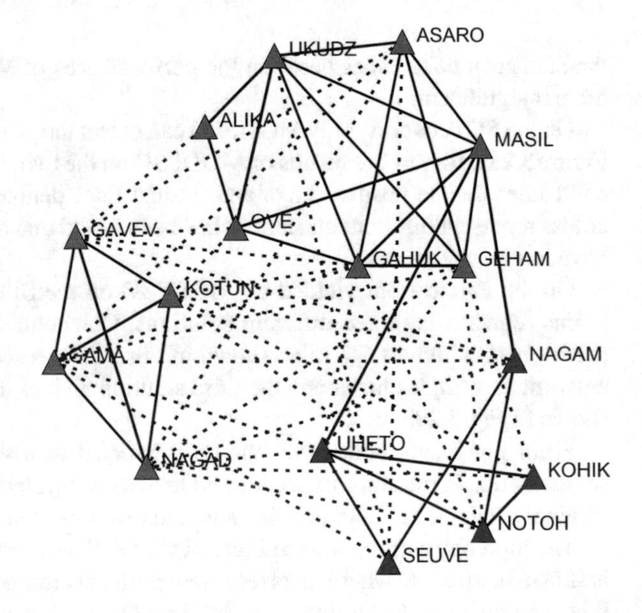

Fig. 3.30 Topology structures recognized by MODPSO on the GGS network

3.4.11 Conclusions

In this section, a novel discrete MOPSO algorithm (MODPSO) for complex network community discovery was introduced. The MODPSO algorithm first adopts decomposition mechanism to decompose multi-objective network community discovery problem into a number of scalar problems, then it optimizes them simultaneously using a discrete PSO framework in which the particle's representation and updating rules have been redefined in discrete context. The diversity among these subproblems will naturally lead to diversity in the population. A problem-specific population initialization method based on label propagation is introduced. This method can generate diverse individuals with high clustering efficiency. A neighbor-based turbulence operator is utilized to promote diversity.

MODPSO minimizes two objectives termed as Kernel K-Means and Ratio Cut. Since these two objectives leave out of consideration of the signed features of complex networks, in order to handle signed networks, the objective functions are extended into signed version. Experimental results show that MODPSO is effective on both signed networks and unsigned networks.

3.5 A Multi-objective Evolutionary Algorithm for Community Discovery in Dynamic Networks

[3]In dynamic networks, the communities may evolve over time so that pose more challenging tasks than in static ones. Only a few studies examined the dynamics of communities in evolving social networks.

The discovery of community structure with temporal smoothness can be formulated as a multi-objective optimization problem. The first objective is the maximization of the community quality, which measures how well the community structure found represents the network at the current time. The second objective is the minimization of the temporal cost, which measures the distance between two community structures at consecutive time steps.

In this section, a novel multi-objective immune algorithm with local search, termed as DYN-LSNNIA to solve the community discovery problem in dynamic networks, is introduced.

3.5.1 Multi-objective Optimization for Community Discovery in Dynamic Networks

The dynamic network G is defined as a sequence of networks $G_t(V_t, E_t)$, i.e., $G = \{G_1, G_2, ..., G_t\}$, where V_t is a set of objects, each $v_i \in V_t$ represents an individual and each edge $e_{ij} \in E_t$ denotes the presence of interactions between v_i and v_j. G_t in the graph is used to represent the snapshot of the network N_t at time t. Let $S_t = \{C_t^1, C_t^2, ..., C_t^k\}$ denote the community structure of the network N_t at time t.

[3]Acknowledgement: Reprinted from Journal of Computer Science and Technology, 27(3), Gong, M.G., Zhang, L.J., Ma, J.J., Jiao, L.C., Community detection in dynamic social networks based on multi-objective immune algorithm, 455–467, Copyright (2012), with permission of Springer.

The main loop of the dynamic multi-objective community detection algorithm based on NNIA with local search, termed as DYN-LSNNIA will be given in this section. In order to solve the problem of community discovery in dynamic social networks, it is needed to deal with the network at the initial time step firstly. Because there is no history information at time step 1, that is to say that the temporal cost is zero, thus the network at time Step 1 can be clustered without smoothing. So, it needs to optimize only the first objective function, i.e., modularity, which is equivalent to the problem of single multi-objective optimization. GA-Net proposed by Pizzuti is an effective algorithm to discover communities in social networks by employing genetic algorithm [25]. Thus the new algorithm adopts GA-Net to process the initial network at time step 1, however, the objective function to be optimized is replaced by the modularity which used in DYN-LSNNIA. The main framework of the algorithm is as follows. In DYN-LSNNIA, the representation and initialization are the same as those in MOEA/D-Net and MICD.

Algorithm 15 DYN-LSNNIA

Input: T: (number of the time steps)
$\{G_1, G_2, ..., G_T\}$: (sequence of dynamic network). **Output:** $\{S_1, S_2, ..., S_T\}$: (sequence of community structure found in the dynamic network).

Step.1: Generate the initial clustering $S_1 = \{C_1^1, C_1^2, ..., C_1^k\}$ of the network G_1 with GA-Net.
Step.2: If $ts \geq T$ is satisfied, export the sequence of network $\{S_1, S_2, ..., S_T\}$ as the output, Stop; Otherwise, go to step 3.
Step.3: Use the procedure of the revised NNIA adapted for community detection to process the network G_{ts} at time step ts. During this procedure, select the dominant population D_t in each generation.
Step.4: Perform the local search on the selected individuals in D_t to generate the new dominant population $= D_t'$. Update the dominant population with $= D_t'$. And then finish the other operations according to the steps of the revised NNIA.
Step.5: Select the solution on the Pareto front, which has the maximum Community Score at the end of time step ts. Decode the selected individual to get the community structure $S_{ts} = \{C_{ts}^1, C_{ts}^2, ..., C_{ts}^k\}$ of the network G_{ts}.
Step.6: $ts = ts+1$, and then return to step 2.

3.5.2 Problem Formation

As described previously, the objective function is composed by the two competing objectives. The first objective is the snapshot cost which measures how well a community structure S_t represents the data at time t. And modularity (Eq. 1.6) which not only maximizes the number of connections inside one community and minimizes the number of links between the communities is the right objective function needed.

The second objective function is the temporal cost which measures how similar the community structure S_t is with the previous community structure S_{t-1}. Thus, the NMI (Eq. 2.3) is used as the second objective function to maximize.

In DYN-LSNNIA, these two objectives to be optimized should be maximized simultaneously.

Some special operators, such as proportional cloning, uniform crossover, and mutation used in DYN-LSNNIA will be described in detail in the following sections.

3.5.3 Proportional Cloning

In DYN-LSNNIA, the proportional cloning T^C on the active population $A = \{a_1, a_2, ..., a_{|A|}\}$ is defined as

$$\begin{aligned} &T^c(a_1 + a_2 + \cdots + a_{|A|}) \\ =&T^C(a_1) + T^C(a_2) + \cdots + T^C(a_{|A|}) \\ =&\{a_1^1 + a_1^2 + \cdots + a_1^{q_1}\} + \{a_2^1 + a_2^2 + \cdots + a_2^{q_2}\} \\ &+ \cdots + \{a_{|A|}^1 + a_{|A|}^2 + \cdots + a_{|A|}^{q_{|A|}}\} \end{aligned} \tag{3.18}$$

where

$$\begin{aligned} T^C(a_i) &= \{a_i^1 + a_i^2 + \cdots + a_i^{q_i}\}, \\ a_i^j =&a_i, \quad i = 1, 2, \cdots, |A|, \quad j = 1, 2, \cdots, q_i \end{aligned} \tag{3.19}$$

q_i is a self-adaptive parameter. The representation + is not the arithmetical operator, but only separates the antibodies here. $q_i = 1$ denotes that there is no cloning on antibody a_i. The individual with greater crowding-distance value is reproduced more times, therefore, the individual with greater crowding-distance value has a larger q_i. Because the crowding-distance value of boundary solutions is positive infinity, before computing the value of q_i for each active antibody, the crowding-distance value of the boundary individuals (in objective space) is set to be equal to the double value of the maximum value of active antibodies except the boundary individuals. Then the value of q_i is calculated as

$$q_i = \left\lceil n_C \times \frac{\zeta(a_i, A)}{\sum_{j=1}^{|A|} \zeta(a_j, A)} \right\rceil \tag{3.20}$$

where $\zeta(a_i, A)$ denotes the crowding-distance value of the active antibodies a_j, n_C is an expectant value of the size of the clone population.

Figure 3.31 illustrates the procedure of proportional cloning. All the antibodies in subpopulation $\{a_i^1, a_i^2, ..., a_i^{q_i}\}$ are the result of the cloning on antibody a_i, and have the same property as a_i. In fact, cloning on antibody a_i is to make multiple identical copies of a_i. The aim is that the greater the crowding-distance value of an individual, the more times the individual will be reproduced. So, there exist more chances to search in less-crowded regions of the trade-off front.

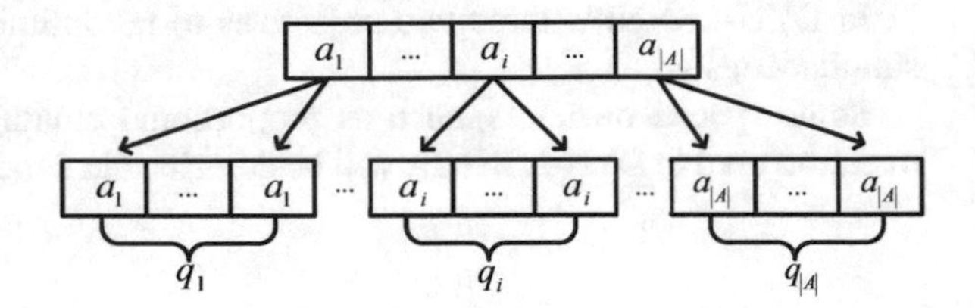

Fig. 3.31 Illustration of the proportional cloning

3.5.4 *Genetic Operators*

Crossover. To guarantee the maintenance of the effective connections of the nodes in the social network in the child individual, the uniform crossover is adopted to replace the recombination in NNIA. Select two arbitrary safe parent individuals, and then produce a random binary vector. If the vector is 1 then select the genes from the first parent, otherwise select the genes from the second parent and combine the genes to form the child. Because of the biased initialization, the child generated from the two safe parents is guaranteed to be safe. That is to say, if a gene i contains a value j, then the edge (i, j) exists.

Mutation. In order to solve the problem of community discovery using NNIA, the static hypermutation which adopted in NNIA is also replaced by mutation operation suited to community discovery. Thus, select the gene of the individual with a certain probability to mutate from the child population. However, the possible value an allele must be one of the replaced genes neighbors, which guarantees the mutated child is also safe as the crossover operation. The uniform crossover and mutation are shown in Fig. 3.32.

3.5.5 *The Local Search Procedure*

When solving community detection problems, it is an effective method to generate a new candidate solution by continuously executing the following three types of operations on current candidate solution, which includes moving single nodes from one community to another, merging multi-communities and splitting single communities [18]. Crossover operator is regarded as a macroscopic operation on individuals, while the mutation operator is regarded as a microcosmic operation on individuals. Thus, if the crossover operator can achieve its global search function by merging and splitting communities, and the mutation operator can achieve its local search function by moving single nodes between communities [20]. Inspired by this idea, the local search algorithm is based on the mutation operator. Because the local search strategy requires a single-objective function, a weighted objective or a Tchebyscheff metric or any other metric which will convert multiple objectives into a single objective can be used. In DYN-LSNNIA, a weighted objective is used:

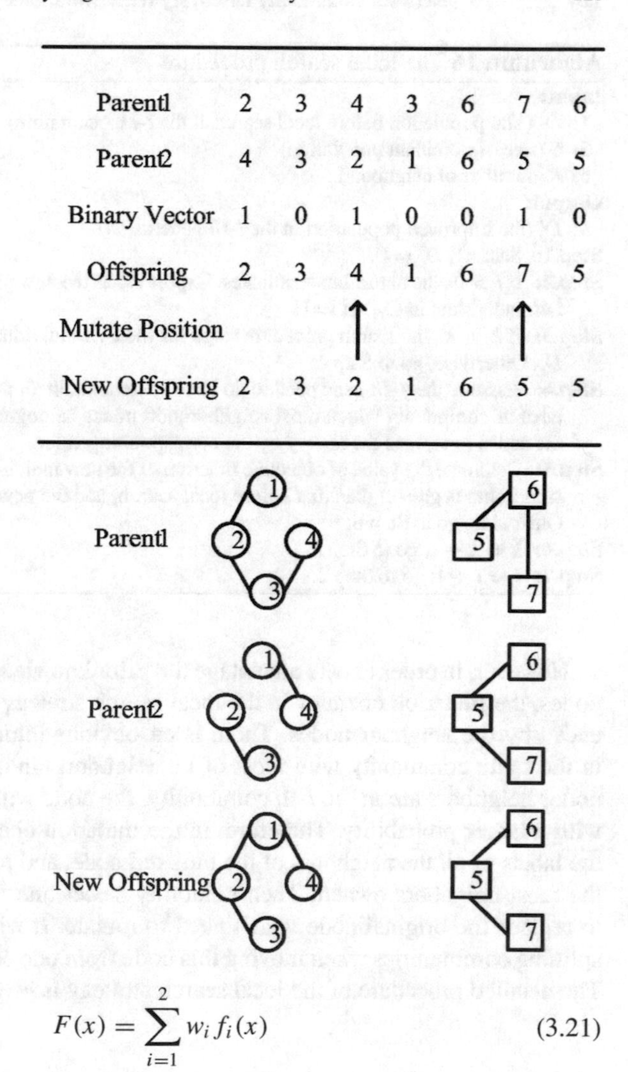

Fig. 3.32 Illustration of the uniform crossover and mutation

$$F(x) = \sum_{i=1}^{2} w_i f_i(x) \tag{3.21}$$

where w_1,w_2 are nonnegative weights for the two objectives, $f_i(x)$ are the objective functions, and the weights are calculated from the obtained set of solutions in a special way. First, the minimum f_i^{min} and maximum f_i^{max} value of each objective function f_i are noted. Thereafter, for any solution x in the obtained set, the weight for each objective function is calculated as follows [6]:

$$w_i = \frac{(f_i(x) - f_i^{\min})/(f_i^{\max} - f_i^{\min})}{\sum_{k=1}^{2} (f_k(x) - f_k^{\min})/(f_k^{\max} - f_k^{\min})} \tag{3.22}$$

where the division of the numerator with the denominator ensures that the calculated weights are normalized or $\sum_{i=1}^{2} w_i = 1$.

Algorithm 16 The local search procedure

Input:

1: D_t (The population before local search at the t-th generation)

2: S (size of dominant population)

3: K (number of neighbors)

Output:

4: D_t' (the improved population in the t-th generation)

Step.1: Set i=1, $D_t' = \emptyset$.

Step.2: If $i > S$, the algorithm terminates. Export D_t' as the new population. Otherwise, select the i-th individual in D_t, set k=1.

Step.3: If $k > K$, the search procedure stops for the i-th individual, add the current individual to D_t'. Otherwise, go to Step 4;

Step.4: Assume the j-th gene need to do local search, attain all the neighbors of node j, find the label of community which most neighborhood nodes belonging to. And then select one from the nodes to replace the node j by the corresponding value.

Step.5: Calculate the value of objective function of the new individual according to Equation (10). If its value is greater than that before local search, add the new individual to D_t', go to Step 7. Otherwise, go to Step 6.

Step.6: $k = k + 1$, go to Step 3.

Step.7: $i = i + 1$, go to Step 2.

However, in order to take advantage the prior knowledge about relations between nodes, the mutation operator in the local search strategy is not randomly but influenced by the neighbor nodes. There is an obvious intuition that the node will be in the same community with most of its neighbors. In other words, if most of the nodes neighbors are in the i-th community, the node will be in the i-th community with a larger probability. Therefore, in the mutation operation, it is needed to find the labels of all the neighbors of the mutated node, and record the node label which the most neighbors owned. Then randomly select one from these neighbor nodes to replace the original node which need to mutate. It will not result in merging or splitting communities when moving this node from one community to the other one. The detailed procedure of the local search strategy is given in the Algorithm 16.

3.5.6 Solution Selection

Actually, the algorithm DYN-LSNNIA returns a set of solutions at the end of each time step, which all contained on the Pareto front. Each of these solutions corresponds to a different trade-off between the two objectives and thus to diverse partitioning of the network consisting of a various number of clusters. The problem is how to select one best solution which denotes the optimal partitioning of the current network at each time step. A criterion should be established to automatically select one solution with respect to another. Unfortunately, there is still no effective selection method in current literature so far.

In DYN-LSNNIA, the Community Score introduced in [25] is used as the selection rule. The Community Score takes into account both the fraction of interconnections among the nodes and the number of interconnections contained in the module. It is defined as $CS = \sum_{i=1}^{k} score(C_i)$, where

$$score(C_i) = \frac{\sum_{i \in C} \mu_i}{|C|} \times \sum_{i,j \in C} A_{ij} \quad (3.23)$$

where $\mu_i = \frac{1}{|C|} \sum_{j \in C} A_{ij}$ denotes the fraction of edges connecting each node i of C to the nodes in the same community C. The community score gives a global measure of the network division in communities by summing up the local score of each module found. The larger community score indicates the community structure is stronger. Thus the best solution selected has the maximum Community Score in the set of solutions.

3.5.7 *Experimental Results*

In this section, the effectiveness and efficiency of DYN-LSNNIA are tested on four artificial datasets and two real-world networks. The compared algorithms include DYN-MOGA which is the only dynamic multi-objective community detection algorithm proposed by Folino et al. [8], and the DYN-LSNNIA without the local search strategy (termed as DYN-NNIA).

The parameter settings are as follows, in DYN-MOGA, crossover rate $P_c = 0.8$, mutation rate $p_M =, 0.2$, elite reproduction equals 10% of the population size, and roulette selection function. And the population size is 100, the number of generation is 300. For DYN-NNIA and DYN-LSNNIA, the maximum size of dominant population n_D=,100, the maximum size of active population $n_A = 20$, and the size of clone population $n_C = 100$, the crossover rate, mutation rate and the number of generation keep the same as DYN-MOGA. In the following experiments, the reported data are the statistical results based on 30 independent runs on each data set.

3.5.7.1 Experiments on Artificial Generated Datasets

The artificial GN datasets are used to test the performance of DYN-LSNNIA. In order to control the noise level in the dynamic networks, a parameter z_{out}, which represents the mean number of edges from a node to nodes in other communities, is introduced to describe the synthetic datasets. In experiments, the datasets are generated under four different noise levels by setting $z_{out} = 3, 4, 5, 6$. In order to introduce dynamics into the network, the community structure of the network evolves in the following way, at each time step after time step 1 and 10% of the nodes are selected to leave their original community and join the other three communities at random. The network with community evolution is generated in this way for 10 time steps.

To measure the accuracy of the community discovery results. NMI, which is described in Sect. 2.2 is used. In order to evaluate the results dependably, the standard error of the NMI is used at each time step to describe the stability of the algorithms. The standard error of a statistic is the standard deviation of the sampling distribution of that statistic. Standard errors are important because they reflect how much sampling fluctuation a statistic will show. The inferential statistics involved in the construction of confidence intervals and significance testing are based on standard errors. The standard error of a statistic depends on the sample size. In general, the larger the sample size, the smaller the standard error. In the experiments, the sample size is 30.

Figures 3.33 shows the statistical average value of NMI with respect to the ground truth over the 10 time steps when $z_{out} = 3$. It can be seen that the average value of NMI at each time step obtained by both DYN-LSNNIA and DYN-NNIA equal to 1, which illustrate these two algorithms can detect the true community structure at each time step. However, DYN-MOGA cannot always get the value 1 at each time step. In addition, the standard error obtained by DYN-MOGA at each time step is larger than both DYN-NNIA and DYN-LSNNIA. It can be known that the results got by DYN-MOGA are not steady enough compared to the other two algorithms.

Figure 3.34 presents the community score obtained by three algorithms at each time step. The larger community score is obtained, which indicates the corresponding network is densely connected within each sub-network. It can be seen that the community score got by these three algorithms at each time step almost the same, because the generated network can be detected effortlessly when $z_{out} = 3$.

Figures 3.35, 3.37, and 3.39 illustrate the statistical average value of NMI over the 10 networks for the 10 time steps, when $z_{out} = 4, 5, 6$. It can be seen that the

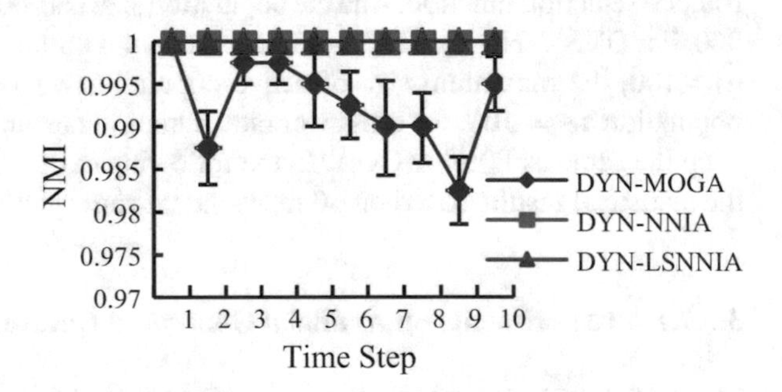

Fig. 3.33 NMI when $z_{out} = 3$

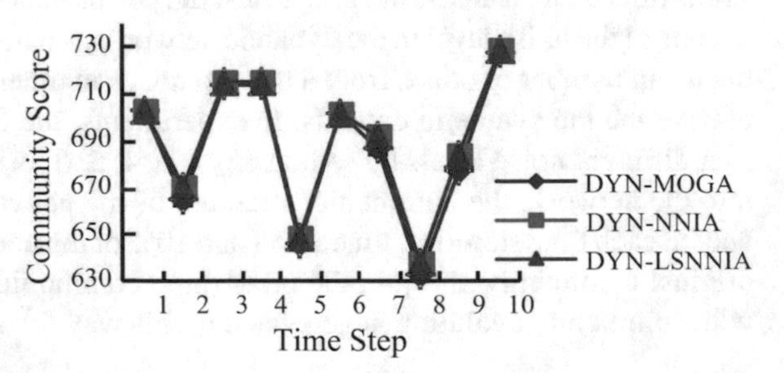

Fig. 3.34 Community score when $z_{out} = 3$

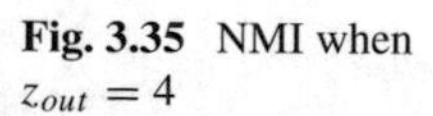

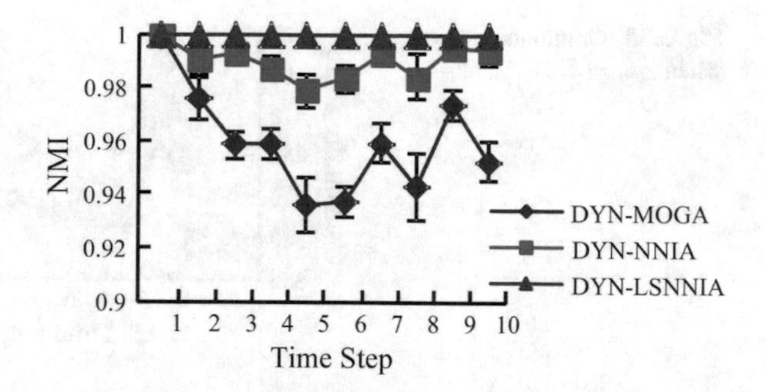

Fig. 3.35 NMI when $z_{out} = 4$

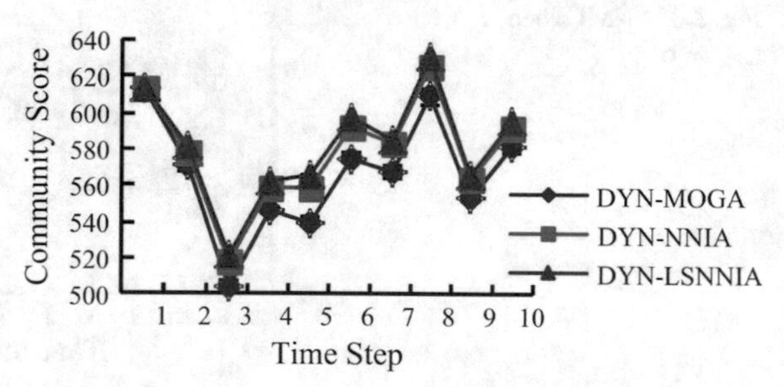

Fig. 3.36 Community score when $z_{out} = 4$

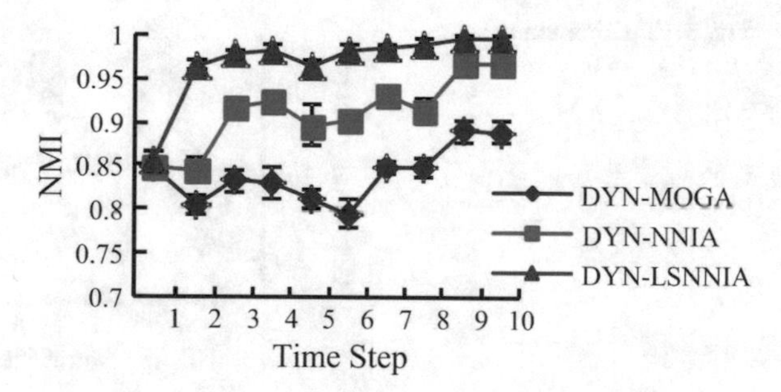

Fig. 3.37 NMI when $z_{out} = 5$

algorithm DYN-LSNNIA can still achieve very high accuracy compared to the other two algorithms when the noise level becomes high. However, it is worth noticing that the value got by three algorithms at time step 1 is basically equal. This is because that GA-Net is used to handle with the initial network in these three algorithms. Thus, the results at time step 1 almost the same. The same phenomena can be seen in all these experiments.

From Fig. 3.35, we can see that only the DYN-LSNNIA can find the true community structure, while the other two algorithms fail. As can be seen from Figs. 3.35, 3.37, and 3.39, with the variation of noise level, the average value of NMI becomes smaller, which demonstrate that the network becomes too complex to detect. Even so, DYN-LSNNIA can still get the better results than the other algorithms. Moreover, the algorithm DYN-LSNNIA is the most steady of all three algorithms, which can be seen from the standard error of the NMI at each time step.

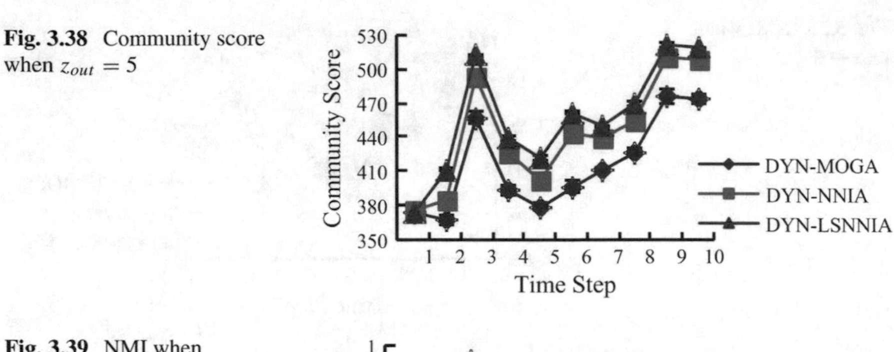

Fig. 3.38 Community score when $z_{out} = 5$

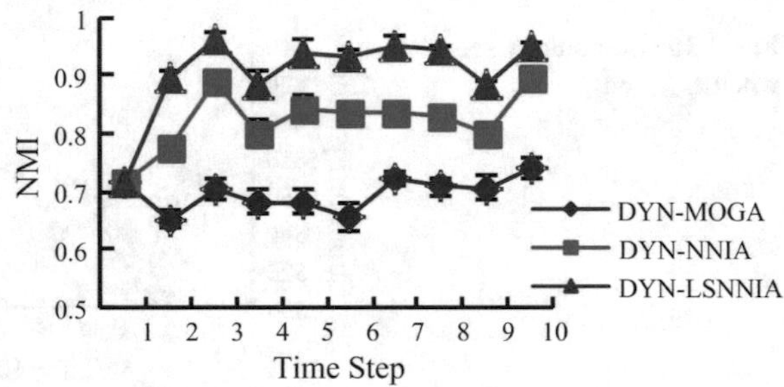

Fig. 3.39 NMI when $z_{out} = 6$

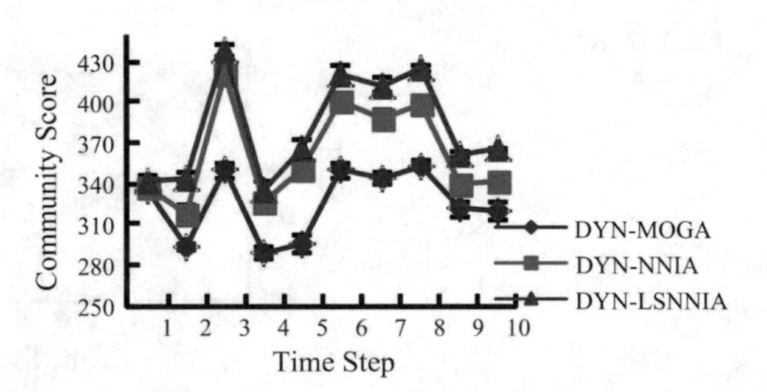

Fig. 3.40 Community score when $z_{out} = 6$

The Community Score obtained by three algorithms when $z_{out} = 4, 5, 6$ are shown in Figs. 3.36, 3.38 and 3.40. It is found that algorithm DYN-LSNNIA still outperforms the other two algorithms. That is to say, the solutions selected by DYN-LSNNIA denote the results approaching to the true community structure.

3.5.7.2 Experiments on Real-World Networks

In this section, experiments are tested on two real-world datasets: football and VAST.

In the football dataset, the year 2005–2009 is selected to evaluate the algorithm DYN-LSNNIA, each year as one time step, where the number of conferences is 12 and the number of teams is 120.

Figure 3.41 shows the statistical average value of NMI with respect to the ground truth over the 5 time steps. It also presents the better performance of DYN-LSNNIA

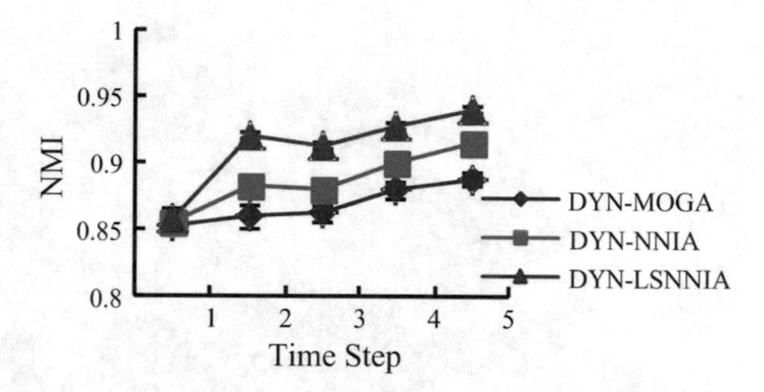

Fig. 3.41 NMI of the football dataset

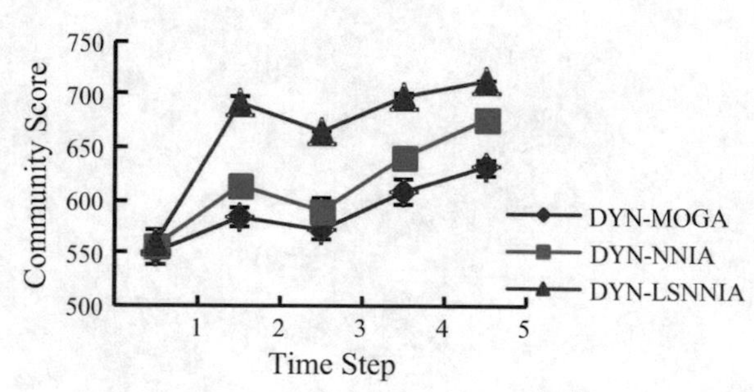

Fig. 3.42 Community score of the football dataset

compared to the other two algorithms. The average value of NMI obtained by DYN-LSNNIA is over 0.9 at each time step except time step 1, which illustrates that DYN-LSNNIA can discover the nearly true community structure at each time step. The community scores obtained by three algorithms are shown in Fig. 3.42.

From Figs. 3.41 and 3.42, it is seen that the results got by three algorithms are becoming better gradually over time except time step 3. Through the analysis, this is because the regular season games between these teams in 2007 are more frequent and irregular. However, this situation is improved in years 2008 and 2009, the regular season games between members of the same conference are arranged more. Thus, the community structure found in these two time steps can be more clear and accurate. In addition, it is also concluded as the synthetic datasets, the results obtained by DYN-LSNNIA are the most steady of all three algorithms from the error bars shown in Figs. 3.41 and 3.42.

In order to analyze visually, the communities recognized by DYN-LSNNIA on the Football data for the year 2009 are shown in Fig. 3.43. The small circles with the same color denote the nodes in one community.

From Fig. 3.43, it is seen that DYN-LSNNIA can be able to recognize 11 different communities. Almost all teams are correctly grouped with the other teams in their conference, which is an impossible mission for the other two algorithms. Only several teams are mistakenly divided, which is shown in different colors in the 11 partitions. That is to say, only the conferences Big 12, MAC, MWC, Pac 12, and WAC have the incorrect teams, which are not belonging to these conferences originally. However, there are four independent teams that do not belong to any conference. They tend

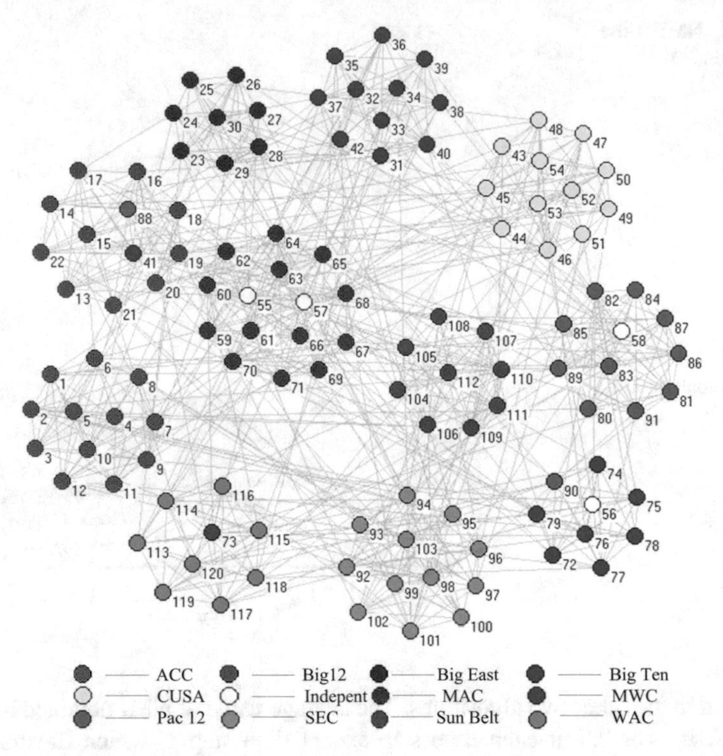

Fig. 3.43 The communities found by DYN-LSNNIA on the football data for the year 2009

to be grouped with the conference with which they are most closely associated. In short, DYN-LSNNIA achieves the best performance of all three algorithms.

The VAST Dataset is a challengeable task from IEEE VAST 2008. However, the experiment is only based on the VAST contest 2008 mini challenge 3, whose primary task is to characterize the Catalno/Vidro social network based on the cell phone call data provided and to characterize the temporal changes in the social structure over the 10-day period.

This dataset consists of information about 9834 calls between 400 cellphones over a 10 day period in June 2006 in the Isla Del Sueno. It includes records with the following fields: identifier for caller, identifier for receiver, time, duration, and call origination cell tower. A call graph G is a pair (V, E), where V is a finite set of vertices (mobile users), and E is a finite set of vertex-pairs from V (mobile calls). So, if user u calls user v, then an edge (u, v) is said to exist in E. The input social network and the corresponding dynamic graph G are converted into 10 different snapshot graphs.

Because the ground truth of the cellphone network is not known, thus the modularity is used to evaluate the network. If the modularity of the network is larger than the other one, it indicates the network connected strongly. Figure 3.44 shows the statistical average value of modularity of Catalno/Vidro social network over 10

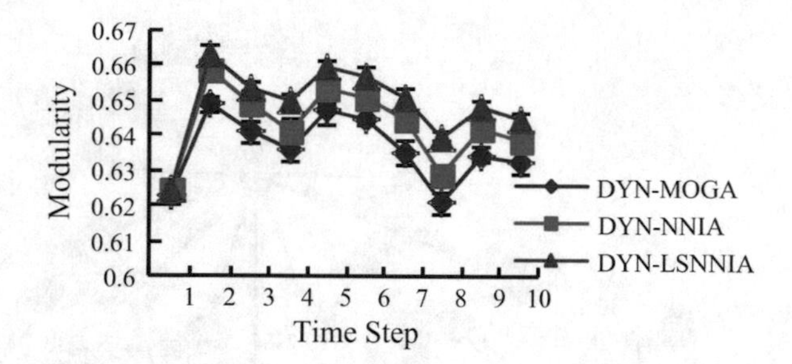

Fig. 3.44 Modularity of VAST dataset

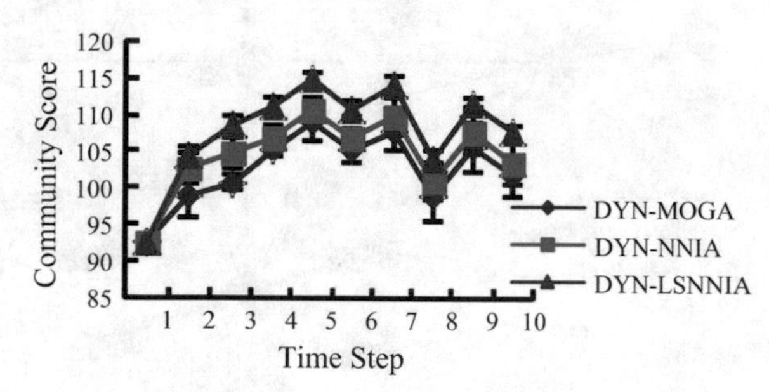

Fig. 3.45 Community score of VAST dataset

time steps. It can be seen that DYN-LSNNIA outperforms the other two algorithms at each time step except time step 1. Similar to the above results, the community structure found by DYN-LSNNIA are not only densely connected, but also more steady at each time step. The community score obtained by three algorithms is shown in Fig. 3.45.

It is known that this is a challenge task from IEEE VAST 2008. Thus, this dataset has been analyzed by many researchers. It has been known that the structure of the cellphone network changed drastically on the 8th day [33]. In other words, there is a significant variation that happened at the high-level leaders during this period. The main structures of the cellphone network at time step 7 are displayed in Fig. 3.46 and time step 8 in Fig. 3.47. From these two figures, it is found that node 200 is the main boss while nodes 1, 2, 3, 5 are important nodes in the Catalano hierarchy at time step 7. While at time step 8 the nodes 300, 306, 309, 360, 397 emerge into prominence. The community structure discovered by DYN-LSNNIA is consistent with the above analysis.

3.5.8 Conclusions

In this section, a novel multi-objective community discovery algorithm was introduced to discover communities and capture community evolutions in dynamic social networks. Experimental results on synthetic datasets and real-world networks demonstrate that DYN-LSNNIA can obtain better performance than the two compared

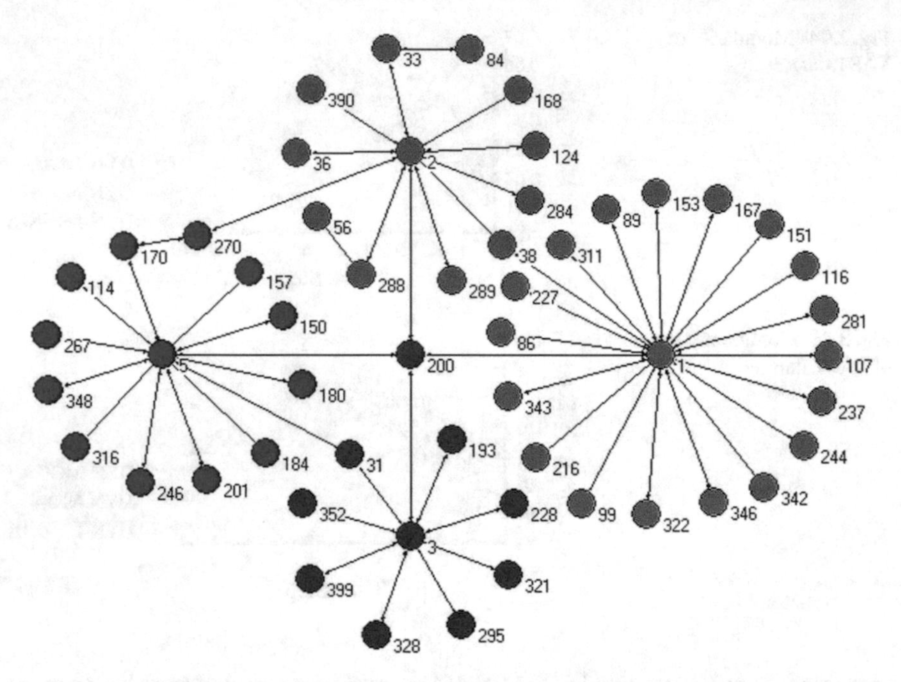

Fig. 3.46 The main community structure of VAST found by DYN-LSNNIA at time step 7

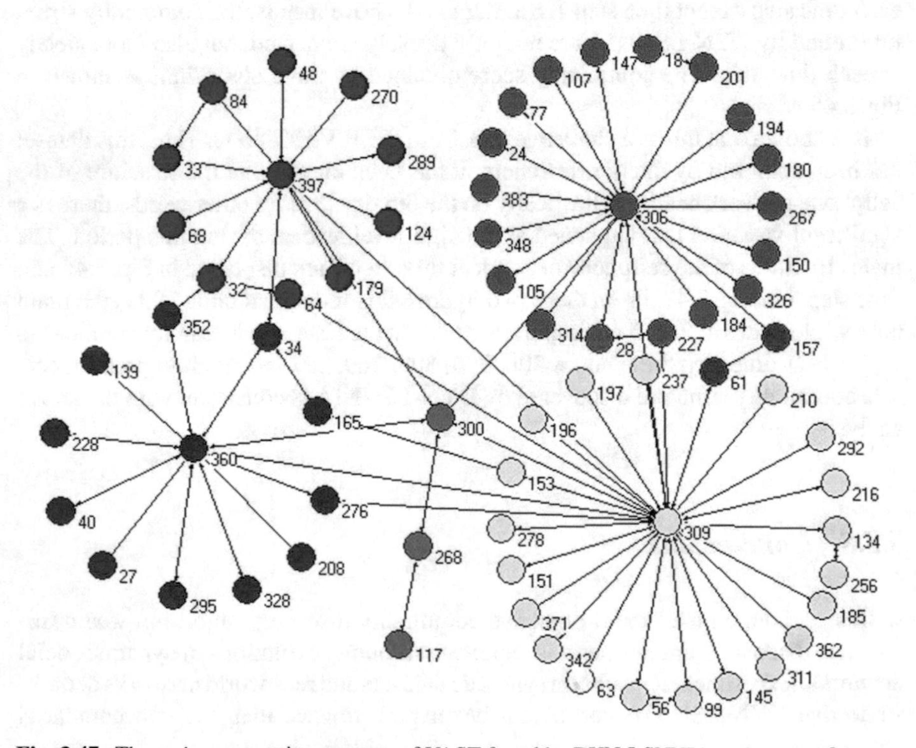

Fig. 3.47 The main community structure of VAST found by DYN-LSNNIA at time step 8

methods. It can achieve better accuracy in community extraction and capture community evolution more faithfully. The results obtained by the algorithm DYN-LSNNIA are not only more accurate but also more steady than the other two algorithms.

References

1. Angelini, L., Boccaletti, S., Marinazzo, D., Pellicoro, M., Stramaglia, S.: Identification of network modules by optimization of ratio association. Chaos Interdisc. J. Nonlinear Sci. **17**(2), 023,114 (2007)
2. Arenas, A., Diaz-Guilera, A., Pérez-Vicente, C.J.: Synchronization reveals topological scales in complex networks. Phys. Rev. Lett. **96**(11), 114,102 (2006)
3. Arenas, A., Fernandez, A., Gomez, S.: Analysis of the structure of complex networks at different resolution levels. New J. Phys. **10**(5), 053,039 (2008)
4. Clauset, A., Newman, M.E., Moore, C.: Finding community structure in very large networks. Phys. Rev. E **70**(6), 066,111 (2004)
5. Coello, C.A.C., Pulido, G.T., Lechuga, M.S.: Handling multiple objectives with particle swarm optimization. IEEE Trans. Evol. Comput. **8**(3), 256–279 (2004)
6. Deb, K., Goel, T.: A hybrid multi-objective evolutionary approach to engineering shape design. In: International Conference on Evolutionary Multi-criterion Optimization, pp. 385–399. Springer (2001)
7. Deb, K., Pratap, A., Agarwal, S., Meyarivan, T.: A fast and elitist multiobjective genetic algorithm: Nsga-ii. IEEE Trans. Evol. Comput. **6**(2), 182–197 (2002)
8. Folino, F., Pizzuti, C.: Multiobjective evolutionary community detection for dynamic networks. In: Proceedings of the 12th Annual Conference on Genetic and Evolutionary Computation, pp. 535–536. ACM (2010)
9. Fortunato, S.: Community detection in graphs. Phys. Rep. **486**(3), 75–174 (2010)
10. Fortunato, S., Barthelemy, M.: Resolution limit in community detection. Proc. Natl. Acad. Sci. **104**(1), 36–41 (2007)
11. Girvan, M., Newman, M.E.: Community structure in social and biological networks. Proc. Natl. Acad. Sci. **99**(12), 7821–7826 (2002)
12. Gong, M., Cai, Q., Li, Y., Ma, J.: An improved memetic algorithm for community detection in complex networks. In: 2012 IEEE Congress on Evolutionary Computation (CEC), pp. 1–8. IEEE (2012)
13. Gong, M., Fu, B., Jiao, L., Du, H.: Memetic algorithm for community detection in networks. Phys. Rev. E **84**(5), 056,101 (2011)
14. Gong, M., Jiao, L., Du, H., Bo, L.: Multiobjective immune algorithm with nondominated neighbor-based selection. Evol. Comput. **16**(2), 225–255 (2008)
15. Gong, M., Ma, L., Zhang, Q., Jiao, L.: Community detection in networks by using multiobjective evolutionary algorithm with decomposition. Phys. A: Stat. Mech. Appl. **391**(15), 4050–4060 (2012)
16. Gong, M., Chen, X., Ma, L., Zhang, Q., Jiao, L.: Identification of multi-resolution network structures with multi-objective immune algorithm. Appl. Soft Comput. **13**(4), 1705–1717 (2013)
17. Gong, M., Cai, Q., Chen, X., Ma, L.: Complex network clustering by multiobjective discrete particle swarm optimization based on decomposition. IEEE Trans. Evol. Comput. **18**(1), 82–97 (2014)
18. Guimera, R., Amaral, L.A.N.: Functional cartography of complex metabolic networks. Nature **433**(7028), 895–900 (2005)
19. Handl, J., Knowles, J.: An evolutionary approach to multiobjective clustering. IEEE Trans. Evol. Comput. **11**(1), 56–76 (2007)

20. Jin, D., He, D., Liu, D., Baquero, C.: Genetic algorithm with local search for community mining in complex networks. In: 22nd IEEE International Conference on Tools with Artificial Intelligence (ICTAI), vol. 1, pp. 105–112. IEEE (2010)
21. Lancichinetti, A., Fortunato, S., Kertész, J.: Detecting the overlapping and hierarchical community structure in complex networks. New J. Phys. **11**(3), 033,015 (2009)
22. Lancichinetti, A., Fortunato, S., Radicchi, F.: Benchmark graphs for testing community detection algorithms. Phys. Rev. E **78**(4), 046,110 (2008)
23. Mostaghim, S., Teich, J.: Strategies for finding good local guides in multi-objective particle swarm optimization (MOPSO). In: Proceedings of the 2003 IEEE Swarm Intelligence Symposium, pp. 26–33 (2003)
24. Palermo, G., Silvano, C., Zaccaria, V.: Discrete particle swarm optimization for multi-objective design space exploration. In: 11th EUROMICRO Conference on Digital System Design Architectures, Methods and Tools, pp. 641–644 (2008)
25. Pizzuti, C.: Ga-net: A genetic algorithm for community detection in social networks. In: Parallel Problem Solving from Nature (PPSN), vol. 5199, pp. 1081–1090. Springer (2008)
26. Pizzuti, C.: A multiobjective genetic algorithm to find communities in complex networks. IEEE Trans. Evol. Comput. **16**(3), 418–430 (2012)
27. Ravasz, E., Barabási, A.L.: Hierarchical organization in complex networks. Phys. Rev. E **67**(2), 026,112 (2003)
28. Rosvall, M., Bergstrom, C.T.: Maps of random walks on complex networks reveal community structure. Proc. Natl. Acad. Sci. USA **105**(4), 1118–1123 (2008)
29. Shi, C., Yu, P.S., Cai, Y., Yan, Z., Wu, B.: On selection of objective functions in multi-objective community detection. In: Proceedings of the 20th ACM International Conference on Information and Knowledge Management, pp. 2301–2304. ACM (2011)
30. Shi, C., Yan, Z., Cai, Y., Wu, B.: Multi-objective community detection in complex networks. Appl. Soft Comput. **12**(2), 850–859 (2012)
31. Villalobos-Arias, M., Pulido, G., Coello Coello, C.: A proposal to use stripes to maintain diversity in a multi-objective particle swarm optimizer. In: Proceedings of the 2005 IEEE Swarm Intelligence Symposium, pp. 22–29 (2005)
32. Wei, Y.C., Cheng, C.K.: Ratio cut partitioning for hierarchical designs. IEEE Trans. Comput. Aided Des. Integr. Circuits Syst. **10**(7), 911–921 (1991)
33. Ye, Q., Zhu, T., Hu, D., Wu, B., Du, N., Wang, B.: Cell phone mini challenge award: social network accuracyłexploring temporal communication in mobile call graphs. In: IEEE Symposium on Visual Analytics Science and Technology, 2008. VAST'08, pp. 207–208. IEEE (2008)
34. Zhang, Q., Li, H.: Moea/d: a multiobjective evolutionary algorithm based on decomposition. IEEE Trans. Evol. Comput. **11**(6), 712–731 (2007)
35. Zitzler, E., Laumanns, M., Thiele, L., et al.: Spea2: improving the strength pareto evolutionary algorithm (2001)

Chapter 4
Network Structure Balance Analytics with Evolutionary Optimization

Abstract Structural balance enables a comprehensive understanding of the potential tensions and conflicts beneath signed networks, and its computation and transformation have attracted increasing attention in recent years. The balance computation aims at evaluating the distance from an unbalanced network to a balanced one, and the balance transformation is to convert an unbalanced network into a balanced one. This chapter focuses on evolutionary algorithms to solve network structure balance problem. First, this chapter overviews recent works on the evolutionary computations for structure balance computation and transformation in signed networks. Then, two representative memetic algorithm for the computation of structure balance in a strong definition are introduced. Next, a multilevel learning based memetic algorithm for the balance computation and the balance transformation of signed networks in a weak definition are presented. Finally, a two-step method based on evolutionary multi-objective optimization for weak structure balance are presented.

4.1 Review on The State of the Art

Social interaction involves friendly and hostile relationships in many complex systems, including Wikipedia, online communities and information recommendation systems [9, 31]. In these systems, the entities with friendly relationships are friends or memberships in a group, while those with hostile relationships are enemies or memberships in different groups. These systems can be represented as signed networks in which nodes represent social entities and positive/negative edges correspond to friendly/hostile relationships.

Structural balance, one of the most popular properties in ensembles of signed networks, reflects the origin of tensions and conflicts, and its computation and transformation have received much attention from physicist, sociologist, economist, ecologist, ecologist, and mathematician [11, 21, 24, 25, 32]. The structural balance theory introduced by Heider states that the relations "the friend of my friend is my enemy" and "the enemy of my enemy is my enemy" are unbalanced in the strong definition, which is based on the statistical analysis of the balance of signed triads from the perspective of social psychology [17]. There are broad applications of the compu-

M. Gong et al., *Computational Intelligence for Network Structure Analytics*,
DOI 10.1007/978-981-10-4558-5_4

tation of structural balance, due to the ubiquity of the multi-relational organization of modern systems in a variety of disciplines. Furthermore, the pursuit of balance is desirable in many real-world signed systems. For instance, the pursuit of balance in international relationship networks can reduce military, economic, and culture conflicts. The pursuit of balance in information systems can improve the authenticity of collected information and accelerate opinion diffusion [20].

The computation of structural balance aims at calculating the least imbalances of signed networks [11]. There are mainly two issues in the computation of structural balance: (i) how to verify imbalances and (ii) how to compute the least imbalances. There have been recent efforts in addressing these issues. The frustration index [25] evaluates the imbalances of signed networks as the number of negative cycles which have an odd number of negative links. The energy function proposed by Facchetti et al. [11, 18] measures the imbalances as the number of unbalanced links in signed networks, and both the gauge transformation [11, 18] and a memetic algorithm [30] are utilized to minimize the energy function. Note that, in many cases these studies are constraint to the strong definition of structural balance, and they are worthy to be applied to the weak definition of structural balance. In the weak definition, "the friend of my friend is my enemy" is unbalanced while "the enemy of my enemy is my enemy" is balanced.

The transformation of structural balance focuses on how to convert an unbalanced network into a balanced one. Classical transformation models are divided into two categories: discrete-time dynamic models and continuous-time dynamic models. Local triad dynamic and constrained triad dynamic are two representative discrete-time models [1, 2]. With these models, an unbalanced network is finally evolved towards a balanced or a jammed state by changing signs of edges [25]. Another classical discrete-time model is based on transforming the least number of unbalanced links in signed networks [8]. Differential dynamic presented by Kulakowski et al. [19] is a classical continuous-time model, and it exhibits the evolutionary process from an unbalanced network to a balanced one. In this model, the final evolutionary state (i.e., conflict or harmony) of social networks is determined by the total amount of positive links [24]. These models can effectively transform an unbalanced network into a balanced one, and they are worth improving if they take into account the cost of balance transformation.

Recent studies have demonstrated that the computation and transformation of structural balance in signed networks are nondeterministic polynomial-time hard (NP-hard) problems [11, 12]. An extended energy function was designed by introducing the cost of edge change [33] and a memetic algorithm was proposed to optimize this extended energy function. Ma et al. proposed an extended energy function with cost of balance transformation based on weak balance [22] and an effective memetic algorithm is proposed to optimize this extended energy function. Cai et al. [4] formulated transformation of structural balance as a multi-objective optimization problem and a two-step algorithm based on evolutionary multi-objective optimization was proposed.

This chapter is organized into five sections. Section 4.2 introduces a memetic algorithm to compute global structural balance. Section 4.3 introduces a memetic algorithm for the computation of transformation of structure balance in strong balance. Section 4.4 introduces a memetic algorithm for the computation of transformation of structure balance in weak balance. Section 4.5 introduces an evolutionary multi-objective algorithm for the computation of transformation of structure balance in weak balance.

4.2 Computing Global Structural Balance Based on Memetic Algorithm

In many cases, in order to understand the structure of signed networks, it is needed to measure the unbalanced degree (i.e., measure a distance to exact balance). Computation of global structural balance offers an approach to solve this problem. More generally, computing global structural balance corresponds to computing the ground state of a (nonplanar) Ising spin glass [3], which is a well-known nondeterministic polynomial-time hard (NP-hard) problem.

Facchetti et al. [11] proposed an energy function to compute a distance to exact balance. Optimizing this energy function is a nondeterministic polynomial-time hard (NP-hard) problem. In this section, an approach which tries to optimize the energy function by employing Memetic Algorithm [26] is introduced.[1] The approach, named as Meme-SB, combines Genetic Algorithm and a greedy strategy as the local search procedure.

4.2.1 Memetic Algorithm for Computing Global Structural Balance

Meme-SB aims to optimize energy function (Eq. 1.8). The framework of Meme-SB is shown in Algorithm 17. The GenerateInitialPopulation() function is used to create the initial population. The Selection() procedure is responsible for selecting parental population for mating in GA. Here, tournament selection is used. The GeneticOperation() function is used to perform crossover and mutation operation. The UpdatePopulation() procedure is used to reconstruct the current population. Here, the current population is constructed taken the best S_{pop} individuals from $\mathbf{P} \bigcup \mathbf{P}_{new}$. The TerminationCriterion() function is used to terminate the algorithm, which can be defined as setting a limit of the total number of iterations, reaching a maximum number of iterations without improvement, etc.

---

[1] Acknowledgement: Reprinted from Physica A: Statistical Mechanics and its Applications, 415, Sun, Y., Du, H., Gong, M., Ma, L., Wang, S., Fast computing global structural balance in signed networks based on memetic algorithm, 261–272, Copyright(2014), with permission from Elsevier.

Algorithm 17 The algorithm framework of Meme-SB.

Input: Maximum number of generations: G_{max}; Population size: S_{pop}; Size of mating pool: S_{pool}; Tournament size: S_{tour}; Crossover probability: P_c; Mutation probability: P_m. **Output**: the fittest chromosome in **P**.

1: $\mathbf{P} \leftarrow$ GenerateInitialPopulation(S_{pop});
2: **repeat**
3: $\mathbf{P}_{parent} \leftarrow$ Selection($\mathbf{P}$, S_{pool}, S_{tour});
4: $\mathbf{P}_{child} \leftarrow$ GeneticOperation($\mathbf{P}_{parent}$, P_c, P_m);
5: $\mathbf{P}_{new} \leftarrow$ LocalSearch($\mathbf{P}_{child}$);
6: $\mathbf{P} \leftarrow$ UpdatePopulation($\mathbf{P}$, $\mathbf{P}_{new}$);
7: **until** TerminationCriterion(G_{max})

4.2.2 Representation and Initialization

Each chromosome in the population consists of genes, in which each gene corresponds to a node in the network. A chromosome is encoded as a string

$$\mathbf{X} = \{x^1 \; x^2 \; \cdots \; x^n\},$$

where n is the number of nodes, and x^i is the sign of node s_i, which can be set to $+1$ or -1 in order to calculate the energy function. Figure 4.1 displays an example of the representation.

The population initialization procedure is given as Algorithm 18. Initially, each chromosome of the population is generated randomly. However, all of these solutions are of low quality. In GAs, it is common to initialize a high-quality population to speed up the convergence. Here, a simple heuristic is employed: the goal is to minimize the energy function, so, if the adjacent pair of nodes have a positive/negative edge, they should have the same/opposite signs. For each chromosome, a simple greedy strategy is employed: select a node (or gene) randomly and reassign a value (or sign) to it to ensure that it makes no contribution to the energy function with another random node which is adjacent to it. This operation is repeated for n times, and at each time a node which has never been selected before is randomly selected.

4.2.3 Genetic Operators

Crossover. Here, in order to improve the search capacity, the two-point crossover is used because this method is easy and effective. The crossing over procedure is defined as follows. Given two parents $\mathbf{X}_a$ and $\mathbf{X}_b$ at random, first randomly select two points i and j (i.e., $1 \leq i \leq j \leq n$), and then everything between the two points of the parents is swapped (i.e., $x_a^k \leftrightarrow x_b^k$, $\forall k \in \{k \,|\, i \leq k \leq j\}$). This procedure returns two new chromosomes $\mathbf{X}_c$ and $\mathbf{X}_d$. An example of the operation of two-point crossover on the encoding step is shown in Fig. 4.2.

Algorithm 18 The Population Initialization Procedure.

Input: Population size: S_{pop}. Generate a population **P** randomly, each chromosome $\mathbf{X}_k$ of which is a string consisting of +1 and -1, where $k = 1, 2, \cdots, S_{pop}$.
Output: Population **P**.
1: **Step 1) Initialization**
2: Generate a population **P** randomly, each chromosome $\mathbf{X}_k$ of which is a string consisting of +1 and -1, where $k = 1, 2, \cdots, S_{pop}$.
3: **for** each chromosome $\mathbf{X}_k$ **do**
4: Generate a random sequence (i.e., $\{r_1\ r_2\ \cdots\ r_n\}$);
5: $t_{counter} \leftarrow 1$;
6: **repeat**
7: select a node s_i, where $i = r_{t_{counter}}$;
8: **repeat**
9: randomly select a node s_j;
10: **until** there is an edge J_{ij} between s_i and s_j
11: **if** $J_{ij} = +1$ **then**
12: $x_k^i \leftarrow x_k^j$;
13: **else**
14: $x_k^i \leftarrow -x_k^j$;
15: **end if**
16: $t_{counter} \leftarrow t_{counter} + 1$;
17: **until** $t_{counter} = n$ (i.e., all the nodes are selected)
18: **end for**

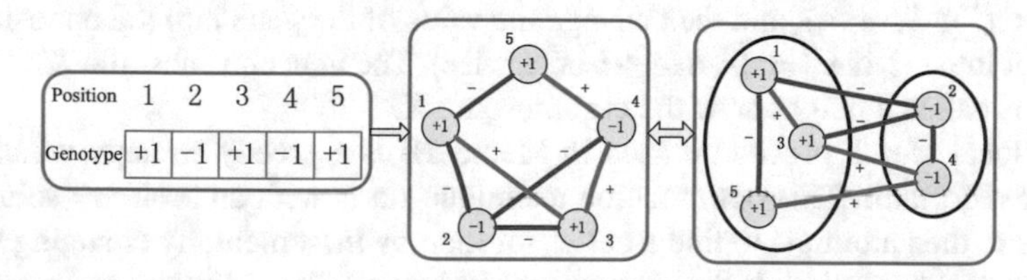

Fig. 4.1 Illustration of the representation. *Left* one possible genotype. *Middle* translate the genotype into the graph. *Right* classification result

Mutation. In this mutation, randomly pick a chromosome from $\mathbf{P}_{parent}$ to be mutated. Here one-point mutation on this chromosome is used: a node is picked randomly on the chromosome, then a value is reassigned to it to ensure that it makes no contribution to the energy function with another random node which is adjacent to it. This procedure follows the simple heuristic mentioned above. This operation is repeated n times on the chromosome. The specialized mutation operator has the abilities of strengthening the local search and keeping the diversity of the population.

4.2.4 The Local Search Procedure

In Meme-SB, a greedy strategy is devised to realize the local search procedure.

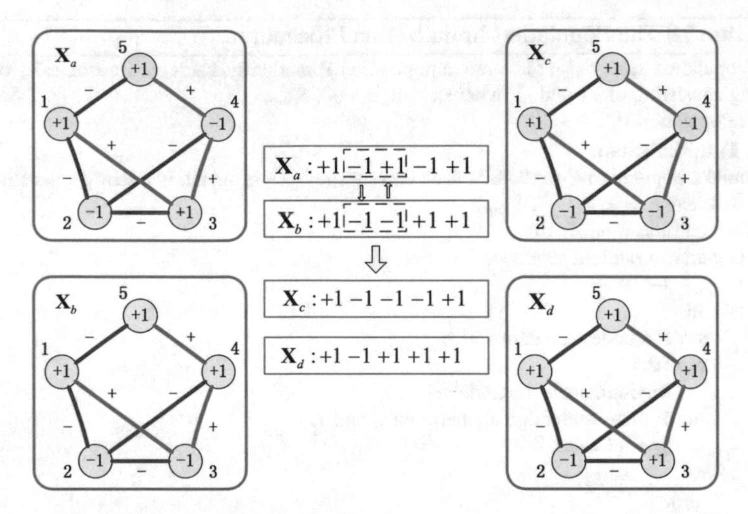

Fig. 4.2 Illustration of the two-point crossover

Before describing the local search procedure, it is needed to define the neighbors of a chromosome. Given a chromosome $\mathbf{X} = \{x^1\ x^2\ \cdots\ x^n\}$, first randomly select a gene x^i $(i \in 1, \cdots n)$, and then change the value of the gene into the opposite (i.e., change it into -1 if $x^i = +1$ and $+1$ otherwise). The new chromosome $\mathbf{X}'$ after this change is called a neighbor of the chromosome $\mathbf{X}$.

The local search procedure used in Meme-SB is a greedy strategy which is an iterative algorithm. This optimization technique starts with an arbitrary solution to a problem, then attempts to find a better solution by incrementally changing several elements of the solution. If the change can produce a better solution, an incremental change is made to the new solution, repeating until no further improvement can be made. The detailed implementation is given as Algorithm 19. Here, this technique is applied to $\mathbf{P}_{child}$, which is the population after crossover and mutation. It is only to find the fittest chromosome in $\mathbf{P}_{child}$ and perform local search on it, until no improvement can be found. In Algorithm 19, the FindBest() function is used to evaluate the fitness of each chromosome in the input population, and return the chromosome having maximum fitness (or minimum energy), on which the local search procedure will be performed. The Energy() function is responsible for evaluating the energy of a solution. The FindBestNeighbor() function is used to find the best neighbor of a chromosome, which can be done easily according to the definition of the neighbors of a chromosome. Here, the procedure of the FindBestNeighbor() function is explained: given a chromosome, select a gene randomly, and then change the value of the gene into the opposite. If this change is an improvement, this change is made to the new chromosome; if this change cannot produce a better solution, this change will be repeated. This operation is repeated for n times, and at each time a gene which has never been selected before is randomly selected.

Algorithm 19 The Local Search Procedure.

Input: $\mathbf{P}_{child}$. **Output**: $\mathbf{P}_{new}$.
  $\mathbf{N}_{current} \leftarrow$ FindBest($\mathbf{P}_{child}$);
  islocal $\leftarrow$ FALSE;
  repeat
    $\mathbf{N}_{next} \leftarrow$ FindBestNeighbor($\mathbf{N}_{current}$);
    if Energy($\mathbf{N}_{next}$)<Energy($\mathbf{N}_{current}$) **then**
      $\mathbf{N}_{current} \leftarrow \mathbf{N}_{next}$;
    else
      islocal $\leftarrow$ TRUE;
    end if
  until *islocal* is TRUE

Table 4.1 Some parameters in the algorithm

Parameter	Meaning	Value
G_{max}	The number of iterations	50
S_{pop}	Population size	500
S_{pool}	Size of the mating pool	150
S_{tour}	Tournament size	2
P_c	Crossover probability	0.9
P_m	Mutation probability	0.1

4.2.5 Experimental Results

Meme-SB is tested on four social networks and four biological networks. The compared algorithm is a GA vision of this algorithm, termed as GA-SB, which is simply developed by removing the local search procedure. The results obtained by the algorithm (described in detail in [11, 18], and termed as HRT-SB here) introduced by Facchetti et al. are also given for comparison.

Because the algorithm Meme-SB is not parameter-free, it is needed to set some values to them in advance such as population size, number of generation, crossover probability, mutation probability, etc. Some parameters are given in Table 4.1. In the following experiments, the reported data are the statistical results based on 30 independent runs on each dataset.

4.2.5.1 Results on Social Networks

In this subsection, Meme-SB is tested on four small-scale social networks: SPP, GGS, SC, and CP. For each social network, the exactly minimum value of the energy function is unknown beforehand. Numbers of nodes, edges, positive edges, and negative edges are provided in Table 1.2.

For all the networks, Fig. 4.3 displays the values of energy function obtained by Meme-SB and GA-SB in one run, when the number of generations increases from 1 to 50. In this experiment, the GA version algorithm is tested on the social networks

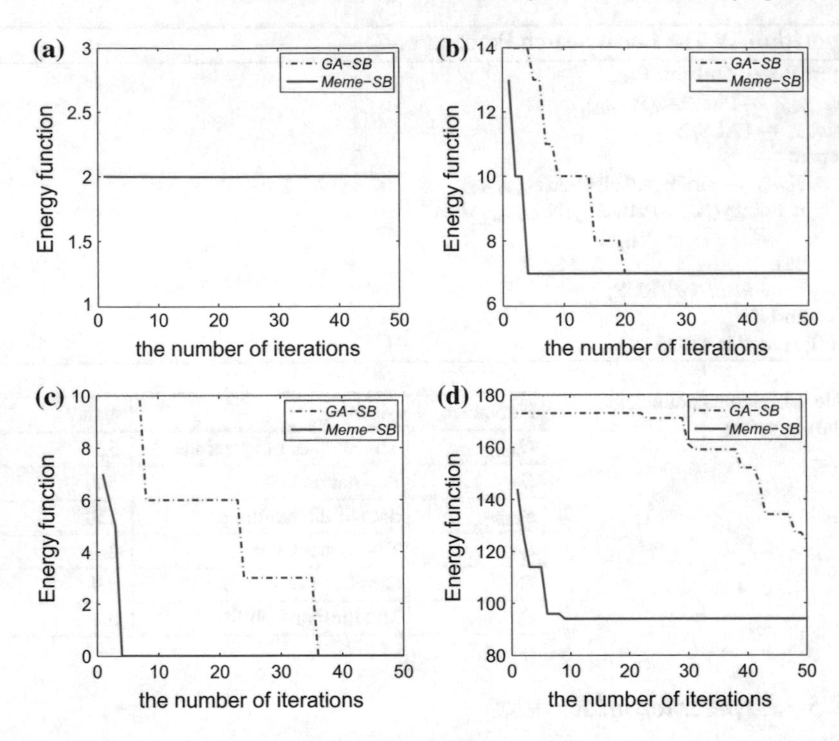

Fig. 4.3 Comparison between Meme-SB and GA-SB on social networks. **a** SPP. **b** GGS. **c** SC. **d** CP

with the same parameters as in Meme-SB. As shown in Fig. 4.3, for each dataset, Meme-SB algorithm can find the minimum value of the energy function within 10 generations. In particular, for SPP, Meme-SB finds a minimum energy function value of 2 in just one generation. Meanwhile, for GGS and SC, Meme-SB only takes less than five generations to find the minimum value. In contrast, GA-SB needs more generations. From this comparison, the local search procedure plays a very important role in Meme-SB. Without the local search procedure, it becomes harder to find the exactly optimum value. Moreover, the local search procedure also speeds up the convergence of Meme-SB. It is obviously shown in Fig. (d), for CP, Meme-SB finds the optimum value of 94 in less than 10 generations. However, without the local search procedure, the GA-SB algorithm does not find the optimum value within 50 generations.

According to the experimental results, the nodes classifications of networks SPP and SC are given, which are shown in Figs. 4.4 and 4.5. For SPP, Meme-SB obtains the exactly minimum value of 2 which means there are two edges lead to the unbalance of the network. As shown in Fig. 4.4, the two edges are marked by circles and if their signs are changed into opposite, the network will become exactly balanced. For SC, Meme-SB can get the exactly minimum value which equals to 0. It indicates that the network SC itself is exactly balanced. As shown in Fig. 4.5, the classification result of SC does not have any “stain”.

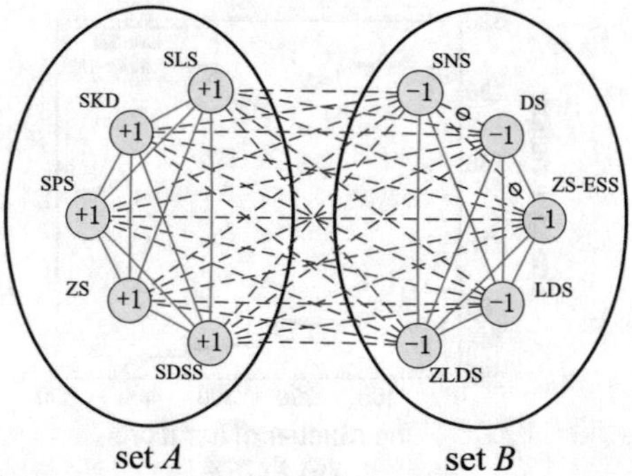

Fig. 4.4 Nodes classification result of SPP. *Solid lines* represent positive relationships, while *dotted lines* represent negative relationships

SLS SNS SKD DS SPS ZS-ESS ZS LDS SDSS ZLDS

set A set B

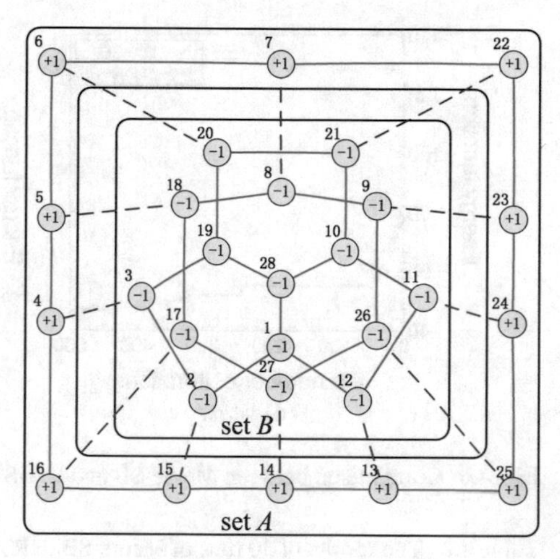

Fig. 4.5 Nodes classification result of SC

This experiment shows Meme-SB is a high valid method for computing the structural balance on the small-scale social networks.

4.2.5.2 Results on Biological Networks

Here, Meme-SB is tested on four biological networks: EGFR, macrophage, yeast, and E.coli. All the networks are symmetric by being previously processed. Here, for each biological network, the energy function's exactly minimum value is unknown. See Table 1.2 for details. The compared algorithm include GA-SB and HRT-SB. Considering the complexity and the large scale of these biological networks, the number of iterations is increased to 500.

For each network, Fig. 4.6 displays the values of energy function obtained by Meme-SB and GA-SB in one run, when the number of generations increases from 1 to 500. It clearly shows that, with the local search procedure, the results obtained by Meme-SB are a lot better than GA-SB. Although the results obtained by Meme-SB may not be the best, at least they are satisfactory. Moreover, as shown in Fig. 4.6b,

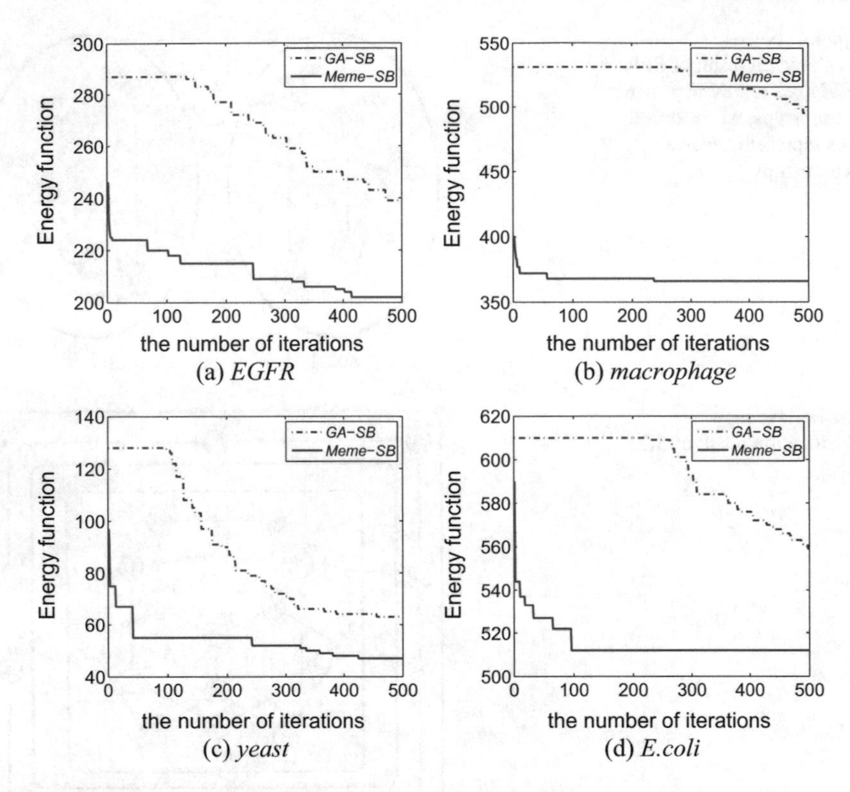

Fig. 4.6 Comparison between Meme-SB and GA-SB on biological networks

Table 4.2 The results of 30 runs of Meme-SB, HRT-SB, and GA-SB on four biological networks

Network	$\delta_{avg}/\delta_{std}$		
	Meme-SB	HRT-SB	GA-SB
EGFR	202.50/1.08	219.30/0.48	240.50/4.14
macrophage	364.50/1.08	374.30/2.79	502.00/12.16
yeast	48.20/1.93	53.20/2.62	62.00/3.65
E.coli	511.10/1.37	517.90/3.14	561.30/6.31

d, the local search procedure can speed up the convergence of Meme-SB obviously. This comparison clearly illustrates the point that, with the help of the local search procedure, Meme-SB performs better than GA-SB.

In the following, for each network, Meme-SB and the two compared algorithms are run 30 times, and the average value and standard deviation of the energy function (δ_{avg} and δ_{std}) are computed over the 30 runs. The results are reported in Table 4.2. Here, it is important to note that considering the time complexity, the process Algorithm 18 of HRT-SB is run. Because Meme-SB has better performances than GA-SB, here, just the experimental results obtained by Meme-SB and HRT-SB are analyzed. As shown in Table 4.2, on *EGFR* network, *macrophage* network, *yeast* network, and *E.coli* network, the average values of the energy function found by Meme-SB are 202.50, 364.50, 48.20, and 511.10, respectively, while the average values found by HRT-SB are 219.30, 374.30, 53.20, and 517.90, respectively. It is seen that, on all these four networks, the results obtained by Meme-SB are better than those by HRT-SB. In the aspect of stability, on *EGFR* network, the standard deviation obtained by Meme-SB is 1.08, while that obtained by HRT-SB is 0.48. It indicates that the stability of HRT-SB is slightly better than Meme-SB on *EGFR* network. However, on the other three networks, the standard deviations obtained by Meme-SB are all smaller than those by HRT-SB. This comparison clearly shows that the results obtained by the algorithm Meme-SB are not only more excellent, but also more steady than the two compared algorithms.

4.2.6 Complexity Analysis

In this part, the time complexity of the Meme-SB is analyzed. Here n and m are used to denote the node and edge numbers of the network, respectively. At each generation, first, it is needed to perform the crossover operator $\lfloor S_{pool}/2 \rfloor$ times and the mutation operator S_{pool} times at most, where S_{pool} is the size of the mating pool. The time complexity of the calculation of energy function is $O(m)$. Therefore, the time complexity of the genetic operator is $O(S_{pool}(n+m))$. Second, it is needed to perform the local search procedure which requires $log\ n$ steps to reach a local optimum in the search space. Moreover, at each step of the local search procedure, it needs to consider $log\ n$ neighbors of each node. Therefore, the time complexity of the local search procedure is $O(n(log\ n)^2)$. Finally, it is needed to reconstruct the population, which needs $O(S_{pop} + S_{pool})$ basic operations, where S_{pop} is the size of the population. In practical applications, $O(S_{pop} + S_{pool}) \leq O(S_{pool}(n+m)) \leq O(n(log\ n)^2)$. Therefore, the time complexity of the proposed Meme-SB algorithm at each generation is $O(n(log\ n)^2)$.

Through the similar complexity analysis with Meme-SB, the computation complexities of GA-SB and HRT-SB are $O(S_{pool}(n+m))$ and $O(n+m)$, respectively. By comparing Meme-SB with GA-SB and HRT-SB, it can be seen that the time complexity of Meme-SB is higher than those of the two compared algorithms. And in the aspect of computation time, Meme-SB need slightly more time to perform its procedure. The reason why Meme-SB is defective in computation time is that it performs the local search procedure. However, with the help of the local search procedure, Meme-SB has better performances on computing global structural balance than GA-SB and HRT-SB. Its perfect performances make it possible for exactly computing global structural balance in a reasonable time.

4.2.7 Conclusions

In this section, Meme-SB was proposed to optimize the energy function for computing global structural balance in signed networks. The proposed algorithm combines GAs and a greedy strategy as the local search procedure. The experimental results show that Meme-SB has better performances than the two compared algorithms. The comparative results on both social and biological networks demonstrate the effectiveness and efficiency of Meme-SB on computing global structural balance. In addition, according to the minimum value of the energy function, optimization know how many edges (or pairwise relationships) which lead to the unbalance of the network exist. By changing the signs of these edges into the opposite, an exactly balanced network can be realized.

4.3 Optimizing Dynamical Changes of Structural Balance Based on Memetic Algorithm

The transformation of structural balance focuses on how to convert an unbalanced network into a balanced one.[2] Computing the energy function (Eq. 1.8) can get the least imbalances of signed networks. Unbalanced network is finally evolved towards a balanced or a jammed state by changing signs of these imbalanced edges in the balance computation of signed networks. It is suggested that there be a certain bias towards flipping positive or flipping negative signs.

This section introduces an algorithm (Meme-DB) to compute the least number of sign changes in the evolution of structural balance, in which the transformation cost is taken into consideration. The advantages of this algorithm are as following. The problem optimizing dynamics of structural balance is formulated as an optimization problem. Instead of using a total number of flipped signs, there is a certain bias towards flipping positive or flipping negative signs and the bias varies with a parameter. Finally, a memetic algorithm with local search procedure is used to optimize the objective function designed. The framework of Meme-DB is similar to that of Meme-SB, which is described in Algorithm 17.

4.3.1 Problem Formation

This section aims to compute the minimum number of sign changes in dynamical evolution of signed networks, and then a criterion is needed to test whether signed networks evolve to balanced ones. As described above, when energy function (Eq. 1.8)

---

[2]Acknowledgement: Reprinted from Social Networks, 44, Wang, S., Gong, M., Du, H., Ma, L., Miao, Q., Du, W., Optimizing dynamical changes of structural balance in signed network based on memetic algorithm, 64–73, Copyright(2016), with permission from Elsevier.

of a signed network is zero, the network is balanced, so that equation (Eq. 1.8) is adopted as the criterion.

It is suggested that, there is a certain bias or probability (transformation cost) towards flipping positive or flipping negative signs in the dynamical evolution of structural balance.

Then, optimizing dynamical evolution of structural balance is mathematically formulated as:

$$\begin{aligned} \textit{minimize} \quad & F(n) = \lambda N_{c+} + (1-\lambda) N_{c-} \\ s.t. \quad & H(s) = 0 \end{aligned}$$

where parameter λ ranges from 0 to 1; N_{c+} is the number of sign-changed positive edges and N_{c-} is the number of sign-changed negative edges; $H(s)$ is the equation (1.8).

From a computational point of view, parameter λ is added to the function metric (1.8) to show that positive and negative edges are changed of sign with different probabilities. By varying the value of λ, different number of sign changes can be got and different evolution processes can be achieved.

From a social point of view, λ could also denote positive or negative force. In the light of people relations, positive links denote friendship and negative ones hostility. The unbalanced network on the left of Fig. 4.7 is taken as an example. When the sign of one edge adjacent to k is flipped, there are two kinds of balanced states, as indicated in Fig. 4.7. In the unbalanced situation, there is pressure on j to update the relationship with k (edge jk) negatively, since k would be negatively recommended by i to j, and similarly j would be recommended positively by i to the nemesis of k, so it would also have a negative impact. Similarly, there is a pressure on i and k to make the relationship ik positive: both are positively recommended by their common friend j. λ then controls what pressure is stronger: the positive force (i.e., make ik positive) or the negative force (i.e., make jk negative). Then this works also can mean how unbalanced networks evolve to balanced ones with minimal pressure.

Meme-DB is to optimize the objective function described in Eq. 4.1. In traditional unconstrained optimization problem, penalty methods are usually used to convert it into an unconstrained problem. A parameter μ is used. Given the number of sign-changed positive edges N_{c+} and the number of sign-changed negative edges N_{c-}, a function is:

$$\min \quad \mathscr{F}(n) = \lambda N_{c+} + (1-\lambda) N_{c-} + \mu H(s) \tag{4.1}$$

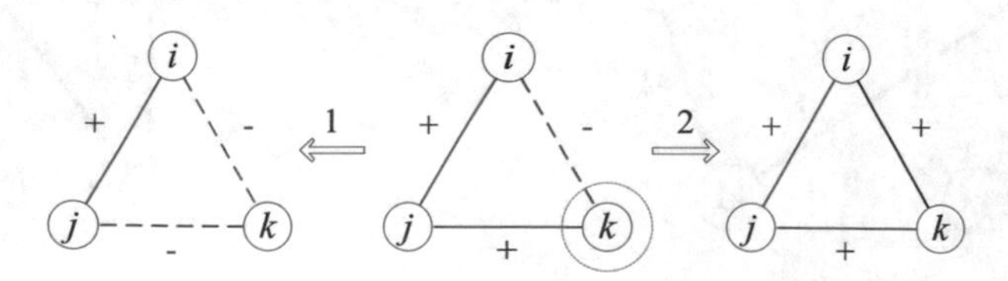

Fig. 4.7 An illustration of evolution of unbalanced triangle to balanced ones with different flipped signs

Then this equality constrained problem is converted to an unconstrained optimization problem. μ is the weighting parameter.

4.3.2 Representation and Initialization

In Meme-DB, a simple genetic representation is used for evolution of unbalanced network. Let N_+, N_- be the number of positive and negative edges, respectively and V be the number of nodes in signed network. A network evolution from unbalanced situation to balanced situation is encoding in a string: $\{X_+, X_-, S_{node}\}$. X_+ denotes a bipolar N_+-vector in $\{1, -1\}^{N_+}$ of positive edges, which is used to judge whether sign of positive edges should be changed. X_- denotes a bipolar N_--vector in $\{1, -1\}^{N_-}$ of negative edges, which is used to judge whether sign of negative edges should be changed. S_{node} denotes a bipolar V-vector in $\{1, -1\}^V$ of nodes, which denote the clusters of all the nodes. For every element $x_i \subseteq \{X_+\}$ *or* $\{X_-\}$, $x_i = -1$ means that the sign of corresponding edge need to be flipped, while $x_i = 1$ means that the sign of corresponding edge remains unchanged. For $s_i \subseteq \{S_{node}\}$, s_i is the sign assigned to ith node, which is consistent with Eq. (1.8). An illustration of genetic representation is described in Fig. 4.8. In Fig. 4.8, there are three positive edges, four negative edges and five nodes, such that the length of vector X_+ is 3, the length of vector X_- is 4 and the length of vector S_{node} is 5. All the elements in X_+ are $+1$, such that none sign of positive edges is changed. The first element in X_- is -1 (i.e., node 4 in this network), such that this edge sign is changed. All the elements in S_{node} are the clusters of nodes.

First, a definition of the positive neighborhood of a node in signed network is given.

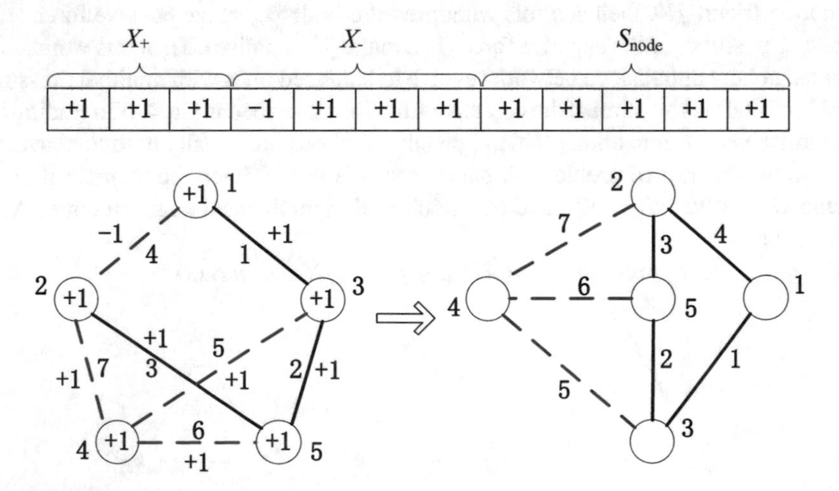

Fig. 4.8 An illustration of genetic representation. The table above is the representation of the network at the *lower left* and the network at the *lower right* is the evolutionary result. In the representation, all the nodes are assigned a $+1$. All the edges are assigned a $+1$, except for the *4*th edge, so that only the sign of *4*th edge needs to be changed

- **Definition of positive neighborhood**: Node N_1 is a positive neighbor of node N_2, if they are connected by a positive edge. the positive neighborhood of node s is defined as the set of nodes connected by positive edges with node s.

Initially, every gene in an individual is assigned a sign +1 or −1 randomly. In order to improve the efficiency and avoid meaningless computation, a preprocess strategy is introduced. For each individual, the allele value j assigned to ith gene is randomly selected from the positive neighborhood of i. The initialization process is described in Algorithm 20.

Algorithm 20 The population initialization procedure.

1: **Input**: the population size: P_{size}, the number of positive edges: N_+, the number of negative edges: N_-, the number of nodes: V
2: Randomly generate the population **P**, each chromosome is a bipolar $(N_+ + N_- + V)$-string.
3: **For** chromosome X_k **do**
4: **For** $i = 1; i < N_+ + N_-; i++$
5: Randomly select an individual $X_k(j)$ from the positive neighborhood of $X_k(i)$;
6: $X_k(i) \leftarrow X_k(j)$;
7: **End for**
8: **output**: the population **P**
[1]

4.3.3 Genetic Operators

Crossover. Crossover used here is two-point crossover, which randomly selects two cut points. The crossover process is defined as followed. Give two chromosomes X^1 and X^2, we randomly select two points in every substring X_+, X_- and S_{node} in the above two chromosome, respectively. In every substring, for example in substring X_+, two points i and j are selected(i.e., $1 \leq i \leq j \leq N_+$), and then all the genes between i and j are swapped between chromosome X^1 and X^2.
Mutation. In this process, a chromosome is randomly selected to be mutated. Then one-point mutation is employed on the chromosome: a gene in the chromosome is randomly picked, and for nodes, sign of the gene is altered to the sign of one positive neighbor of this node. For edges, sign of the gene is shifted to opposite sign. The times of mutation m is selected randomly, where $m \in [1, V+N]$. The neighbor-based mutation decreases meaningless operation.

4.3.4 The Local Search Procedure

Given a individual noted $X = \{x_1, x_2, \cdots, x_n\}$, randomly select a gene x_i in this chromosome. If the change of x_i can result in decrease of the function, then the sign of x_i is changed. In this local search procedure, every gene is searched. The pseudocode of local search is illustrated in the Algorithm 21.

In Meme-DB, local search procedure is just performed on one of the best chromosomes in population $\mathbf{X}_{child}$ after crossover and mutation. In Meme-DB, local search procedure will just be performed on the chromosomes with optimal fitness.

Algorithm 21 The local search procedure on chromosome.

1: **Input**: $\mathbf{X}_{parent}$.
2: *islocal* $\leftarrow$ FALSE;
3: $\mathbf{X}_{child} = \mathbf{X}_{parent}$
4: **For** $i = 1; i \leq N_+ + N_- + V; i++$ **do**
5: $\mathbf{X}_{child}[i] = -1 * \mathbf{X}_{parent}[i]$;
6: **if eva**$(\mathbf{X}_{child}) >$ **eva**$(\mathbf{X}_{parent})$;
7: $\mathbf{X}_{child}[i] = \mathbf{X}_{parent}[i]$;
8: **else**
9: *islocal* $\leftarrow$ TRUE;
10: **End if**
11: **End for**
12: **End**
13: **Output**: $\mathbf{X}_{child}$.
[1]

4.3.5 Transformation

An algorithm based on the application of equivalence transformations is proposed in [18]. This algorithm preserves the overall energy function in Eq. 1.8. Meme-DB is to have the least number of sign changes by optimizing objective function Eq. 4.1. Inspired by the algorithm in [18], an improved transformation procedure is designed, which also preserves the overall energy function. In this transformation, network is still balanced while the results of objective function Eq. 4.1 are better. For a node v_i, $i = 1, 2, ..., V$, pseudo code of transformation procedure is illustrated in Algorithm 22.

Algorithm 22 The transformation procedure on chromosome.

Parameters: the number of sign-changed positive edges ended in node v_i: N_{c+}; the number of sign-changed negative edges ended in node v_i: N_{c-}; the number of sign-unchanged positive edges ended in node v_i: N_{u+}; the number of sign-unchanged negative edges ended in node v_i: N_{u-}.
1: **Input**: chromosome.
2: **For** $i = 1; i \leq V; i++$ **do**
3: **if** $\lambda * N_{c+} + (1-\lambda) * N_{c-} \geq \lambda * N_{u+} + (1-\lambda) * N_{u-}$;
4: change sign to all the edges ended in v_i and sign assigned to node v_i in chromosome.
5: **End if**
6: **End for**
7: **End**
8: **Output**: chromosome.

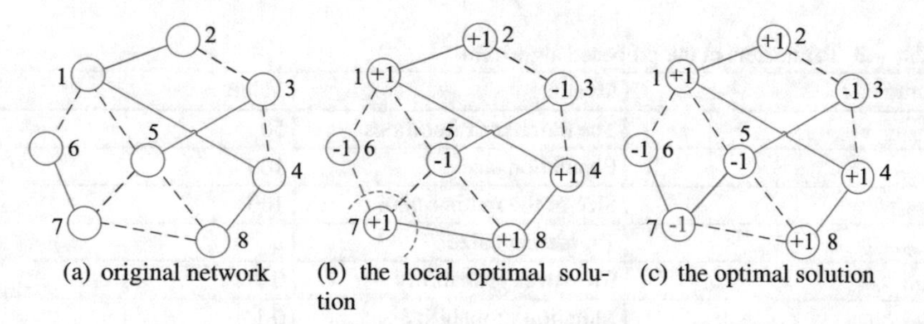

(a) original network (b) the local optimal solution (c) the optimal solution

Fig. 4.9 An illustration of the transformation procedure. **a** The original network; **b** The solution trapped in local optimal solution; **c** The optimal solution

The 8-node network shown in Fig. 4.9 is used in order to explain the transformation procedure introduced in this paper. In this algorithm, solid lines represent positive edges and dashed lines denote negative ones. Green lines represent sign-changed edges. Figure 4.9a is the original network which is unbalanced. The energy function (Eq. 1.8) of Fig. 4.9b equals zero and Fig. 4.9b is partitioned into two clusters. Though Fig. 4.9b is balanced, this result is not the optimal solution of the problem to be solved in this paper. In Fig. 4.9b, for the node 7, $N_{c+} = 1$, $N_{c-} = 1$, $N_{u+} = 0$ and $N_{u-} = 1$. When $\lambda = 0.5$, $\lambda * N_{c+} + (1-\lambda) * N_{c-} > \lambda * N_{u+} + (1-\lambda) * N_{u-}$. Then all the edges adjacent to node 7 will have a sign flip, and sign assigned to this node will also be switched. Figure 4.9c which is achieved by applying the transformation procedure to Fig. 4.9b is balanced. Figure 4.9c is partitioned into two clusters where all the nodes within the same cluster are friends while nodes from different clusters are foes.

4.3.6 *Experimental Results*

Meme-DB is tested on some signed networks. To demonstrate the necessity and effectiveness of local search procedure, GA-DB is chosen to be compared with Meme-DB, which is developed by removing the local search procedure. Moveover, the algorithm (named GT here) proposed by Facchetti et al. [18], the method named Meme-SB proposed by [30] and constrained triad dynamics (CTD) proposed by Antal et al. [1] are also given for comparison.

In Meme-DB, there are various parameters. The parameters are given in Table 4.3.

In experiments, Meme-Net is tested on two illustrative signed networks: SN, SC(IN), and two real signed networks: EGFR (EN) and Macrophage (MN). The properties of these networks are given in Table 1.2.

In the objective function, there are two undetermined parameters. In experiments, the variations of objective function with both of parameters μ and λ will be known. First, influence of μ on objective function is tested. The algorithm without the transformation procedure is run 30 times on four networks for parameter $\lambda = 0.2$, 0.5 and 0.8. In Tables 4.4, 4.5 and 4.6, the averages of Meme-DB for $\mu = 0.1$, 1, 5, 10,

Table 4.3 Parameters in the proposed algorithm

Parameter	Meaning	Value
G_{max}	The number of iterations	50
S_{pop}	Population size	100
S_{pool}	Size of the mating pool	100
S_{tour}	Tournament size	2
P_c	Crossover probability	0.9
P_m	Mutation probability	0.1

Table 4.4 Results of 30 runs of our algorithm on four networks for $\lambda = 0.2$

	Index	μ				
		0.1	1	5	10	100
SN	N_{c+}	0	0	2.0	2.4	2.6
	N_{c-}	0	2	4.5	5.1	5.0
	OF	0.2	**1.6**	4.0	4.6	4.5
	EF	**2**	0	0	0	0
IN	N_{c+}	0	4.2	8.3	8.2	8.9
	N_{c-}	0	0.9	2.6	2.8	2.9
	OF	0.1	**1.6**	3.7	3.9	4.1
	EF	**1.1**	0	0	0	0
EN	N_{c+}	145.0	207.6	209.8	204.5	203.9
	N_{c-}	57.5	127.5	128.2	130.7	131.3
	OF	104.6	**143.5**	144.5	145.5	145.8
	EF	**296.1**	0	0	0	0
MN	N_{c+}	319.8	399.0	398.0	392.1	411.1
	N_{c-}	137.5	239.8	239.4	239.5	238.0
	OF	228.2	271.6	271.1	**270.0**	272.6
	EF	**542.2**	0	0	0	0

100 are given. In these tables, "N_{c+}", "N_{c-}" denote the numbers of sign-changed positive and negative edges, respectively, and "OF" denotes the values of objective function.

$\mu = 1$ is set as the default value and test the variation of different variables with parameter λ. For each network, The algorithm is run 30 times for different λ. In the experiments, objective function is optimized with the mixing parameter λ increasing from 0 to 1 with an interval of 0.1. The averages and standard deviations of three variables over 30 times are given in Table 4.7, where N_{c+} is the number of sign-changed positive edges, N_{c-} is the number of sign-changed negative edges and OF is the result of objective function.

On SN network, except for $\lambda = 0.9$ and 1, averages of N_{c+} and N_{c-} remain unchanged with the increase of λ, for the number of nodes and edges in SN network

Table 4.5 Results of 30 runs of our algorithm on four networks for $\lambda = 0.5$

	Index	μ				
		0.1	1	5	10	100
SN	N_{c+}	0	0.1	2.2	1.9	2.4
	N_{c-}	0	2.2	5.0	4.6	5.0
	OF	0.2	**1.2**	3.6	3.2	3.7
	EF	**2**	0	0	0	0
IN	N_{c+}	0	2.6	6.8	6.0	6.8
	N_{c-}	0	1.8	3.8	4.1	3.7
	OF	0.0	**2.2**	5.3	5.1	5.2
	EF	**0.2**	0	0	0	0
EN	N_{c+}	131.3	188.9	192.4	192.0	188.8
	N_{c-}	70.8	136.2	134.5	138.4	139.7
	OF	131.2	**162.6**	163.5	165.2	164.3
	EF	**301.1**	0	0	0	0
MN	N_{c+}	299.9	362.5	364.4	363.2	359.0
	N_{c-}	155.8	258.1	257.8	257.7	261.3
	OF	282.4	310.3	300.1	310.5	310.1
	EF	**545.6**	0	0	0	0

Table 4.6 Results of 30 runs of Meme-DB on four networks for $\lambda = 0.8$

	Index	μ				
		0.1	1	5	10	100
SN	N_{c+}	0	0.2	2.1	2.6	1.0
	N_{c-}	0	2.3	5.9	7.0	6.3
	OF	0.2	**0.6**	2.9	3.4	2.1
	EF	**2**	0	0	0	0
IN	N_{c+}	0	1.8	4.5	5.7	5.5
	N_{c-}	0	3.3	5.7	5.0	5.1
	OF	0.2	**2.1**	4.7	5.6	5.4
	EF	**2.4**	0	0	0	0
EN	N_{c+}	127.5	182.3	187.5	185.0	181.7
	N_{c-}	98.5	144.7	144.1	143.5	147.0
	OF	150.4	174.8	178.8	176.7	174.8
	EF	**287.2**	0	0	0	0
MN	N_{c+}	305.1	350.7	349.0	356.3	357.9
	N_{c-}	234.7	269.6	272.8	270.3	268.5
	OF	318.7	334.5	333.8	339.1	340.0
	EF	**277.5**	0	0	0	0

Table 4.7 The average and standard deviation of results over 30 runs on four networks for different λ. "N_c+", "N_c-" denote the number of sign-changed positive edges and sign-changed negative edges and "OF"denotes the values of the objective function

	Index	0	0.1	0.2	0.3	0.4	0.5	0.6	0.7	0.8	0.9	1.0
SN	N_{c+ave}	0	0	0	0	0	0	0	0	0	0	0
	N_{c+std}	0	0	0	0	0	0	0	0	0	0	0
	N_{c-ave}	2	2	2	2	2	2	2	2	2	10.3	15.3
	N_{c-std}	0	0	0	0	0	0	0	0	0	12.0	12.7
	OF_{ave}	2	1.8	1.6	1.4	1.2	1.0	0.8	0.6	0.4	1.0	0
	OF_{std}	0	0	0	0	0	0	0	0	0	1.2	0
IN	N_{c+ave}	11.2	7.9	6.4	4.7	1.6	2.0	1.7	1.4	1.5	1.4	0.1
	N_{c+std}	2.7	2.2	3.6	3.7	2.1	2.3	2.1	2.0	2.1	1.8	0.4
	N_{c-ave}	0	0	0	0	1.0	1.4	1.40	1.2	2.0	1.4	2.7
	N_{c-std}	0	0	0	0	1.7	2.0	2.1	2.0	2.8	2.2	3.4
	OF_{ave}	0	0.8	1.3	1.4	1.2	1.7	1.6	1.4	1.6	1.4	0.1
	OF_{std}	0	0.2	0.7	1.1	1.5	1.4	1.5	1.6	1.8	1.7	0.4
EN	N_{c+ave}	248.7	207.0	205.7	200.9	171.7	133.9	88.6	65.7	58.6	53.6	35.9
	N_{c+std}	4.3	4.6	5.6	5.4	5.8	8.7	8.9	12.6	16.5	11.2	8.1
	N_{c-ave}	21.0	25.7	26.5	27.6	41.0	69.9	131.0	172.1	194.3	200.8	223.3
	N_{c-std}	2.3	3.4	2.9	3.1	4.3	8.0	9.7	12.7	17.6	12.7	9.0
	OF_{ave}	21.0	43.9	62.3	79.6	93.3	101.9	105.6	97.6	85.7	68.3	35.9
	OF_{std}	2.3	2.9	2.2	1.9	1.5	2.0	2.9	5.4	9.7	8.9	8.1
MN	N_{c+ave}	460.0	336.7	331.5	320.3	274.2	216.9	143.9	116.6	99.2	96.0	61.0
	N_{c+std}	9.5	6.4	6.2	7.8	5.6	5.9	9.3	13.6	15.3	15.2	10.5
	N_{c-ave}	61.7	65.9	66.6	68.7	90.9	132.3	230.5	280.2	327.8	351.4	410.0
	N_{c-std}	2.1	2.2	3.9	3.3	3.7	6.15	10.3	15.4	19.8	20.2	12.6
	OF_{ave}	61.7	93.0	119.6	144.2	164.2	174.6	178.5	165.7	144.9	121.5	61.0
	OF_{std}	2.1	1.9	3.1	2.6	1.9	2.2	3.0	5.5	8.7	11.9	10.5

are small and Meme-DB could get the optimal solution. For $\lambda = 0.9$ and 1, all the negative edges within SN network sometimes are changed of their signs and all the nodes in this network are friends. The IN network is originally structurally balanced, i.e., no edge flips are needed. Meme-DB cannot always find the optimal solution and sometimes get the near-optimal solution.

On EN and GN networks, the numbers of sign-changed negative edges increase with the increase of λ. The numbers of sign-flipped positive edges decrease with the increase of λ. Moreover, the results of objective function increase first and then decrease. To show the phenomenon more clearly, except for N_{c+}, N_{c-}, and OF, variation of averages of the overall numbers of sign-changed edges N_c with λ on four networks is also illustrated in Fig. 4.10. Contrary to OF, the averages of N_c decrease first and then increase and reach the minimum for $\lambda = 0.5$ (i.e., in this case there is no bias towards flipping positive or flipping negative signs). That is,

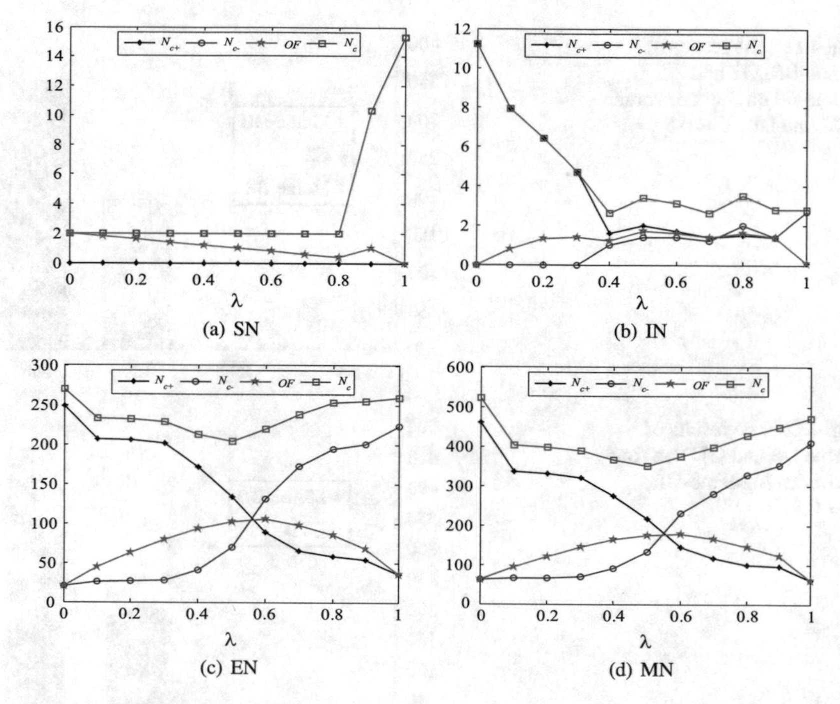

Fig. 4.10 The variation of N_{c+}, N_{c-}, OF and N_c of Meme-DB with λ on four networks. **a** SN Network; **b** IN network; **c** EN network; **d** MN network

when there is a bias towards flipping positive or negative flips, more edges should be changed of sign.

When there is no bias towards flipping positive or flipping negative signs, results achieved by optimizing objective function (Eq. 4.1) should be equal to the minimum result of energy function (Eq. 1.8). Then to show the performance of Meme-DB, the comparison of results of Meme-DB on equation (Eq. 4.1) for $\lambda = 0.5$ is given with the results of the method proposed by [18] and algorithm by [30] on equation (Eq. 1.8). As shown in Fig. 4.11, though except for SN network, the results of Meme-DB are slightly less than that of GT on other three networks, the results of our algorithm are acceptable. On the network MN, our algorithm is better than algorithm Meme-SB, while on network EN, Meme-DB is a little bit worse.

Many discrete dynamics models for structural balance have been proposed. Here, constrained triad dynamics (CTD) proposed by Antal et al. [1] is used to make a comparison with Meme-DB. First, a definition of "jammed state" for incomplete network is given because"jammed state" is originally defined for complete networks. **Definition of jammed state for incomplete networks**: an algorithm is struck in a jammed state, when no change of edge signs can decrease the number of imbalanced triads.

As shown in Fig. 4.12, CTD can give the optimal solutions on networks SN and IN, while the total numbers of sign changes of CTD are higher than those of Meme-DB on another two networks. CTD on incomplete networks is easily trapped in "jammed

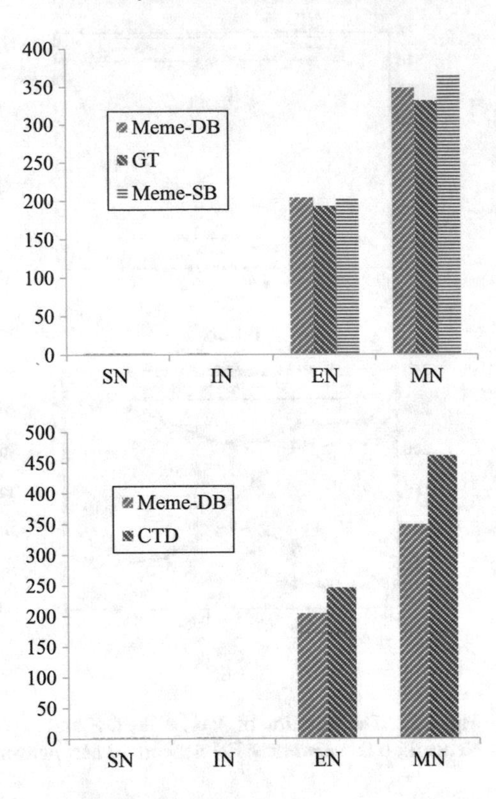

Fig. 4.11 The results of Meme-DB, GT and Meme-SB on four networks. In Meme-DB, $\lambda = 0.5$

Fig. 4.12 The results of Meme-DB and CTD on four networks. In Meme-DB, $\lambda = 0.5$

states". Meme-DB gives the least number of sign changes of evolution, while CTD evolves following update rules and does not give the least number of sign changes, which causes that the results of CTD is worse than those of Meme-DB.

To show the necessity of procedures of local search and preprocess method in Meme-DB, three algorithms named as Meme-DB1, Meme-DB2, and GA-DB are chosen to make a comparison. In Meme-DB, a preprocess method is introduced. Meme-DB1 and Meme-DB2 are introduced to show that the preprocess procedure in Meme-DB can speed up the algorithm convergence and contribute to find the global optima. Meme-DB1 is a method with the random initialization method without preprocessed procedure. In Meme-DB2, a random number is produced for each gene in the chromosome intended for preprocess. If the number is above 0.5, then the gene will be preprocessed. Otherwise, the gene is without preprocess procedure. To demonstrate the necessity of local search procedure, GA-DB is introduced, which is achieved by only deleting the local search procedure of our algorithm. The proposed algorithm Meme-DB and other three algorithms above are run 30 times without the transformation procedure. The averages, standard deviations, and minimum values of four algorithms on four networks are illustrated in Table 4.8. As Table 4.8 shown, all four algorithms can get the global optimal solutions on networks SN and IN, while Meme-DB has the minimum average on network IN. On the network EN, both of the average and minimum of Meme-DB are the best among four algorithms. On

Table 4.8 Results of 30 runs of four algorithms on four networks for different values when $\lambda = 0.5, \mu = 1$

		Meme-DB	Meme-DB1	Meme-DB2	GA-DB
SN	ave	1	1.3	1	1.6
	std	0	1.0	0	1.6
	min	1	1	1	1
IN	ave	**2.5**	4.3	3.0	3.5
	std	1.5	1.6	1.3	1.9
	min	0	0	0	0
EN	ave	**163.8**	182.0	168.1	419.4
	std	3.8	4.5	4.7	14.6
	min	**157**	174	161	379
MN	ave	**306.2**	338.8	314.5	860.7
	std	8.5	4.7	10.7	17.7
	min	286.5	328.5	284.5	807

the network MN, the average of Meme-DB is minimum, while the minimum value of algorithm Meme-DB2 is optimal.

To explain the results of four algorithms further, the convergence trajectories of optimal values of four algorithms on four signed networks in experiments above are given in Fig. 4.13. All four algorithms can find the optimal solutions of networks SN and IN. Meme-DB, Meme-DB2, and GA-net all can find the global solution of SN in the 1th iteration. On the network IN, algorithm Meme-DB1 has the best performance, which may be because of a good random initialization. Meme-DB has a better performance than another two algorithms. On the network EN, Meme-DB has the best performance among the four algorithms. On the network MN, Meme-DB has the best initial value, while the speed of convergence is worse than Meme-DB2. Algorithm GA-DB cannot converge to the optimal solutions on networks EN and MN within 50 iterations. It is concluded that the preprocessing procedure and local search procedure in Meme-DB are necessary.

4.3.7 Conclusions

This section introduced a work which aims to optimize dynamical evolution of structural balance in signed networks. Instead of using a total number of sign changes, it is suggested that there be a certain bias towards flipping positive or flipping negative signs. Dynamical evolution of structural balance is formulated as an optimization problem. An algorithm based on memetic algorithm named Meme-DB is proposed to optimize the proposed objective function. The results of Meme-DB on four networks

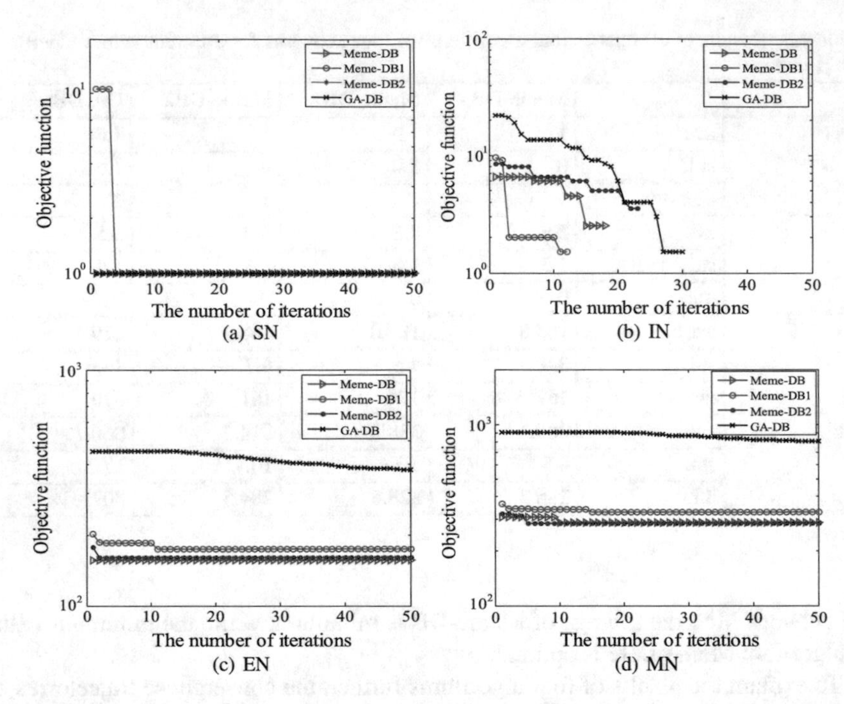

Fig. 4.13 Comparison of the results of four algorithms on four networks. **a** SN Network; **b** IN network; **c** EN network; **d** MN network

show the variation of objective function with parameter in the objective function. Further experiments show the efficiency of the preprocessing procedure and local search procedure.

4.4 Computing and Transforming Structural Balance Based on Memetic Algorithm

Section 4.2 introduced a memetic algorithm to compute global structural balance in signed networks by optimizing Eq. 1.8.[3] Section 4.3 introduces a memetic algorithm to optimize the extended objective function Eq. 4.1. These two objective functions are just applicable to the strong definition of structural balance.

This section introduces a fast memetic algorithm, referred to as MLMSB, to compute and transform structural balance of signed networks. The advantages of MLMSB are as follows. First, the energy function criterion (Eq. 1.8) is extended to the weak definition of structural balance, and the computation of structural balance is mod-

[3] Acknowledgement: Reprinted from Knowledge-Based Systems, 85, Ma, L., Gong, M., Du, H., Shen, B., Jiao, L., A memetic algorithm for computing and transforming structural balance in signed networks, 196–209, Copyright(2015), with permission from Elsevier.

eled as the optimization of the extended energy function H. Second, by introducing a cost coefficient parameter, a more general energy function H_w is presented to evaluate the balance transformation cost, and the transformation of structural balance is modeled as the optimization of H_w. Finally, a multilevel learning based local search is integrated into a population-based genetic algorithm (GA) to solve the modeled optimization problems. For the proposed MLMSB algorithm to converge fast, the network-specific knowledge such as the interactions of nodes, clusters, and partitions are adopted.

4.4.1 *Optimization Models*

4.4.1.1 Optimization Model for the Computation of Structural Balance in Signed Networks

The computation of structural balance of signed networks in the weak definition can be modeled as follows:

$$\min H = \sum_{(i,j)} [\frac{(1 - J_{ij}\Theta(x_i, x_j))}{2}] \tag{4.2}$$

where H is the extended energy function and $\Theta(x_i, x_j)$ is a sign function whose value is 1 if $x_i = x_j$ and -1 otherwise. Here, the x_i value is an integer value in the range of $[1, N]$, where N is the number of nodes. The minimization of H is to find the minimum number of unbalanced links by dividing the nodes of a signed network into $k \geq 2$ clusters.

It is worthwhile to note that the energy function in Eq. 1.8 is a special case of H with k fixed at 2, and the function H can cater both the strong and weak definitions of the structural balance of signed networks. For instance, the social relation $- - -$ can be classified as an unbalanced (balanced) structure with $H = 1$ ($H = 0$) by dividing the nodes into two (three) clusters. It is necessary to set $k = 2$ in advance when we use H to compute the structural balance of signed networks in the strong definition. An illustration of the computation of structural balance on a toy network G_2 is shown in Fig. 4.14.

4.4.1.2 Optimization Model for Structural Balance Transformation in Signed Networks

The unbalanced entities and connections of complex systems are the potential threats to their functionality. For complex systems to improve the functional security, it is necessary to alleviate their potential imbalances.

The extended energy function H can be further expressed as follows:

$$H = -\sum_{(i,j),x_j \neq x_i} [\frac{J_{ij}^{+}\Theta(x_i, x_j)}{2}] - \sum_{(i,p),x_p = x_i} [\frac{J_{ip}^{-}\Theta(x_i, x_p)}{2}]$$
$$= \sum_{(i,j),x_j \neq x_i} |J_{ij}^{+}|/2 + \sum_{(i,p),x_p = x_i} |J_{ip}^{-}|/2, \tag{4.3}$$

where J_{ij}^{-} (J_{ij}^{+}) represents the negative (positive) links between nodes v_i and v_j. $\sum_{(i,p),x_i = x_p} |J_{ip}^{-}|/2$ and $\sum_{(i,j),x_j \neq x_i} |J_{ij}^{+}|/2$ indicate the number of the negative links within the same cluster and the positive links across different clusters, respectively.

A balanced network will be achieved if the signs of the unbalanced links found by the optimization of H are changed.

In reality, the cost of changing a positive link may be different from that of a negative link. Thus, a more general energy function H_w is introduced with cost coefficient parameter w evaluating the different cost of changing positive and negative links for balance transformation. The corresponding optimization model for the balance transformation with minimum cost is as follows:

$$\min H_w = w \cdot \sum_{(i,j),x_j \neq x_i} |J_{ij}^{+}|/2 + (1 - w) \cdot \sum_{(i,p),x_p = x_i} |J_{ip}^{-}|/2, \tag{4.4}$$

where $w \in \{0, 1\}$ evaluates the emphasis on either positive links or negative ones. If $0 \leq w < 0.5$, changing a negative link is more costly compared with changing a positive one; whereas if $0.5 < w \leq 1$, positive links affect more. An illustration of balance transformation on the toy network G_2 is shown in Fig. 4.15.

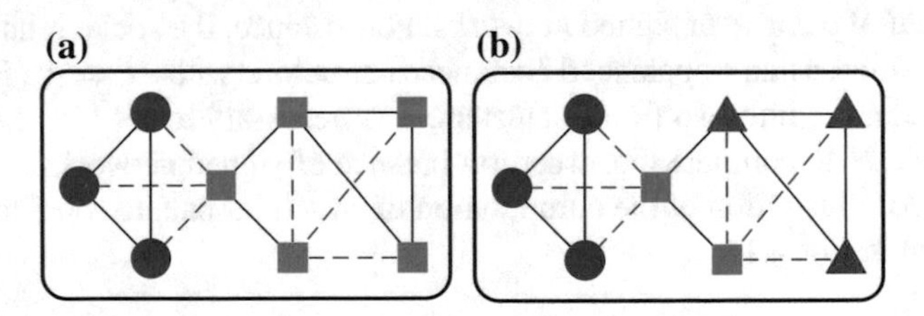

Fig. 4.14 Two balance transformations on the toy network G_2 under different values of w. The unbalanced toy network is transformed into a balanced one at minimum cost under **a** $w = 0.9$ and **b** $w = 0.1$. For the unbalanced toy network to be balanced, three negative edges in (**a**) and one positive edge in (**b**) are necessary to be changed of signs

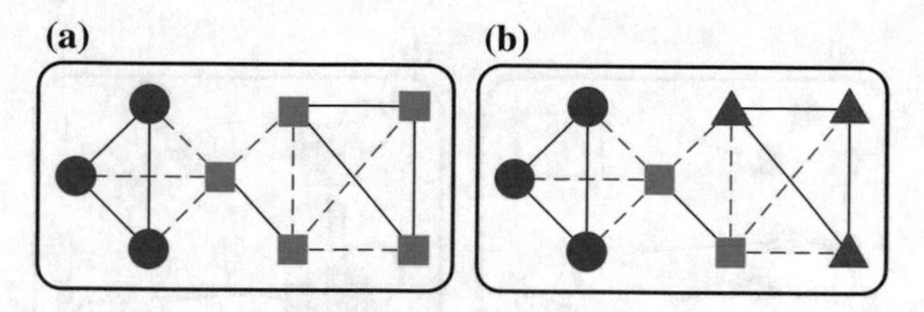

Fig. 4.15 Illustration of the computation of structural balance on a toy network G_2. G_2 is divided into **a** two clusters with $H = 4$ and **b** three clusters with $H = 0$

4.4.2 *Memetic Algorithm for the Computation and Transformation of Structural Balance in Signed Networks*

MLMSB is first initialized with a population of solutions, each of which represents a possible balance transformation. The population of solutions is then evolved via global search conducted by Genetic Algorithm (GA) and a network-specific local search iteratively. The network-specific knowledge, such as the neighborhoods of node, cluster and partition, is incorporated into the GA and local search.

4.4.2.1 Definitions of Neighborhoods

A signed network can be viewed as a microscopic graph $G = \{V, E\}$, with a set of nodes $V = \{v_1, v_2, \ldots, v_N\}$ and positive/negative links E. And the links between nodes can be represented as an adjacency matrix J.

In the meantime, from a macroscopic viewpoint, a signed network can also be divided into a set of clusters $S = \{s_1, s_2, \ldots, s_k\}$ with positive and negative connections E' between clusters, in summary $G' = \{S, E'\}$. Here, weighted adjacency matrices A^+ A^- are used to measure the effect of the positive and negative links, respectively, and their items are defined as follows: $A^+_{ij} = \sum_{u \in s_i} \sum_{q \in s_j} J^+_{uq}$ and $A^-_{ij} = \sum_{u \in s_i} \sum_{q \in s_j} J^-_{uq}$. Especially, A^+_{ii}/A^-_{ii} indicates the number of internal positive/negative links within the cluster s_i; and A^+_{ij}/A^-_{ij} $(j \neq i)$ is the number of external positive/negative links between clusters s_i and s_j.

Figure 4.16 demonstrates both the microscopic and the macroscopic viewpoints of a toy network.

Based on the knowledge that negative links have little influences on the formation of communities [10] and that the final balance state of signed networks is determined by the number of positive links [24], the neighborhoods of node, cluster, and partition are defined as follows.

Definition of the neighborhood of node. The neighborhood of a node v_i is a set of nodes that have a positive link with v_i.

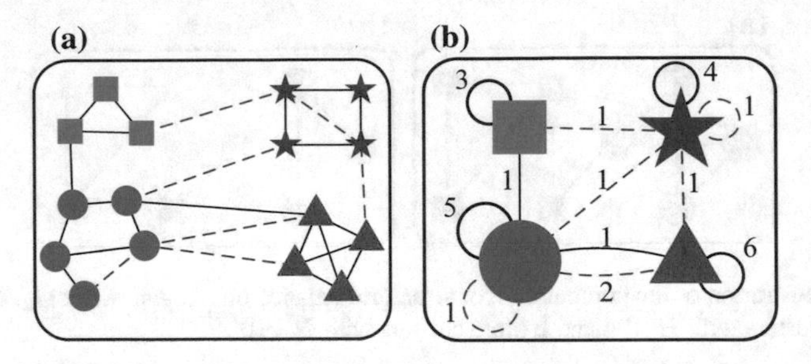

Fig. 4.16 Illustration of the representations of the toy network G_3 under the microscopic and the macroscopic viewpoints. **a** In a microscopic viewpoint, G_3 can be represented by a microscopic graph with *16* nodes and *20* positive and *7* negative links. **b** In a macroscopic viewpoint, G_3 is composed of *4* clusters, *18* internal positive and *2* internal negative links and *2* external positive and *5* external negative links. The self-loop represents the internal links between nodes in the same community. Nodes are in different clusters if they are plotted by different *colors* and shapes

Definition of the neighborhood of cluster. The neighborhood of a cluster s_i is a set of clusters that have external positive links with s_i.

Definition of the neighborhood of partition. The neighborhood of a partition S_i is a set of partitions that can be transformed into S_i by merging any of their two neighboring nodes or clusters.

4.4.2.2 Representation and Initialization

In MLMSB, each chromosome (or solution) $\mathbf{x}_i$ is expressed as an integer vector.

$$\mathbf{x}_i = \{x_{i1}, x_{i2}, \ldots, x_{iN}\}, \tag{4.5}$$

where $x_{ij} \in \{1, 2, ..., N\}$ is the cluster identifier of node v_j. For instance, $x_{ij} = 10$ indicates that the node v_j of ith chromosome is classified into the 10th cluster. With this representation, each solution represents a possible clustering division of this network. Thereafter, the nodes are divided into several clusters and the ones with the same cluster identifier are grouped in the same cluster.

In the transformation of structural balance, it is necessary to change signs of positive edges across different clusters and negative edges within the same cluster. Figure 4.17 illustrates the integer vector representation on a toy network G_4 with 10 nodes. It can be clearly seen that the number of clusters is automatically determined by the integer vector $\mathbf{x}_i$.

Initialization corresponds to **Step 1** of MLMSB. In general, the convergence of MAs can be accelerated by a population of initialized solutions with high quality and good diversity. For the initialization to generate a population of high-quality solutions, it is necessary to incorporate network-specific knowledge. Here, the neighborhood of node is used, and a population of solutions is generated by assigning

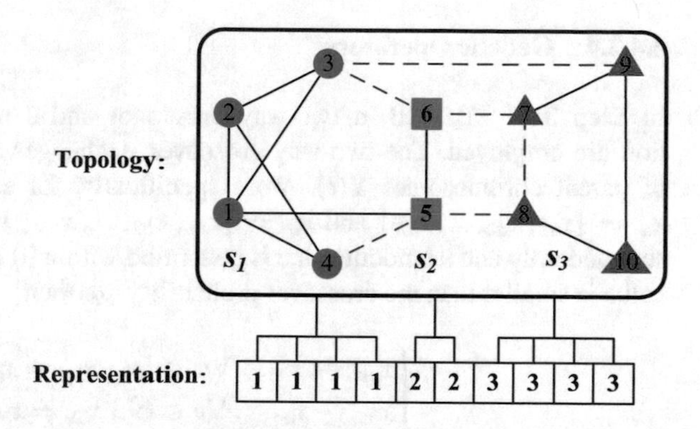

Fig. 4.17 Illustration of the integer vector representation on a toy network G_4. The toy network is divided into three clusters, $s_1 = \{1, 2, 3, 4\}$, $s_2 = \{5, 6\}$ and $s_3 = \{7, 8, 9, 10\}$. The integer vector $\mathbf{x}_i$ can be represented as $\mathbf{x}_i = \{1, 1, 1, 1, 2, 2, 2, 3, 3, 3, 3\}$

the cluster identifiers of nodes to those of their neighbors. In order to maintain the diversity of the initial solutions, a random sequence strategy is adopted to update the cluster identifiers of nodes. Algorithm 23 lists the pseudocode of the network-specific initialization.

Algorithm 23 Network-specific initialization.

1: **Input**: The size of population N_P.
2: Generate N_P solutions $\mathbf{X}(0) = \{\mathbf{x}_1, \mathbf{x}_2, \ldots, \mathbf{x}_{N_P}\}$, where $x_{ij} \leftarrow j, 1 \leq i \leq N_P, 1 \leq j \leq N$.
3: **for** each solution $\mathbf{x}_i$ of $\mathbf{X}(0)$ **do**
4: Generate a random sequence (e.g., $\{r_1, r_2, \ldots, r_N\}$).
5: **for** each chosen node v_{r_j} of G **do**
6: $x_{iu} \leftarrow x_{ir_j}, \forall u \in \{u \mid J_{ur_j} = 1\}$.
7: **end for**
8: **end for**
9: **Output**: $\mathbf{X}(0) = \{\mathbf{x}_1, \mathbf{x}_2, \ldots, \mathbf{x}_{N_P}\}$.

In the end, the network-specific initialization will generate a population of unsupervised solutions $\mathbf{X}(0)$.

4.4.2.3 Tournament Selection

In **Step 2** of MLMSB, N_m solutions are selected from the population as the parent chromosomes (solutions). Here, the classical tournament selection is used to choose the parent chromosomes. The tournament selection firstly generates N_m tournaments. Each tournament consists of N_o solutions chosen from the population randomly, where N_o is the size of the tournament. Then, the solution of each tournament with minimum H_w is chosen as the parent chromosomes $\mathbf{Y}(t)$, where t is the running number of generations.

4.4.2.4 Genetic Operators

In **Step 3** of MLMSB, a two-way crossover and a neighborhood-based mutation are employed. The two-way crossover exchanges the clustering information of parent chromosomes $\mathbf{Y}(t)$. More specifically, for each pair of chromosomes $\mathbf{x}_a = \{x_{a1}, x_{a2}, ..., x_{aN}\}$ and $\mathbf{x}_b = \{x_{b1}, x_{b2}, ..., x_{bN}\}$ in $\mathbf{Y}(t)$, a node v_q is chosen randomly and a random value is generated within [0 1]. If the generated random value is smaller than the crossover probability p_c, then

$$\begin{cases} x_{a\mu} \leftarrow x_{bq}, & \forall \mu \in \{\mu \mid x_{b\mu} = x_{bq}\} \\ x_{b\nu} \leftarrow x_{aq}, & \forall \nu \in \{\nu \mid x_{a\nu} = x_{aq}\} \end{cases}$$

The two-way crossover can inherit useful clustering divisions from their parents [15]. Figure 4.18 gives an illustration of the two-way crossover on two parent chromosomes $\mathbf{x}_a$ and $\mathbf{x}_b$. It shows that the offspring $\mathbf{x}_c$ inherits the cluster $\{1, 2, 3\}$ and $\{4, 5, 6, 7\}$ from $\mathbf{x}_b$ and $\mathbf{x}_a$, respectively.

In order to reduce useless explorations and generate dependable offsprings, a neighborhood-based mutation is used which mutates the cluster identifiers of nodes with those of their neighbors. In the mutation processes, when a node has more than one neighbor and these neighbors are in different clusters, its cluster identifier is randomly replaced with one of its neighbors. The neighborhood-based mutation performs on each solution $\mathbf{x}_i$ generated by the crossover, and its processes can be

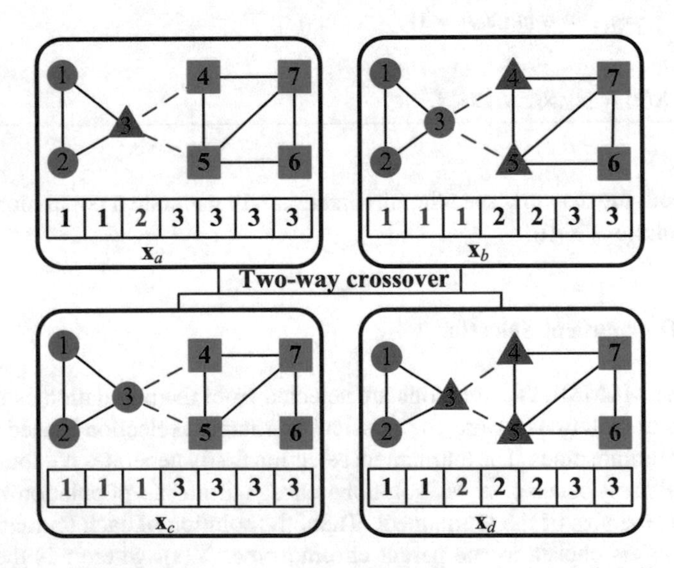

Fig. 4.18 Illustration of the two-way crossover. $\mathbf{x}_a = \{1, 1, 2, 3, 3, 3, 3\}$ and $\mathbf{x}_b = \{1, 1, 1, 2, 2, 3, 3\}$ are two parent chromosomes. Assumed that v_3 is chosen, nodes v_1, v_2 and v_3 in $\mathbf{x}_a$ are assigned with the cluster identifier of v_3 of $\mathbf{x}_b$ and node v_3 in $\mathbf{x}_b$ with the cluster identifier of v_3 of $\mathbf{x}_a$ after crossover

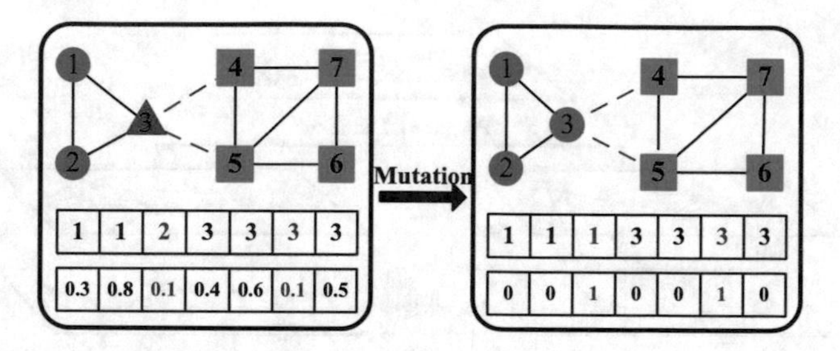

Fig. 4.19 Illustration of mutation operation with $p_m = 0.2$

expressed as follows. For each node v_j in $\mathbf{x}_i$, $1 \leq j \leq N$, generate a random value within [0, 1] and choose at random a node v_u from the neighbors of v_j. If the generated random value is smaller than the mutation probability p_m, then $x_{ij} \leftarrow x_{iu}$.

Figure 4.19 gives an illustration of the neighborhood-based mutation. The solutions generated by the mutation form the offspring population $\mathbf{Z}(t)$.

4.4.2.5 Multilevel Learning Based Local Search

In MLMSB, GA can explore and exploit a few promising regions in the solution space. However, it cannot converge to the optimal solutions of promising regions within a few generations. A local search technique is designed to accelerate the convergence of GA to the optimal solutions [5]. Generally, starting from an initial solution, and then a local search technique is adopted to repeatedly search for an improved solution by local changes [5]. A local search can effectively reduce useless explorations by incorporating problem-specific knowledge.

In this study, the neighborhoods of node, cluster, and partition are incorporated into the multilevel learning technique which consists of a microscopic-level node learning, a macroscopic-level clustering learning and a partition learning. In **Step 4** of MLMSB, the multilevel learning is employed on the best solution $\mathbf{x}_l$ in $\mathbf{Z}(t)$. The visualization of the multilevel learning is shown in Fig. 4.20. In the following, detailed descriptions of the node learning, cluster learning, and partition learning are given.

Node Learning. In the node learning, each node v_{r_i} in $\mathbf{x}_l$ is reassigned to the cluster of its neighbor v_{r_u}, which will result in the maximum decrement in H_w. More specifically, for each node v_{r_i} chosen in a random sequence $R_1 = \{r_1, r_2, ..., r_N\}$, its cluster identifier x_{lr_i} is updated as follows.

$$x_{lr_i} = \arg\max_{x_{lu}} \left(-\Delta H_w(\mathbf{x}_l \mid_{x_{lr_i} \leftarrow x_{lu}}) \right), \; u \in \{u \mid J_{r_i u} = 1\},$$

where $\Delta H_w(\mathbf{x}_l \mid_{x_{lr_i} \leftarrow x_{lu}})$ represents the decrement of H_w when node v_{r_i} is reassigned to the cluster of v_u. The node learning ends when the assignment of each node to the clusters of its neighbors cannot result in the decrement of H_w.

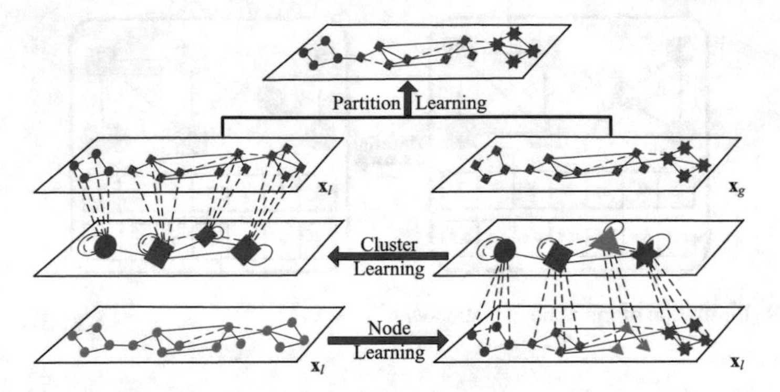

Fig. 4.20 Visualization of the multilevel learning on the best solution $\mathbf{x}_l$ in $\mathbf{Z}(t)$. The node learning technique updates $\mathbf{x}_l$ by merging any two neighboring nodes. The cluster learning technique consists of two phases: *1* build a new network whose nodes are the found clusters and *2* update $\mathbf{x}_l$ by merging any neighboring nodes in the new network. The partition learning technique evolves a consensus neighbor of two solutions $\mathbf{x}_l$ and $\mathbf{x}_g$ towards a better solution. The *red dash lines* are negative edges while the *black dash lines* represent mappings between nodes in a microscopic viewpoint and super-nodes in a macroscopic viewpoint

The node-level learning can converge to a good solution quickly. However, the node learning technique falls into a local optimum easily because it cannot traverse all solutions in the promising regions. By incorporating the neighborhood structure of cluster, a cluster learning technique is used to find a better solution around the solution $\mathbf{x}_l = \{s_1, s_2, \ldots, s_k\}$ generated by the node learning technique, where $s_i = \{v_{i_1}, v_{i_2}, \ldots, v_{i_n}\}$ is a cluster with n nodes.

Cluster Learning. In the cluster learning, each cluster s_i updates its cluster identifier with that of its neighbor s_q, which will lead to the maximum decrement in H_w. More specifically, for each cluster s_i chosen in a random order $R_2 = \{r_1, r_2, ..., r_k\}$, its cluster identifier x_{ls_i} is updated as follows.

$$x_{ls_i} = \arg\max_{x_{ls_q}} \left(-\Delta H_w(\mathbf{x}_l \mid_{x_{ls_i} \leftarrow x_{ls_q}}) \right),\ \forall q \in \{q \mid A^+_{s_i s_q} \geq 1\}.$$

The updating process ends when the cluster identifier of each cluster keeps unchanged.

The cluster learning considers different neighborhood structures compared with the node learning, which makes it possible to find a better solution around the solution $\mathbf{x}_l$. However, the cluster learning is still easy to trap into a local optimal clustering division because it is hard to separate the merged nodes and communities again [23].

Partition Learning. In order to overcome the constraint of cluster learning, the partition learning is designed. The partition learning performs on two solutions $\mathbf{x}_l = \{x_{l1}, x_{l2}, \ldots, x_{lN}\}$ and $\mathbf{x}_g = \{x_{g1}, x_{g2}, \ldots, x_{gN}\}$, where $\mathbf{x}_l$ is the solution generated by the cluster learning and $\mathbf{x}_g$ is the best solution in the population, and it consists of two steps. The first step is to find the consensus neighbor $\mathbf{x}_f$ of $\mathbf{x}_g$ and $\mathbf{x}_l$. At this step, the nodes are divided into the same cluster if they have the same identifiers in both $\mathbf{x}_g$ and $\mathbf{x}_l$, but separated into different clusters if they have the same identifiers in $\mathbf{x}_g$

while different identifiers in $\mathbf{x}_l$. An illustration of this step is shown in Fig. 4.20. The second step uses the node and cluster learning to optimize the consensus neighbor $\mathbf{x}_f$, and generates a solution $\mathbf{x}_l^{'}$.

The partition learning can learn the skeletal structure from $\mathbf{x}_l$ and $\mathbf{x}_g$ and evolve the consensus neighbor towards $\mathbf{x}_l$ or $\mathbf{x}_g$, or even a better solution.

Computation of ΔH_w. The ΔH_w of merging a node v_i from its cluster s_i into its neighboring cluster s_j can be quickly computed as

$$\Delta H_w(\mathbf{x}_l\,|_{x_{li}\leftarrow \mathbf{x}_{ls_j}}) = w \cdot \sum_{(i,q),q\in s_i} |J_{iq}^{+}| + (1-w)\cdot \sum_{(i,q),q\in s_i} |J_{iq}^{-}| - \ldots$$
$$w \cdot \sum_{(i,p),p\in s_j} |J_{ip}^{+}| - (1-w)\cdot \sum_{(i,p),p\in s_j} |J_{ip}^{-}| \tag{4.6}$$

The computational complexity of ΔH_w is $O(\bar{k})$, where $\bar{k}$ is the averaged degree of a signed network.

The reasons why the multilevel learning technique works are that the multilevel learning technique incorporates the network-specific knowledge into the search rules. Moreover, the node learning and the cluster learning can quickly converge to the optimal solutions in the promising regions discovered by GA. In addition, the partition learning overcomes the constraint of cluster learning. Finally, extensive experiments demonstrate the effectiveness of the multilevel learning technique.

4.4.2.6 Complexity Analysis

It is time-consuming to implement the genetic operations and multilevel learning of MLMSB. In genetic operations, the two-way crossover and the mutation are implemented for $N_c/2$ and N_c independent times, respectively, where N_c represents the size of chosen solutions. The computational complexities of the crossover and the mutation are $O(\bar{k})$ and $O(\bar{k}+M^{+})$, respectively, where $\bar{k}$ is the averaged degree and M^{+} represents the number of positive edges. In the node and cluster learning, ΔH_w is computed M^{+} times. Moreover, the formation of clusters needs $\bar{k}$ basic operations in the cluster learning. In addition, the identification of consensus neighbor needs N basic operations in the partition learning. Therefore, the computational complexities of the node learning, cluster learning and partition learning are $O(\bar{k}M^{+})$, $O(M\bar{k}^2)$ and $O(N+M^{+}\bar{k}+M\bar{k}^2)$, respectively. For MLMSB, its computational complexity is $O(\bar{k}^2 M g_{max})$, where g_{max} is the maximum number of iterations.

4.4.2.7 Comparisons Between MLMSB and Previously Introduced Works

MLMSB and Meme-Net (in Sect. 2.2), MLCD (in Sect. 2.3) and Meme-SB (in Sect. 4.2) model the issue of complex systems as a network optimization problem, and use the network-specific MA to solve the modeled problem. However, they are different in the motivation, optimization model, GA operation, network-specific knowledge, and local search.

As for the motivation, Meme-Net focuses on detecting the multi-resolution community structures of undirected networks. However, it is impossible for Meme-Net to detect communities in large-scale networks because of its high computational complexity. MLMCD is devised to identify communities in large-scale networks. Meme-SB aims at computing the structural balance of signed networks in the strong definition. MLMSB is designed for the transformation of structural balance of signed networks.

In the optimization model, Meme-Net models the community detection in undirected networks as the optimization of modularity density (Eq. 1.7). Multi-resolution communities can be detected by maximizing Eq. 1.7 with different λ. MLCD models the community detection of networks as the optimization of modularity (Q) (Eq. 1.6). The maximization of Q can detect communities with massive internal links and few external links. Meme-SB models the computation of structural balance as the optimization of energy function h while MLMSB solves the transformation of structural balance by optimizing a more general energy function H_w.

In the adopted GA operation, Meme-Net, MLCD, and MLMSB have different representations from Meme-SB. Meme-Net, MLCD, and MLMSB adopt an integer vector representation while Meme-SB uses a bit vector representation. Moreover, Meme-Net and MLMSB have different initializations from Meme-SB and MLCD. In the initialization, Meme-Net and MLMSB assign the cluster identifiers of nodes to those of their neighbors while Meme-SB and MLCD update the cluster identifiers of nodes with those of their neighbors. In addition, Meme-Net, MLCD and MLMSB have different crossover operators from Meme-SB. Meme-Net, MLCD, and MLMSB use a two-way crossover while Meme-SB adopts a two-point crossover.

In the incorporated network-specific knowledge, Meme-Net uses the undirected links among nodes while MLCD adopts both the undirected connections of node and community. Meme-SB and MLMSB are used for tackling signed networks. Meme-SB uses the neighborhood of node while MLMSB adopts the neighborhoods of node, cluster, and partition.

In the devised local search, Meme-Net and Meme-SB adopt a hill climbing and node learning as the local search, respectively. The hill climbing and the node learning optimize the clustering division of networks by updating the cluster identifier of nodes iteratively. MLCD and MLMSB adopt multilevel learning techniques as the local search. However, the detailed operations of the multilevel learning are different because MLCD and MLMSB tackle different optimization models and types of networks.

Table 4.9 Parameters settings and computational complexity of the algorithms

Algorithm	g_{max}	N_p	N_m	p_c	p_m	N_o	Complexity
MLMSB	100	100	10	0.9	0.1	2	$O(\bar{k}^2 M g_{max})$
GA	100	100	10	0.9	0.1	2	$O(N_m M g_{max})$
OLMSB	100	100	10	0.9	0.1	2	$O(\bar{k} M g_{max})$
TLMSB	100	100	10	0.9	0.1	2	$O(\bar{k}^2 M g_{max})$
SLPAm	100	—	—	—	—	—	$O(\bar{k} M g_{max})$
SBGLL	100	—	—	—	—	—	$O(\bar{k}^2 M g_{max})$
MODPSO	100	100	10	—	0.1	2	$O(N_P M g_{max})$
SN2013	300	—	—	—	—	—	$O(\bar{k} M g_{max})$
PRE2014	—	—	—	—	—	—	$O(\bar{k}^2 M)$
FEC	—	—	—	—	—	—	$O(M)$

4.4.3 Experimental Results

In this section, MLMSB is tested on five real-world signed networks and compare it with nine classical algorithms.

Five signed networks, including two small-scale international relationship networks, two large-scale online vote networks and a large-scale online blog network, are chosen in experiments. The chosen networks are GGS, War, WikiElec (Wikiele network), Wiki-rfa network and Slashdot network. The main properties of these networks are shown in Table 1.2.

$$J'_{ij} = \begin{cases} J_{ij} & \text{if } J_{ij} = J_{ji} \vee J_{ij} \neq 0 \wedge J_{ji} = 0, \\ J_{ji} & \text{if } J_{ij} = 0 \wedge J_{ji} \neq 0, \\ 0 & \text{if } J_{ij} = -J_{ji}. \end{cases} \quad (4.7)$$

In the experiments, MLMSB is compared with nine classical algorithms, namely GA, OLMSB, TLMSB, SLPAm, SBGLL, MODPSO, SN2013, PRE2014, and FEC. Their computational complexities are recorded in Table 4.9.

GA, OLMSB, and TLMSB are the variant versions of MLMSB without the local search, the cluster, and partition learning and the partition learning, respectively. Comparisons between MLMSB, GA, OLMSB, and TLMSB are made to demonstrate the effectiveness of the devised node, cluster, and partition learning techniques on the acceleration of MLMSB to an optimal solution.

SLPAm and SBGLL are the extended versions of classical greedy algorithms LPAm and BGLL, respectively. LPAm detects the communities of unsigned networks by a modularity-specific label propagation while SLPAm computes the imbalances of signed networks by a H_w-specific label propagation. BGLL is a fast two-phase community detection algorithm based on the optimization of modularity [27, 28]. In the first (second) phase, the node (community) chosen from the first one to the last one is assigned to its neighboring community when this change can increase modularity maximally. SBGLL focuses on the imbalance computation of signed networks based on the optimization of H_w.

FEC [34] adopts an agent-based technique to identify the communities in signed networks. In FEC, communities are composed of a few nodes with dense positive but sparse negative links. The positive links across different communities and the negative links within the same community are considered as unbalanced links. A comparison between MLMSB and FEC is made to analyze the relations between the structural balance and the communities under different definitions in signed networks.

MODPSO is the algorithm introduced in Sect. 3.4. Comparison between MLMSB and MODPSO is made to illustrate the super performance of the proposed algorithm in terms of accuracy and stability compared with the population-based evolutionary algorithm.

PRE2014 [10] and SN2013 [7] are two relevant works on the computation of imbalance in signed networks. PRE2014 adopts a classical community detection algorithm Infomap [29] to detect communities in signed networks. SN2013 updates the cluster information of each node by: (i) the movement of the node from its cluster to another cluster and (ii) the interchange of two nodes between two clusters. PRE2014 and SN2013 evaluate the imbalance of signed networks by computing the number of positive links across the found communities and negative links within the same community.

4.4.3.1 Experiments for the Computation of Structural Balance

Table 4.10 records the statistical results over 50 independent trials of all algorithms with parameters settings shown in Table 4.9. In Table 4.10, the H_{min}, $\overline{H}$, and H_{std} indicate the minimum, mean, and standard deviation of H values, respectively. Moreover, the P-value is a statistical hypothesis (i.e., t-test) value, and it is the probability of getting the observed data when the null hypothesis is actually true. Here, the null hypothesis is expressed as that where there is no difference between MLMSB and its comparison algorithm on the computation of structural balance, and the observed data are the statistical H values of the comparison algorithm over 50 trials. If the P-value is less than a significance level $\alpha = 0.05$, it indicates that the observed data are inconsistent with the assumption that the null hypothesis is true. It is noticed that a small P-value cannot imply that there is a meaningful difference of MLMSB with its comparison algorithm on the computation of structural balance. In addition, the symbols ‘>’, ‘<’ and ‘=’ denote that MLMSB has better significance than, lower significance than and similar significance to its comparison algorithm, respectively, and they are generated based on the P-value of the comparison between MLMSB and the others algorithms and their $\overline{H}$ values. More specifically, if the P-value is smaller than 0.05 and MLMSB has larger (smaller) $\overline{H}$ value than its comparison algorithm, then the significance is labeled as “>” (“<”). If the P-value is larger than 0.05 and the $\overline{H}$ value of MLMSB is far larger than or far smaller than that of its comparison algorithm, then the significance is labeled as “>” or “<”. In the case of values with similarity, we label the significance as “=”.

Table 4.10 Statistic results of H over 50 independent trials on the six real-world signed networks. "—" represents that the algorithm cannot tackle it. "×" indicates that the observed data are the same as the tested data

Networks	Indexes	MLMSB	GA	OLMSB	TLMSB	SLPAm	SBGLL	MODPSO	SN2013	PRE2014	FEC
GGS	H_{min}	**2**	**2**	**2**	**2**	**2**	**2**	**2**	**2**	**2**	4
	$\overline{H}$	**2**	**2**	**2**	**2**	2.720	**2**	2.620	2.560	**2**	4
	H_{std}	**0**	**0**	**0**	**0**	0.9648	**0**	1.638	1.057	**0**	**0**
	P-value	×	×	×	×	$3.3e^{-11}$	∞	0.0003	$7.12e^{-7}$	×	0
	Significance	=	=	=	=	>	=	>	>	=	>
War	H_{min}	**40**	43	**40**	**40**	43	44	44	44	103.0	59
	$\overline{H}$	**41.23**	59.99	45.91	41.53	62.38	44	49.25	80.38	103.0	59
	H_{std}	0.6678	5.994	3.638	0.7582	5.299	**0**	4.693	16.42	**0**	**0**
	P_{value}	×	$9.9e^{-53}$	$1.2e^{-21}$	0.0887	$2.8e^{-63}$	$4.5e^{-52}$	$1.9e^{-29}$	$2.1e^{-42}$	$6e^{-185}$	$3e^{-131}$
	Significance	=	>	>	=	>	>	>	>	>	>
Wiki-ele	H_{min}	**13838**	26355	14310	14012	13969	13902	23829	14077	34215	18786
	$\overline{H}$	**13841**	29658	14456	14097	16919	13902	33113	14104	34215	18786
	H_{std}	2.201	2868	92.23	40.92	4954	**0**	7625	14.73	**0**	**0**
	P_{value}	×	$2.61e^{-8}$	$5.70e^{-9}$	$9.16e^{-9}$	0.0081	$1.6e^{-14}$	$2.23e^{-5}$	$2.3e^{-12}$	$3.2e^{-37}$	$1.1e^{-31}$
	Significance	=	>	>	>	>	>	>	>	>	>
Wiki-rfa	H_{min}	**25972**	54345	26796	26283	26205	26026	49967	26328	62516	33761
	$\overline{H}$	**25974**	63081	27143	26370	33072	26026	49967	26363	62516	33761
	H_{std}	1.429	6105	788.6	56.07	4310	**0**	**0**	12.98	**0**	**0**
	P_{value}	×	$1.29e^{-8}$	0.0011	$3.26e^{-9}$	0.0006	$3.6e^{-41}$	$1.5e^{-39}$	$6.8e^{-15}$	$3.5e^{-41}$	$3.8e^{-35}$
	Significance	=	>	>	>	>	>	>	>	>	>
Slashdot	H_{min}	**62117**	185421	62433	62290	62321	62338	—	62530	70843	70628
	$\overline{H}$	**62120**	189225	62604	62310	62370	62338	—	62534	70843	70628
	H_{std}	1.398	2577	113.6	16.41	36.59	**0**	—	4.243	**0**	**0**
	P_{value}	×	$9.3e^{-17}$	$2.81e^{-7}$	$2.4e^{-11}$	$8.5e^{-20}$	$3.0e^{-21}$	—	$2.1e^{-42}$	$1.1e^{-35}$	$1.4e^{-35}$
	Significance	=	>	>	>	>	>	—	>	>	>

Table 4.10 shows/demonstrates that MLMSB can achieve the smallest H_{min} and $\overline{H}$ for all networks. Though some algorithms, e.g., OLMSB and TLMSB, can also yield the smallest H_{min} and $\overline{H}$ for the small-scale networks such as GGS and War, none of the comparison algorithms can effectively find the solutions with the smallest H_{min} and $\overline{H}$ values in the large-scale networks such as Wiki-ele, Wiki-rfa and Slashdot.

It is worthwhile to note that SBGLL, PRE2014 and FEC are deterministic algorithms which produce the same output for a particular input, whereas MLMSB, GA, OLMSB, TLMSB, SPLAm, MODPSO, and SN2013 are nondeterministic algorithms whose output varies. Of the nondeterministic algorithms, MLMSB has the smallest H_{std} for the GGS, War, Wiki-ele, Wiki-rfa, and Slashdot networks and GA, OLMSB, and TLMSB have the smallest H_{std} values for the GGS network. The comparisons of H_{std} between MLMSB and its comparison algorithms demonstrate the stability of MLMSB on the computation of H.

From Table 4.10, it is seen that the P-value of most comparisons are smaller than the significance level $\alpha = 0.05$. Table 4.10 also shows that the significance symbols of the comparisons between MLMSB, GA, OLMSB, TLMSB, SBGLL and PRE2014 are labeled as '=' for the small-scale GGS network. Furthermore, the significance symbols of the comparisons between MLMSB, GA, OLMSB, TLMSB, SLPAm, SBGLL, SN2013 and PRE2014 are labeled as '>' for the large-scale Wiki-ele, Wiki-rfa and Slashdot networks.

From Table 4.10, the effectiveness of the proposed learning techniques is verified easily. For the small-scale GGS network, GA, OLMSB, TLMSB and MLMSB have the same H_{min}, $\overline{H}$ and H_{std} values. However, for the small-scale War network and the large-scale Wiki-ele, Wiki-rfa and Slashdot networks, the H_{min}, $\overline{H}$ and H_{std} values found by GA are the largest. OLMSB which incorporates the node learning into GA has smaller H_{min}, $\overline{H}$ and H_{std} values than GA; TLMSB which incorporates the cluster learning into OLMSB has smaller H_{min}, $\overline{H}$ and H_{std} values than OLMSB; and MLMSB which incorporates the partition learning into TLMSB has smaller H_{min}, $\overline{H}$ and H_{std} values than TLMSB. It is noticed that the improvements of these algorithms are achieved at the cost of an acceptable increase in computational complexity.

Figures 4.21 and 4.22 exhibit the clustering results of the GGS network and the War network with the smallest H, respectively. The clustering results reflect the potential community structures and the unbalanced links obviously. The GGS network is divided into three clusters, and it has two unbalanced links (e.g., the relations between "MASIL" and "NAGAM", "MASIL" and "UHETO"). The War network is mainly classified into two clusters, and it has 40 unbalanced links highlighted with black lines.

Figure 4.23 shows the variations of the H values obtained by the iterative optimization algorithms MLMSB, GA, OLMSB, TLMSB, SLPAm, SBGLL, MODPSO and SN2013 with the number of iterations on the War and Wiki-ele networks. The results illustrate that MLMSB, OLMSB, TLMSB, SLPAm, SBGLL and MODPSO converge within 100 generations while SN2013 converge within 300 generations. The results also show that the global search algorithm GA cannot converge within 100 generations, and that MLMSB, TLMSB and SBGLL can quickly converge to good solutions with low H values.

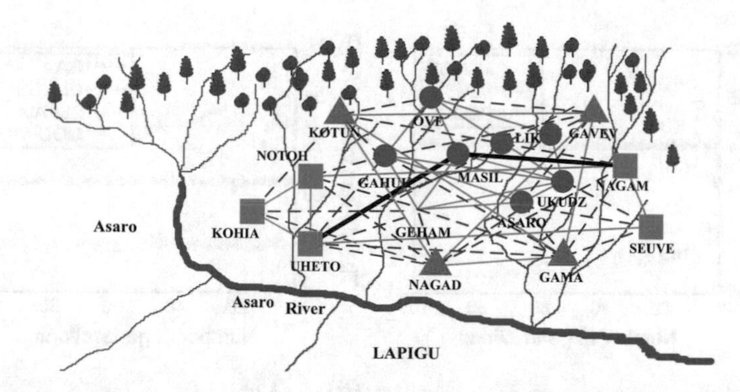

Fig. 4.21 Clustering result of the GGS network with $H = 2$. The unbalanced edges are highlighted in *black*

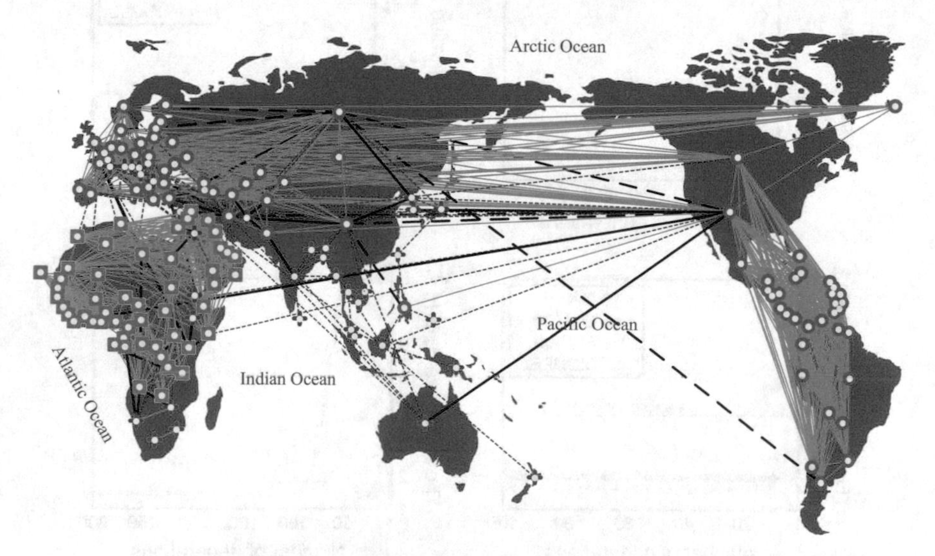

Fig. 4.22 Clustering result of the war network with $H = 40$. The nodes in the clusters with less than *3* nodes are marked with cross, and the unbalanced edges are highlighted in *black*

4.4.3.2 Experiments for the Transformation of Structural Balance

The mean value of H_w of all algorithms is listed in Table 4.11, where H_w denotes the balance transformation cost.

In this subsection, multiple balance transformations at minimum cost by varying the cost coefficient w from 0 to 1 with interval 0.1 are investigated. Note that, only situations where $0 < w < 1$ are investigated since the cases where $w = 0$ and $w = 1$ hardly take place in practical applications.

The results in Table 4.11 show that for the small-scale GGS and War networks, MLMSB, OLMSB, TLMSB, and SBGLL have smaller H_w values than SLPAm,

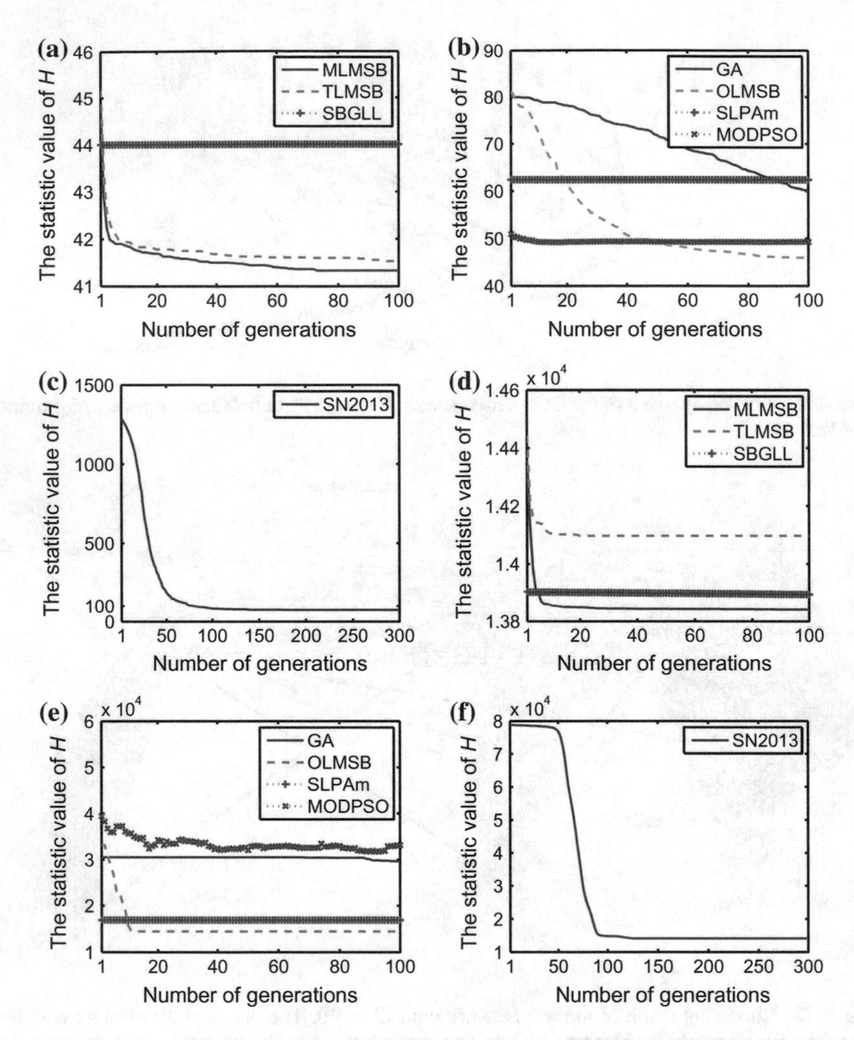

Fig. 4.23 Variation of H obtained by **a** MLMSB, TLMSB, and SBGLL, **b** GA, OLMSB, SLPAm, and MODPSO and **c** SN2013 with the number of generations on the war network and that by **d** MLMSB, TLMSB, and SBGLL, **e** GA, OLMSB, SLPAm, and MODPSO and **f** SN2013 with the number of generations on the Wiki-ele network

MODPSO, SN2013, PRE2014, and FEC. For the large-scale Wiki-ele, Wiki-rfa, and Slashdot networks, MLMSB, SBGLL, and SN2013 have smaller H_w values than GA, OLMSB, TLMSB, SLPAm, MODPSO, PRE2014, and FEC. Furthermore, MLMSB has the smallest H_w values for all networks under different w values but the Wiki-rfa network under $w = 0.1$.

From Table 4.11, it is seen that MLMSB and its three variations can transform the unbalanced GGS network into balanced ones at minimum cost. For the War, Wiki-ele, Wiki-rfa, and Slashdot networks, GA has the largest H_w values. OLMSB

Table 4.11 Comparison results of H_w between MLMSB and the others algorithms on the five real-world networks

Network	Alg.	$w = 0$	$w = 0.1$	$w = 0.2$	$w = 0.3$	$w = 0.4$	$w = 0.5$	$w = 0.6$	$w = 0.7$	$w = 0.8$	$w = 0.9$	$w = 1$
GGS	MLMSB	**0**	**0.4**	**0.8**	**1.2**	**1.6**	**2**	**2**	**2.8**	**2.8**	**1.4**	**0**
	GA	**0**	**0.4**	**0.8**	**1.2**	**1.6**	**2**	**2**	**2.8**	**2.8**	**1.4**	**0**
	OLMSB	**0**	**0.4**	**0.8**	**1.2**	**1.6**	**2**	**2**	**2.8**	**2.8**	**1.4**	**0**
	TLMSB	**0**	**0.4**	**0.8**	**1.2**	**1.6**	**2**	**2**	**2.8**	**2.8**	**1.4**	**0**
	SLPAm	**0**	0.5560	1.104	1.632	2.208	2.720	2.920	3.458	4.372	4.794	5.040
	SBGLL	**0**	**0.4**	**0.8**	**1.2**	**1.6**	**2**	**2**	**2.8**	**2.8**	**1.4**	**0**
	MODPSO	1.680	1.868	2.056	2.244	2.432	2.620	2.808	2.996	3.184	3.372	3.560
	SN2013	**0**	0.5560	1.000	1.608	2.328	2.560	2.968	3.738	3.664	3.924	3.800
	PRE2014	**0**	**0.4**	**0.8**	**1.2**	**1.6**	**2**	2.4	2.8	3.2	3.6	4
	FEC	**0**	0.8	1.6	2.4	3.2	4	4.8	5.6	6.4	7.2	8
War	MLMSB	**0**	**13.83**	**24.04**	**30.83**	**36.40**	**41.23**	**41.00**	**36.03**	**28.40**	**16.40**	**0**
	GA	**0**	20.38	24.11	44.68	52.96	59.99	64.55	69.72	71.91	71.81	75.68
	OLMSB	**0**	14.78	25.47	33.80	40.31	45.91	48.85	46.73	42.10	38.88	33.56
	TLMSB	**0**	14.06	24.08	31.04	36.60	41.53	41.08	36.06	**28.40**	**16.40**	**0**
	SLPAm	**0**	17.29	29,95	42.07	52.21	62.38	69.34	76.85	85.06	90.85	95.36
	SBGLL	**0**	16	27.60	36.60	42.40	44	42.80	38.20	30.80	16.40	**0**
	MODPSO	40.24	44.14	46.82	48.03	48.76	49.25	48.65	41.35	33.44	25.42	17.40
	SN2013	**0**	21.29	37.98	51.03	66.04	80.38	86.09	89.73	110.7	112.4	107.4
	PRE2014	24	39.80	55.60	71.40	87.20	103.0	118.8	134.6	150.4	166.2	182.0
	FEC	22	29.40	36.80	44.20	51.60	59.00	66.40	73.80	81.20	88.60	96.00
Wiki-ele	MLMSB	**0**	**7475**	**11706**	**13997**	**14537**	**13841**	**12128**	**9828**	**7022**	**3738**	**0**
	GA	3532	16298	24935	27505	29874	29658	29838	33409	30805	41810	40337
	OLMSB	**0**	7888	12367	14696	15316	14456	12569	10078	7143	3766	7
	TLMSB	**0**	7619	11955	14329	14854	14097	12316	9935	7077	3746	**0**
	SLPAm	**0**	8017	12263	14560	15325	16919	14084	11357	7095	5740	39.40

(continued)

Table 4.11 (continued)

Network	Alg.	$w = 0$	$w = 0.1$	$w = 0.2$	$w = 0.3$	$w = 0.4$	$w = 0.5$	$w = 0.6$	$w = 0.7$	$w = 0.8$	$w = 0.9$	$w = 1$
	SBGLL	**0**	7617	11864	14147	14630	13902	12184	9855	7036	3704	**0**
	MODPSO	10980	20643	29530	31157	32135	33113	34091	35069	36039	36993	37948
	SN2013	**0**	7645	12131	14420	14951	14104	12334	11401	7154	3813	76.00
	PRE2014	20106	22923	25750	28571	31393	34215	37036	39858	42680	45502	48324
	FEC	37520	33773	30026	26279	22533	18786	15039	11292	7545	3799	52
Wiki-rfa	MLMSB	**0**	14263	**22255**	**26537**	**27432**	**25974**	**22709**	**18291**	**12894**	**6762**	**0**
	GA	7030	28554	45581	53161	58627	63081	72800	69620	79809	96478	99156
	OLMSB	0.200	14813	23098	27622	28699	27143	23358	18663	13063	6804	7.400
	TLMSB	**0**	14367	22660	27091	27972	26370	23016	18469	12973	6778	**0**
	SLPAm	**0**	14922	25819	33150	33518	33071	37227	37124	28228	29013	23111
	SBGLL	**0**	**13997**	22374	26657	27559	26026	22760	18316	12920	6766	**0**
	MODPSO	99934	89941	79947	69954	59960	49967	39974	29980	19987	9993	**0**
	SN2013	**0**	14267	23074	28047	28070	26363	24308	20098	14931	6827	52.40
	PRE2014	35708	41070	46431	51792	57154	62516	67886	73239	78601	83962	89324
	FEC	67402	60674	53946	47217	40489	33761	27033	20305	13576	6848	120
Slashdot	MLMSB	**0**	**29346**	**51498**	**63277**	**66153**	**62120**	**52804**	**40985**	**27770**	**13944**	**0**
	GA	1873	41894	81432	118385	155100	189225	226890	262687	295975	334869	370136
	OLMSB	**0**	30126	52094	64241	66989	62604	53035	41059	27778	**13944**	**0**
	TLMSB	**0**	30420	52001	63982	66624	62310	52883	41008	**27770**	**13944**	**0**
	SLPAm	**0**	30859	52370	64138	66745	62370	52925	41030	27778	**13944**	**0**
	SBGLL	**0**	29399	51738	63698	66511	62338	53167	41070	27806	**13944**	**0**
	MODPSO	—	—	—	—	—	—	—	—	—	—	—
	SN2013	**0**	30289	53342	65219	67233	62534	52910	41006	27778	**13944**	**0**
	PRE2014	134982	122154	109326	96499	83671	70843	58015	45187	32360	19532	6704
	FEC	141256	121730	113004	98879	84754	70628	56502	42377	28251	14126	**0**

can effectively reduce the balance transformation cost of GA. More specifically, the maximum decrease of the cost can reach 45.86, 90.99, 92.95, and 95.84% for the War, Wiki-ele, Wiki-rfa, and Slashdot networks when $0 < w < 1$, respectively. Moreover, TLMSB can reduce 57.82, 3.41, 3.01, and 0.98% transformation cost for the War, Wiki-ele, Wiki-rfa, and Slashdot networks, respectively. In addition, MLMSB can reduce the balance transformation cost of TLMSB. More specifically, H_w is maximally decreased by 1.64% for the War network, 2.32% for the Wiki-ele network, 2.04% for the Wiki-rfa network and 3.53% for the Slashdot network.

Figure 4.24 records the variations of the balance transformations found by MLMSB with the cost parameter w. As shown in Fig. 4.24, there are multiple balance transformations for the GGS, War, Wiki-ele, Wiki-rfa, and Slashdot networks. In general, more negative links and less positive links need to be changed of signs with the increase of w. To illustrate the balance transformations intuitively, the clustering results of the GGS network with different w is plotted in Figs. 4.21 and 4.25, and the clustering results of the War network with $w = 0.5$ and $w = 0.4$ in Figs. 4.22 and 4.26.

Figures 4.21 and 4.25 illustrate that it is necessary to transform two unbalanced positive links and seven unbalanced negative links when $0.1 \leq w \leq 0.7$ and $0.8 \leq w < 1$, respectively. As shown in Fig. 4.21, the GGS network is divided into three clusters and the positive relations between "MASIL" and "UHETO", "MASIL" and "NAGAM" need to be transformed. As shown in Fig. 4.25, the GGS network is divided into two clusters and the seven negative links highlighted with black lines need to be transformed.

Figure 4.22 shows that the War network is mainly divided into two clusters with 32 unbalanced positive links and 8 unbalanced negative links when $w = 0.5$. Figure 4.26 illustrates that the War network is mainly divided into three clusters with 35 unbalanced positive links and 7 unbalanced negative links when $w = 0.4$.

It is noticed that in the GGS network, the two clusters drawn with blue box and red circle in Fig. 4.21 are merged into the cluster drawn with blue box in Fig. 4.25. In the War network, the two clusters drawn with red triangle and green circle in Fig. 4.22 mainly split from the cluster drawn with green circle in Fig. 4.26. The interesting phenomenon indicates that different balance transformations may lead to the mergence and the separation of clusters in signed networks.

In practical applications, if the unbalanced factors and the transformation cost are known in advance, the potential conflicts of real systems by the transformation of the unbalanced factors can be reduced. The MLMSB algorithm and transformation optimization models are devised to identify and reduce the imbalances in real-world signed systems.

4.4.3.3 Impacts of Parameter Settings

In MLMSB and its variations, there are six crucial parameters: the population size N_P, the parent population size N_m, the iteration number g_{max}, the pool size N_o, the crossover probability p_c and the mutation probability p_m. Figure 4.27 records the

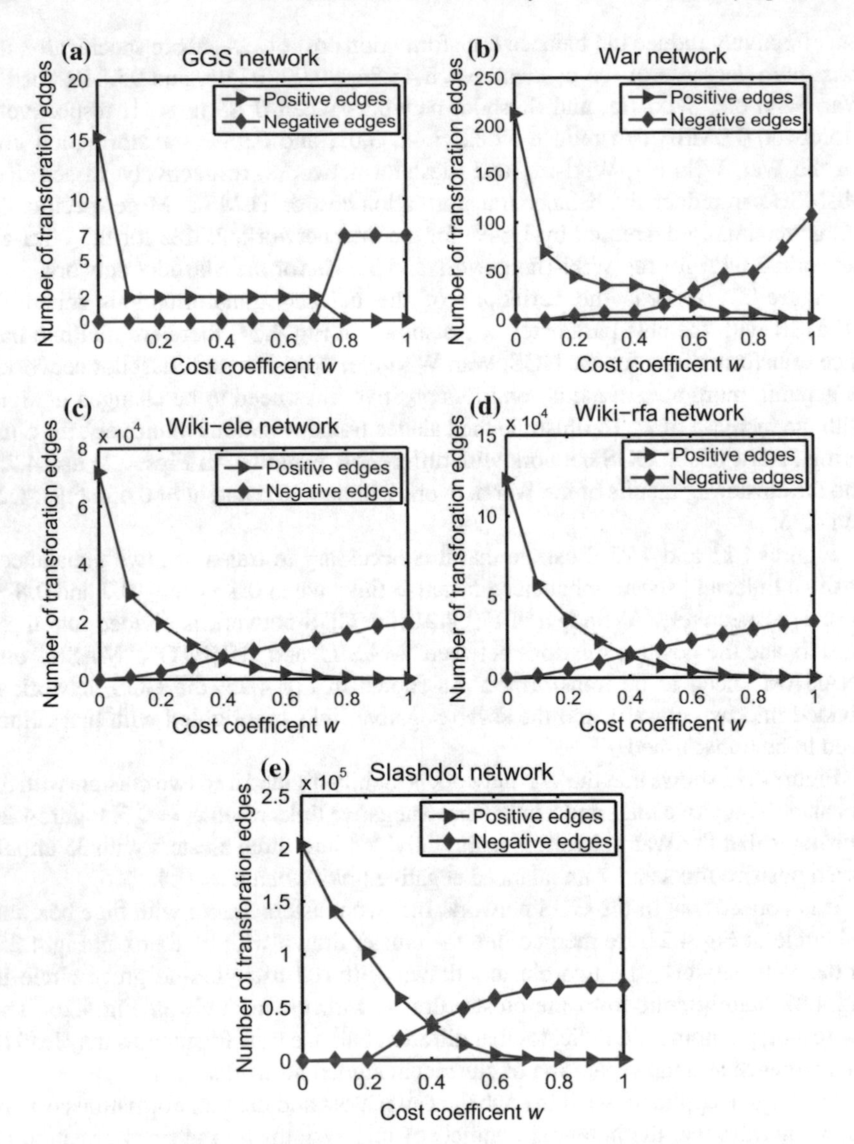

Fig. 4.24 Variation of the balance transformations with the cost parameter w for the **a** GGS, **b** War, **c** Wiki-ele, **d** Wiki-rfa and **e** Slashdot networks

variations of the statistic H values over 50 independent trials with different N_P, N_m, g_{max}, N_o, p_c and p_m in the War network.

The curves in Fig. 4.27 show that MLMSB and TLMSB are more robust than GA and OLMSB to the parameters N_P, N_m, g_{max}, N_o, p_c, and p_m. Generally, the statistic H values of GA decrease as the N_P, N_m, and g_{max} values increase. Note that, the computational complexity of GA is related to the N_P, N_m, and g_{max} values.

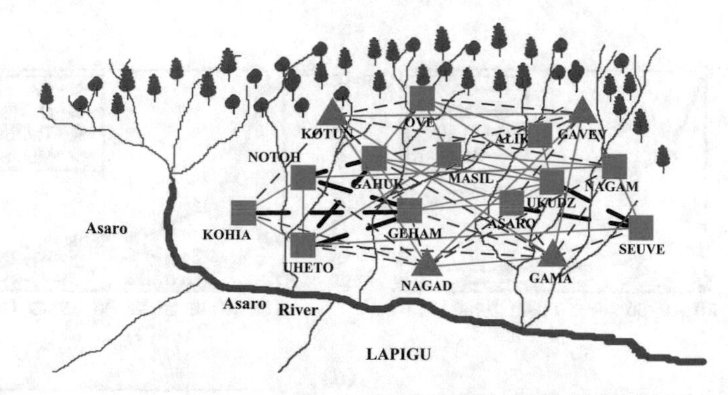

Fig. 4.25 Balance transformation of the War network based on the minimization of H_w with $0.8 \leq w < 1$

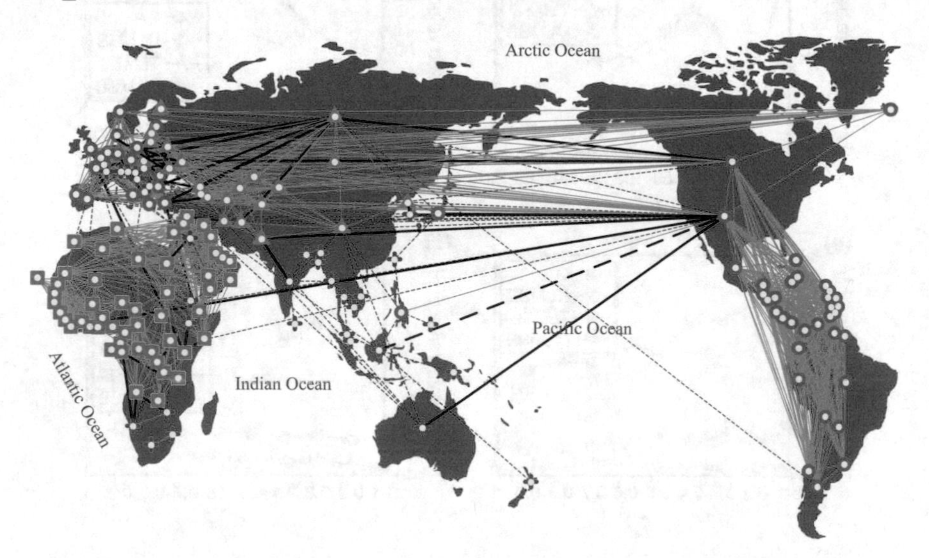

Fig. 4.26 Balance transformation of the War network based on the minimization of H_w with $w = 0.4$. The nodes in the clusters with less than *3* nodes are marked with cross, and the unbalanced edges are highlighted in *black*

OLMSB is less sensitive than GA to N_P, N_m, g_{max}, N_o, p_c, and p_m. However, it may be trapped in local optima and cannot achieve convergence in a few iterations. The algorithms TLMSB and MLMSB are robust to N_P, N_m, g_{max}, N_o, p_c and p_m and they converge to good solutions in a few generations (e.g., $g_{max} = 40$).

Figure 4.27 also illustrates that GA, OLMSB, TLMSB, and MLMSB have low H values when $N_P \geq 60$, $N_m \geq 10$, $g_{max} \geq 80$, $1 \leq N_o \geq 3$, $0.7 \leq p_c \geq 1$ and $0 \leq p_m \geq 0.4$. Considering both the consuming time and the performances, $N_p = 100$, $N_m = 10$, $g_{max} = 100$, $N_o = 2$, $p_c = 0.9$ and $p_m = 0.1$ for MLMSB and its variations. The parameters settings in MLMSB are the same as those in the population-based MODPSO algorithm.

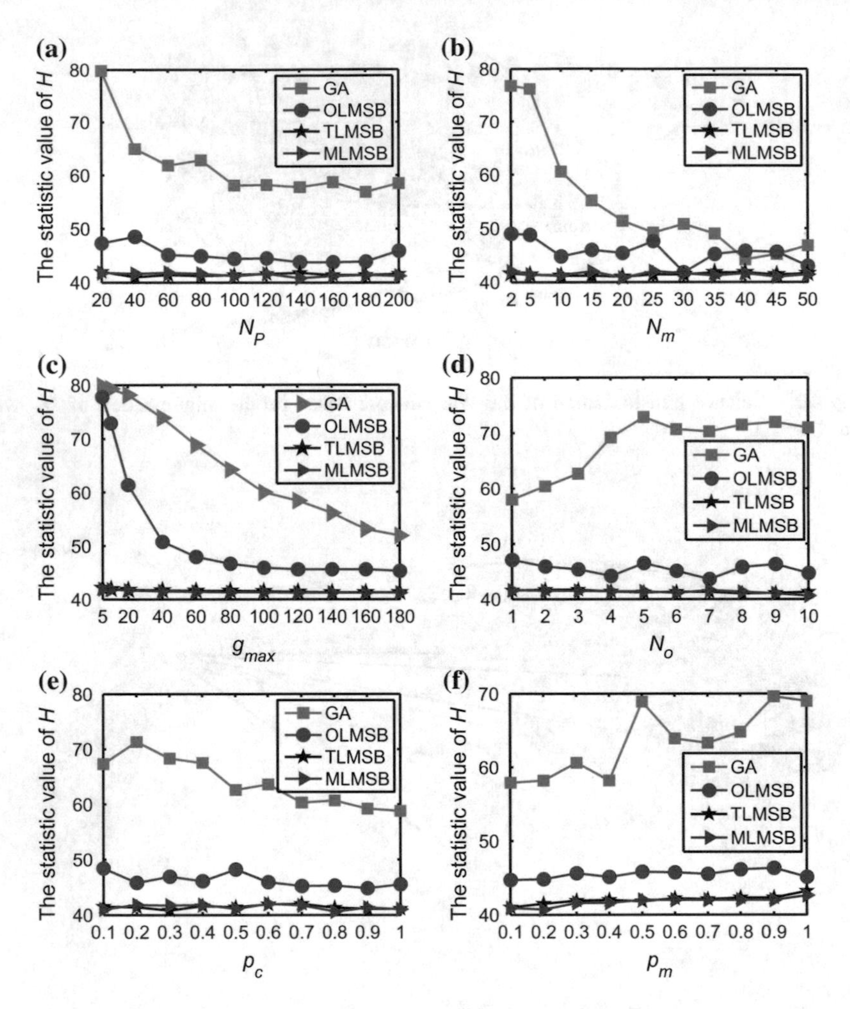

Fig. 4.27 Statistic values of H averaged over *50* independent runs for MLMSB and its three variations with different **a** population size N_P, **b** chosen population size N_m, **c** iteration numbers g_{max}, **d** pool size N_o, **e** crossover probability p_c and **f** mutation probability p_m

4.4.4 Conclusions

In this section, the classical energy function was extended so that it can compute the structural balance of signed networks both in strong and weak definitions. After that, the unbalanced links can be transformed by changing signs of unbalanced edges. Moreover, a more general energy function was introduced to evaluate the transformation cost. In addition, by incorporating the neighborhoods of node, cluster and partition, a fast memetic algorithm was introduced to compute and transform structural balance in signed network. Experimental results have demonstrated the superior performance of MLMSB compared with other state-of-the-art models in five real-world signed networks. The results also have illustrated the effectiveness of MLMSB on the pursuit of balance at minimum cost.

4.5 Computing and Transforming Structural Balance Based on Evolutionary Multi-objective Optimization

In the extant literature, the idea of traditional ways, including the work introduced in Sect. 4.4, to realize the structural balance for an arbitrary network is shown in Fig. 4.28.

In Fig. 4.28, a technique is utilized to divide the network into small clusters, afterward, all the imbalanced edges have been flipped so as to make the network balanced.

However, there are two drawbacks for the traditional methods. For one thing, it may need to set the clusters in advance. For another thing, each single run of the method can only output one solution. It should be pointed out that many social networks have hierarchical structures, i.e., a network can be divided into diverse kind of partitions based on different standards. Thus, one should have plenty of choices to realize the balanced structure of a network. However, each run of the traditional methods can only provide a single choice.

To avoid the drawbacks of traditional methods, a two-step approach (termed as two-step) was proposed in [4]. The basic idea of the approach is graphically illustrated in Fig. 4.29.

The two-step approach involves evolutionary multi-objective optimization, followed by model selection. In the first step, an improved version of the multi-objective discrete particle swarm optimization framework developed in previous section is suggested. The suggested framework is then employed to implement network multi-resolution clustering. In the second step, a problem-specific model selection strategy is devised to select the best Pareto solution from the Pareto front produced by the first

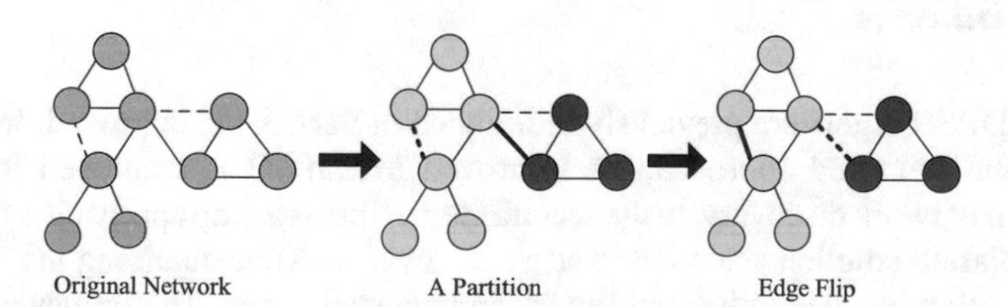

Fig. 4.28 Traditional way to realize the structural balance for an arbitrary network

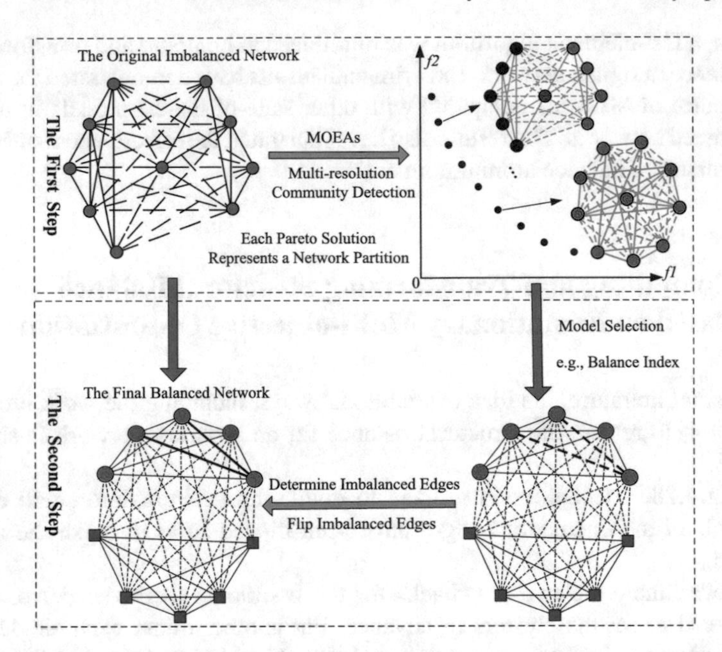

Fig. 4.29 Graphical illustration of our idea for solving the network structural balance problem

step. The best Pareto solution is then decoded into the corresponding network community structure. Based on the discovered community structure, imbalanced edges are determined. Afterward, imbalanced edges are flipped so as to make the network be structurally balanced.

It is noticed from Fig. 4.29 that, compared with traditional methods, the two-step approach actually can give many choices to realize the balance structure of a network. Based on different purposes, one can redesign the second step so as to obtain a preferable network balance structure with desired community topology.

4.5.1 The Two-Step Algorithm for Network Structural Balance

The MODPSO algorithm previously introduced in Sect. 3.4 is improved. In the first step of the proposed approach, the improved MODPSO is employed for multi-resolution network discovery. In the second step of the two-step approach, a problem-specific Pareto solution selection strategy is devised. After obtaining the final network partition, we determine and flip the imbalanced edges. The framework of the proposed two-step approach for addressing network structural balance problem is presented in Algorithm 24.

Algorithm 24 General framework of the proposed two-step approach for addressing network structural balance problem.

Parameters: maximum algorithm iterations: $gmax$, particle swarm size: pop, turbulence probability pm, inertia weight: ω, the learning factors: c_1, c_2.
Input: The adjacency matrix A of an imbalanced network G.
Output: Pareto solutions X^*, each solution corresponds to a partition of a signed network; the balanced network G^*.
The First Step: multi-resolution network discovery.

1.1 Population Initialization:

a) Position initialization: $\boldsymbol{P} = \{\boldsymbol{x}_1, \boldsymbol{x}_2, ..., \boldsymbol{x}_{pop}\}^T$, where $\boldsymbol{x}_i$ is the position vector of the i-th particle.
b) Velocity initialization: $\boldsymbol{V} = \{\boldsymbol{v}_1, \boldsymbol{v}_2, ..., \boldsymbol{v}_{pop}\}^T$, where $\boldsymbol{v}_i$ is the velocity vector of the i-th particle.
c) Generate a uniformly distributed weighted vectors: $\boldsymbol{W} = \{\boldsymbol{w}_1, \boldsymbol{w}_2, ..., \boldsymbol{w}_{pop}\}^T$.
d) Personal best solution initialization: $\boldsymbol{Pbest} = \{\boldsymbol{pb}_1, \boldsymbol{pb}_2, ..., \boldsymbol{pb}_{pop}\}^T$, where $\boldsymbol{pb}_i = \boldsymbol{x}_i$.

1.2 Initialize reference point z^*.
1.3 Initialize neighborhood $B(i)$ based on Euclidean distance.
1.4 Population Update:

a) Set $iter = 0$.
b) for $i = 1, 2, ..., popsize$, do
  i) Randomly select one particle from the N sub-problems as the leader.
  ii) Calculate the new velocity $\boldsymbol{v}_i^{t+1}$ and the new position $\boldsymbol{x}_i^{t+1}$ for the i-th particle. See the work in Sect. 3.4.
  iii) If $iter < gmax \cdot pm$, implement turbulence operation on $\boldsymbol{x}_i^{t+1}$.
  iv) Evaluate $\boldsymbol{x}_i^{t+1}$.
  v) Update neighborhood solutions with $\boldsymbol{x}_i^{t+1}$. See Algorithm 25 for more information.
  vi) Update reference point z^*.
  vii) Update personal best solution $\boldsymbol{pb}_i$.
c) If $iter < gmax$, then $iter$ ++ and go to 1.4b.

The Second Step: structural balance transformation.

2.1 Select the ultimate Pareto solution BS from the PF yielded by the first step, using the proposed model selection strategy.
2.2 Decode BS to a network partition $\Omega = (c_1, c_2, ..., c_k)$.
2.3 Determine the negative edges in c_i, $i = 1, ..., k$.
2.4 Determine the positive edges between c_i and c_j, $i \neq j$.
2.5 Flip all the imbalanced edges.

It should be pointed out that, in Algorithm 24, the initialization step, the two objective functions, i.e., SRA (signed ratio association) and SRC (signed ratio cut), and the particle status update principles are the same as those in MODPSO introduced in Sect. 3.4. Small change has been made to the turbulence operation used in Step 1.4. In MODPSO, the label of a vertex is assigned to all its neighboring vertices. In the two-step algorithm, randomly choose a label from the neighborhood and assign it to a vertex.

To better preserve population diversity, the subproblems update strategy is modified. After a new offspring solution $\boldsymbol{x}_i^{t+1}$ is generated, this solution is used to update only n_r subproblems from the whole population. The pseudocode of the modified strategy is shown in Algorithm 25.

Algorithm 25 Pseudocodes of the improved subproblems update strategy.

1. Set $P = \{1, ..., pop\}$. Set $c = 0$.
2. For each index $j \in P$, if $g^{te}(\boldsymbol{x}_i^{t+1}|\boldsymbol{w}_j, z^*) \leq g^{te}(\boldsymbol{x}_j^{t+1}|\boldsymbol{w}_j, z^*)$, then $c = c + 1$. //$g^{te}(\cdot)$ is the Tchebycheff decomposition approach used in [35].
3. **if** $c \leq n_r$ **then**
   - For each index $j \in P$, if $g^{te}(\boldsymbol{x}_i^{t+1}|\boldsymbol{w}_j, z^*) \leq g^{te}(\boldsymbol{x}_j^{t+1}|\boldsymbol{w}_j, z^*)$, then set $\boldsymbol{x}_j^{t+1} = \boldsymbol{x}_i^{t+1}$ and $F(\boldsymbol{x}_j^{t+1}) = F(\boldsymbol{x}_i^{t+1})$.
4. **else**
   - **for** indexes $j \in P$ with $g^{te}(\boldsymbol{x}_i^{t+1}|\boldsymbol{w}_j, z^*) \leq g^{te}(\boldsymbol{x}_j^{t+1}|\boldsymbol{w}_j, z^*)$ **do**
     - Rank the Euclidean distances between $F(\boldsymbol{x}_i^{t+1})$ and $F(\boldsymbol{x}_j^{t+1})$ with an ascend order.
     - Choose the former n_r solutions, then set $\boldsymbol{x}_j^{t+1} = \boldsymbol{x}_i^{t+1}$ and $F(\boldsymbol{x}_j^{t+1}) = F(\boldsymbol{x}_i^{t+1})$.
   - **end for**
5. **end if**

4.5.2 Model Selection

The first step of the two-step approach can return a set of equally good solutions. Each solution represents a certain network partition. In order to recommend to the decision maker a best solution, in the second step, a problem-specific model selection strategy is developed. The flow chart of the devised model selection strategy is shown in Fig. 4.30.

Given that a network G has a partition $\Omega = (c_1, c_2, ..., c_k)$. $\Gamma(\cdot) = s_i \cdot s_j$. Because, if nodes i and j are in the same group, then $\Gamma(\cdot) = 1$, otherwise, $\Gamma(\cdot) = -1$. Then energy function $H(s)$ (Eq. 1.8) can be rewritten as

$$\begin{aligned} H(s) = & \sum_{m=1}^{k} \sum_{i,j \in c_m} (1 - J_{ij}\Gamma(\cdot))/2 \quad + \\ & \sum_{m=1}^{k} \sum_{n=1, n \neq m}^{k} \sum_{i \in c_m, j \in c_n} (1 - J_{ij}\Gamma(\cdot))/2 \\ = & \sum_{m=1}^{k} \sum_{i,j \in c_m} |J_{ij}^-| + \sum_{m=1}^{k} \sum_{n=1, n \neq m}^{k} \sum_{i \in c_m, j \in c_n} |J_{ij}^+| \\ = & \sum_{m=1}^{k} |e_m^-| + \sum_{m=1}^{k} \sum_{n=1, n \neq m}^{k} |e_{mn}^+| \end{aligned} \tag{4.8}$$

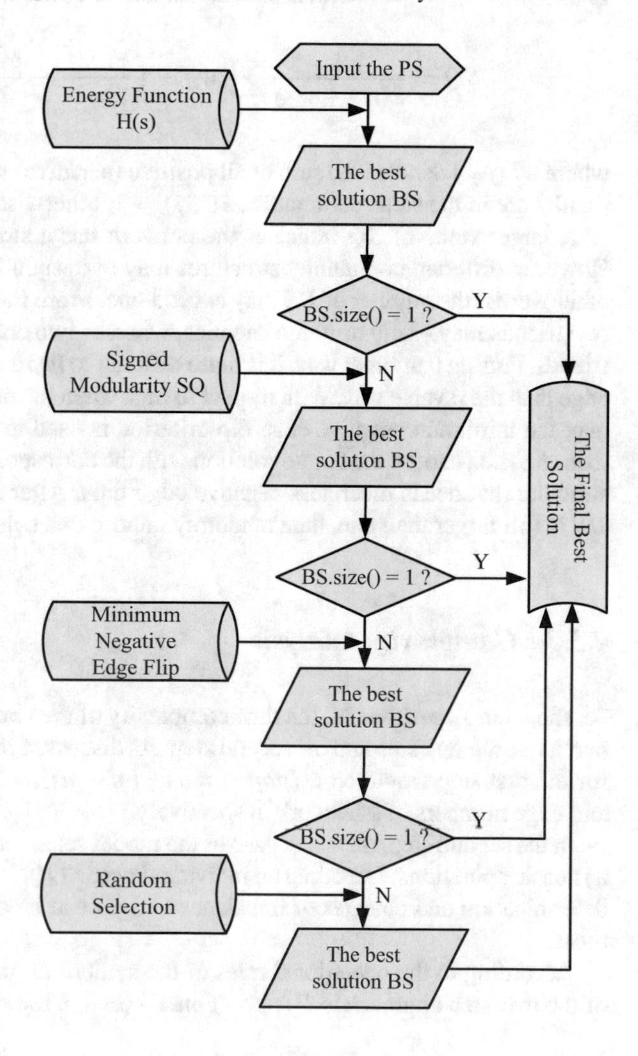

Fig. 4.30 Flow chart of the suggested model selection strategy

where $|e_m^-|$ denotes the number of negative edges within community c_m, and $|e_{mn}^+|$ represents the number of positive edges between community c_m and c_n.

From the above equation, it is noticed that, the value of $H(s)$ equals the number of imbalanced edges of a network. Consequently, if a signed network has smaller energy value, it will need to flip much less edges to reach balance. Therefore, after obtaining the Pareto solutions PS, the $H(s)$ index is utilized to choose the solutions that have minimum $H(s)$ value. Because different Pareto solutions may have the same $H(s)$ value, if there are more than one best solution BS, then the signed modularity index SQ is used to choose the best solutions that have the maximum SQ value. The SQ index is proposed by Gómez et al. in [13]. It is a metric to evaluate the goodness of a network community structure. The SQ index reads:

$$SQ = \frac{1}{2(w^+ + w^-)} \sum_{i,j} \Big[w_{ij} - \Big(\frac{w_i^+ w_j^+}{2w^+} - \frac{w_i^- w_j^-}{2w^-} \Big) \Big] \delta(i, j) \tag{4.9}$$

where w_i^+ (w_i^-) denotes the sum of all positive (negative) weights of node i. If nodes i and j are in the same community, $\delta(i, j) = 1$, otherwise, $\delta(i, j) = 0$.

A larger value of SQ indicates the network has a strong community structure. However, different community structures may correspond to the same SQ value. In other words, the number of BS may exceed one. From the perspective of sociology, two friends may easily turn into enemies, whereas two enemies may hardly become friends. Putting it another way, it is more difficult to flip a negative edge into positive edge than the reverse way. With respect to this, when the number of BS is larger than one, the minimum negative edge flip criterion is used to choose the best solutions from BS, i.e., choose the Pareto solution with the corresponding imbalanced network structure that needs much less negative edge flips. After all these, if the number of BS is still larger than one, then randomly choose one solution as the output.

4.5.3 Complexity Analysis

As shown in Algorithm 24, the time complexity of the two-step approach lies in two parts, i.e., the first step and the second step. As discussed in [14], the time complexity for the first step would be $O(pop \cdot gmax \cdot (m + n))$, where n and m are the node and edge numbers of a network, respectively.

In the second step, the worst case of the model selection step needs $O(pop \cdot (m + n))$ basic operations. Decoding an individual needs $O(n \log n)$ basic operations. The determination and changes of imbalanced edges can be computed in $O(m)$ time at most.

According to the operational rules of the symbol O, the overall time complexity of the two-step approach is $O(pop \cdot gmax \cdot (m + n \log n))$ in the worst case.

4.5.4 Experimental Results

In this section, first the two-step algorithm is tested on the artificial generated signed network: signed LFR benchmark. And then it is tested on several real-world signed networks: SPP, GGS, EGFR, Macrophage, Yeast, Ecoli, WikiElec, and Slashdot. The statistics of these network are given in Chap. 1. And the two-step algorithm is compared with several other MOEAs.

4.5.4.1 Validation Experiments

Because, in two-step algorithm, the output of the first step is the input of the second step. In this subsection, first the abilities of the first step for community detection is checked, then it will check whether the second step can solve the structural balance problem effectively.

Community Detection Validation

The first step of the two-step approach is run for 30 independent trials for each of the 216 generated benchmark networks. Because the ground truth of each benchmark network is known, the *NMI, Normalized Mutual Information* is used (Eq. 2.3) to evaluate the goodness of a network partition.

For each single run, the maximum *NMI* value from the PF is recorded. The statistical results are presented in Fig. 4.31.

As can be seen from Fig. 4.31 that when μ is no bigger than 0.3 and $p-$ is no bigger than 0.2, the ground truths of the networks can be successfully detected. As μ increases, the community structures become vague and it is failed to obtain the right partitions.

It is noticed from the figure that when $p-$ increases, the decline of the *NMI* values is obvious. However, from the viewpoint of $p+$, the decline of the *NMI* values is not obvious. This phenomenon indicates that from the perspective of community detection, and the two-step method is robust to noises of positive edges between communities.

Structural Balance Validation

The above validation experiments indicate that the first step is promising for community detection. Here, the performance of the second step will be validated.

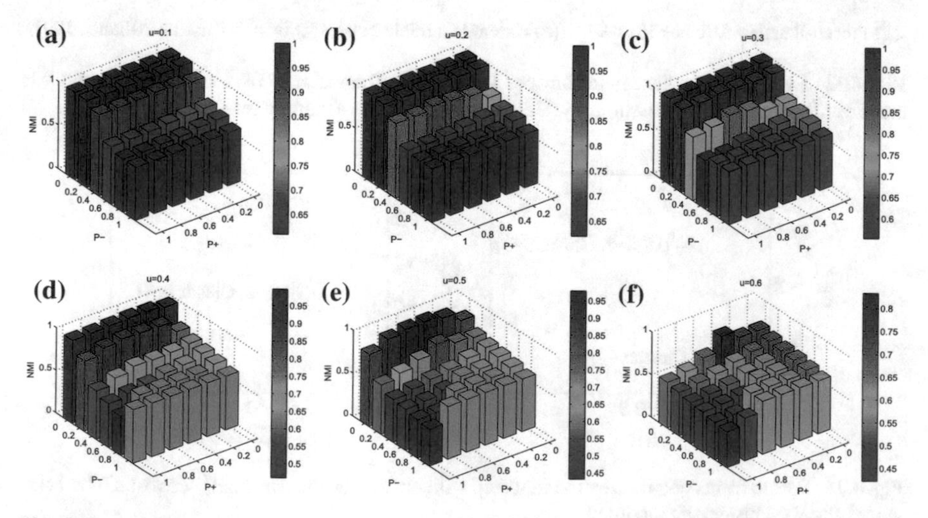

Fig. 4.31 Averaged *NMI* values for the combinations of the three parameters μ, $p-$ and $p+$ when experimenting on the *216* signed benchmark networks

In order to better visualize network structures, the two-step method is tested on the two small networks shown in Fig. 4.32d, e. The PFs obtained by the first step with one run are shown in Fig. 4.33.

It is seen from the PFs that for the first network, the first step obtains two nondominated solutions. The optimal solution with the $H(s)$ value of 0 is got which indicates the two-step algorithm has discovered the network partition that is structurally balanced. For the second network, only one optimal solution with the $H(s)$ value of 1 is obtained which means one edge flip is needed to make the network balanced. The balanced structures of the two networks are presented in Fig. 4.34.

The balance result shown in Fig. 4.34b is interesting. The result suggests that, in order to realize structural balance, the negative edge between nodes "Ru" and "GB"

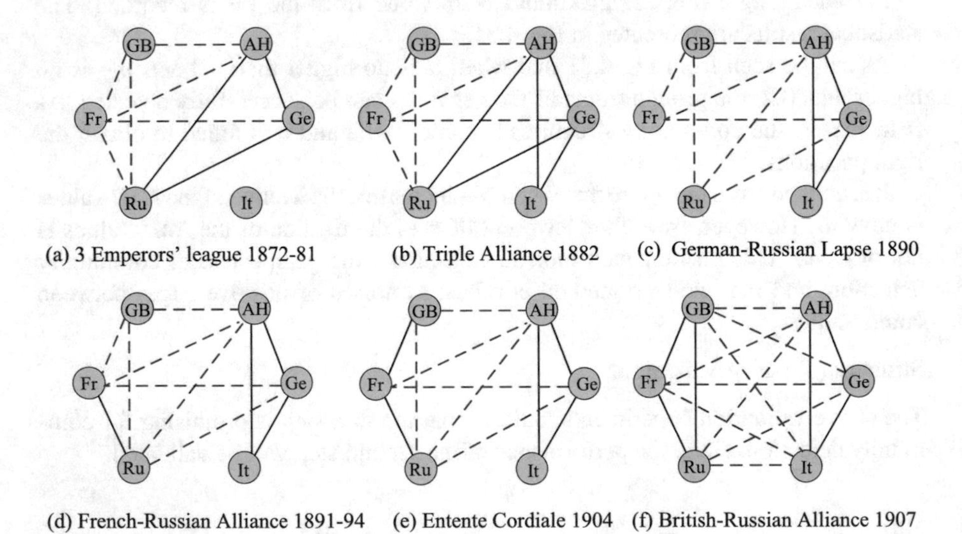

Fig. 4.32 Evolution of alliances in Europe, 1872–1907. The nations GB, Fr, Ru, It, Ge, and AH are Great Britain, France, Russia, Italy, Germany, and Austria-Hungary respectively

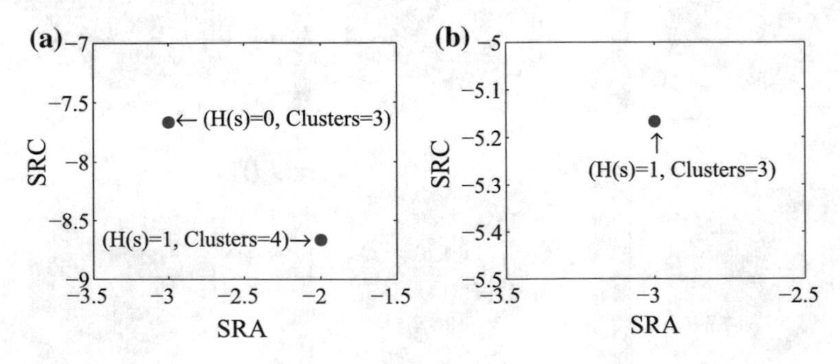

Fig. 4.33 Pareto fronts obtained by the first step with one run on the two small networks. The $H(s)$ values and the clusters are recorded

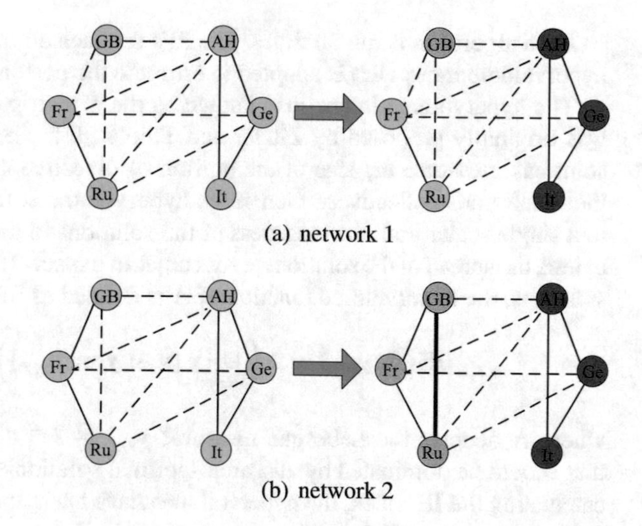

Fig. 4.34 Structural balance results for the two signed networks

should be changed into a positive one. In other words, the result suggests that Great Britain should cease fire with Russia.

According to the history, in 1907, a bipartite agreement between Russia and Great Britain was established, that then bound Russia, Great Britain, and France into the Triple Entente. From this viewpoint, we can see that the second step of our proposed approach is effective. The balance results are helpful for decision-making.

4.5.4.2 Comparisons with Other MOEAs

Here, comparisons between the two-step and several other MOEAs which optimize the same objective functions as two-step does are made. The compared MOEAs are the MODPSO algorithm, the fast and elitist multi-objective genetic algorithm (NSGA-II) proposed in [6], and the multi-objective evolutionary algorithm based on decomposition (MOEA/D) devised in [16]. The parameter settings of the algorithms are listed in Table 4.12.

Table 4.12 Parameter settings of the algorithms

Algorithm	S_{pop}	G_{max}	P_c	P_m	T	n_r	Ref.
Two-step	100	100	—	0.9	10	2	
MODPSO	100	100	—	0.9	40	—	[14]
NSGA-II	100	100	0.9	0.9	—	—	[6]
MOEA/D	100	100	0.9	0.9	40	—	[16]

P_c and P_m are the crossover and mutation possibility, respectively. T is the number of the weight vectors in the neighborhood of each weight vector

Each algorithm is run 30 times. The PFs for each algorithm are recorded and the hypervolume index (IH) is adopted to estimate the performance of the MOEAs.

The hypervolume index, also known as the S metric or the Lebesgue measure, was originally proposed by Zitzler and Thiele [37]. The hypervolume of a set of solutions measures the size of the portion of objective space that is dominated by those solutions collectively. Generally, hypervolume is favored because it captures in a single scalar both the closeness of the solutions to the optimal set and, to some extent, the spread of the solutions across objective space. If A is a set of nondominated solutions, the hypervolume function of A is defined as follows:

$$\mathrm{IH}(A, \boldsymbol{y}_{ref}) = \Lambda\Big(\bigcup_{y \in A} \{\boldsymbol{y}' | \boldsymbol{y} \prec \boldsymbol{y}' \prec \boldsymbol{y}_{ref}\}\Big), \quad A \subseteq \mathbb{R}^m \tag{4.10}$$

where Λ denotes the Lebesgue measure; $\boldsymbol{y}_{ref} \in \mathbb{R}^m$ denotes the reference point that should be dominated by all Pareto-optimal solutions. In the experiments, when calculating the IH index, the hypervolumes have been normalized and the reference point $\boldsymbol{y}_{ref}$ is set to (1.2, 1.2).

A statistical analysis using the Wilcoxon's rank sum test (RST) is performed. In the two-sided rank sum test experiments, the significance level is set to 0.05 with the null hypothesis that the two independent samples (the two IH sets) come from distributions with equal means. The statistical results are summarized in Table 4.13.

It can be seen from the table that the two-step approach outperforms the compared MOEA-based methods. The two-step approach obtains larger IH values on the eight real-world signed networks. The rank sum tests also suggest that the two-step approach works better. It is observed from the table that the two-step approach outperforms the MODPSO method.

Figure 4.35 displays the PFs with the largest IH values obtained by the two-step, MODPSO, MOEA/D, and NSGA-II. It is graphically obvious that the two-step approach has better performance than the compared methods.

An MOEA generally concerns two key components, the reproduction (i.e., generate offspring) and the replacement (i.e., update parents). When integrating PSO into an MOEA, the replacement strategy should be taken into thorough consideration. MODPSO uses the Tchebycheff decomposition method. When applying MODPSO to implement the multi-objective network clustering, several outstanding Pareto solutions which are better than the majority of the solutions in the population appear easily. In this situation, if the neighboring subproblems with these outstanding solutions are updated, population diversity will be lost. However, in the MODPSO, the particle position status update rule is a differential-like operator. Because the leader particle is randomly selected from the neighboring subproblems, even if the newly generated solution is excellent, the population diversity can still be preserved by the differential-like particle position status update rule. This is the reason why the PF obtained by MODPSO is better than that obtained by MOEA/D.

MODPSO simply updates all the neighboring subproblems. When the newly generated solution is better enough, the neighboring subproblems will be replaced by the newly generated solution, and consequently, diversity will be lost. To overcome this

Table 4.13 Hypervolumes of the PFs generated by the two-step, MODPSO, MOEA/D, and NSGA-II after 30 independent runs for the eight signed networks

Network	IH	Two-step	MODPSO	MOEA/D	NSGA-II
SPP	Max	**1.4212**	1.4212	1.4169	0.9558
	Mean	**1.4210**	1.3923	1.3729	0.6313
	Std	0.0013	0.0551	0.0265	0.1271
	RST	×	≈	+	+
GGS	Max	**1.3497**	1.2126	1.2315	0.8081
	Mean	**1.3146**	1.0019	1.0142	0.6894
	Std	0.0148	0.1701	0.0992	0.1394
	RST	×	+	+	+
EGFR	Max	**1.2039**	0.8695	0.6948	0.4461
	Mean	**1.1330**	0.7957	0.5865	0.4114
	Std	0.0318	0.0347	0.0722	0.0185
	RST	×	+	+	+
Macrophage	Max	**1.2552**	0.9815	0.7750	0.4495
	Mean	**1.2065**	0.8653	0.6314	0.4251
	Std	0.0294	0.0472	0.0758	0.0130
	RST	×	+	+	+
Yeast	Max	**1.2138**	1.056	0.9008	0.5482
	Mean	**1.1695**	0.8766	0.6346	0.4967
	Std	0.0321	0.0758	0.1217	0.0263
	RST	×	+	+	+
Ecoli	Max	**1.1923**	0.9200	0.8929	0.5337
	Mean	**1.1257**	0.8054	0.6781	0.4789
	Std	0.0421	0.0705	0.1203	0.0255
	RST	×	+	+	+
WikiElec	Max	0.1468	0.0475	**0.2456**	0.0831
	Mean	0.0980	0.0462	**0.1464**	0.0811
	Std	0.0325	0.0006	0.0595	0.0009
	RST	×	+	−	+
Slashdot	Max	**0.8428**	0.1989	0.4941	0.3510
	Mean	**0.6121**	0.1172	0.4827	0.2979
	Std	0.1336	0.0537	0.0091	0.0748
	RST	×	+	+	+

The symbols "≈", "+" and "−" denote that the performance of the two-step approach is similar to, significantly better than, and worse than that of the compared method, respectively

drawback, the replacement strategy is modified. Instead of updating all the neighboring subproblems, only n_r subproblems from the population are updated. In the MODPSO, each particle represents a subproblem, the personal best solution update is actually an update of the subproblem of itself. If subproblems from the population are updated, it is to avoid the situation that all the neighboring solutions are replaced

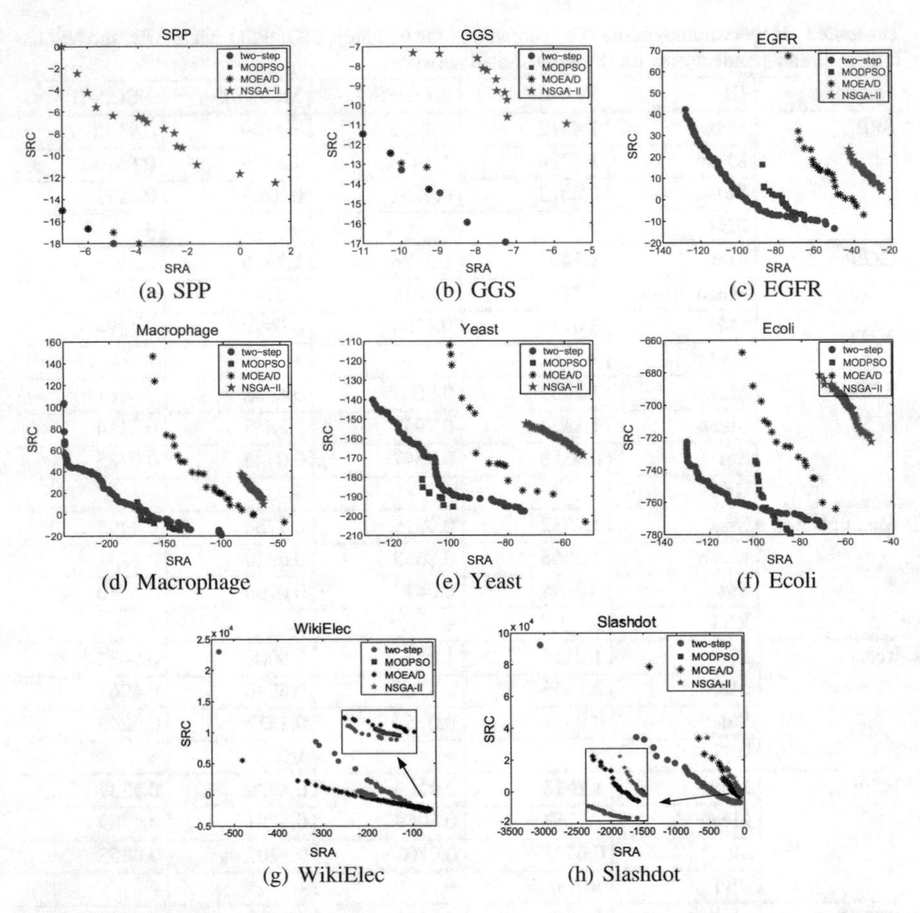

(a) SPP (b) GGS (c) EGFR (d) Macrophage (e) Yeast (f) Ecoli (g) WikiElec (h) Slashdot

Fig. 4.35 Comparisons between the Pareto fronts with largest IH values obtained by the four MOEAs

by one offspring solution, and diversity will be enhanced. It can be seen from both Fig. 4.35 and Table 4.13 that the improved replacement strategy works better.

It is concluded from the above experiments that the improved MODPSO framework is efficient. The effectiveness of the first step ensures the ultimate performance. In the next subsection, The performance of the two-step approach on addressing the structural balance problem is tested.

4.5.4.3 Structural Balance Experiments

For each signed network, only four Pareto solutions on the PF are analyzed. These solutions have the smallest $H(s)$ values. The statistical results are summarized in Table 4.14.

Table 4.14 Structural balance experimental results obtained by our proposed approach

Network	$H(s)$	SQ	Cluster	E^{+-}	E^{-+}	$H_b(s)$
SPP	2	0.4086	3	2	0	0
	5	0.3228	4	5	0	0
	7	0.2685	5	7	0	0
	×	×	×	×	×	×
GGS	4	0.3870	4	4	0	0
	7	0.3148	5	7	0	0
	8	0.3511	5	8	0	0
	9	0.2678	6	9	0	0
EGFR	283	0.2848	73	212	71	0
	289	0.2789	76	219	70	0
	289	0.2779	73	217	72	0
	291	0.2728	73	220	71	0
Macrophage	461	0.3284	74	168	293	0
	462	0.3280	77	173	289	0
	464	0.3269	76	179	285	0
	465	0.3266	81	179	286	0
Yeast	146	0.6065	109	121	25	0
	149	0.6013	108	124	25	0
	154	0.5987	113	130	24	0
	155	0.5934	113	130	25	0
Ecoli	443	0.3989	401	387	56	0
	444	0.3967	401	384	60	0
	474	0.3907	408	415	59	0
	476	0.3873	408	416	60	0
WikiElec	18,753	0.0016	915	18,715	38	0
	18,754	0.0017	914	18,660	94	0
	18,759	0.0017	936	18,636	123	0
	18,870	0.0018	927	18,604	266	0
Slashdot	86,197	0.0035	2,810	98	86,099	0
	86,216	0.0029	2,906	487	85,729	0
	86,309	0.0057	2,887	539	85,770	0
	86,474	0.0073	2,927	997	85,475	0

E^{+-} (or E^{-+}) denotes the number of positive (or negative) edges that need to be flipped into negative (or positive) edges. $H_b(s)$ is the energy function value of the network to which edge flips have been made

When experimenting on the SPP network, the first step of the two-step proposed approach yields only three Pareto solutions, shown in Fig. 4.35a. After the second step, the solution with the smallest $H(s)$ value is chosen as the final result. Figure 4.36 displays the structural balance result of the SPP network.

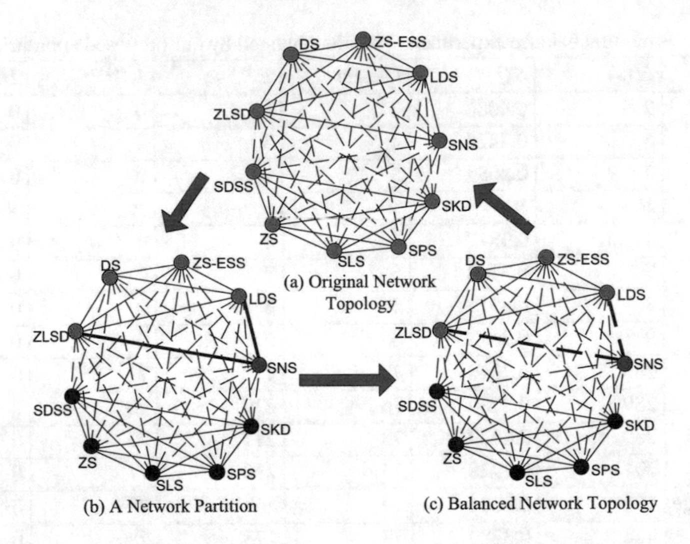

Fig. 4.36 Imbalanced and balanced topology structures of the SPP network

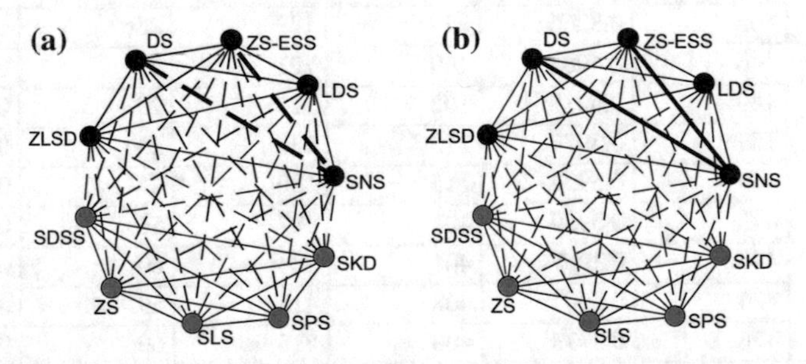

Fig. 4.37 Ground truth partition of the SPP network and the corresponding balanced network structure

For the SPP network, a partition with three clusters is obtained, shown in Fig. 4.36b. The $H(s)$ value is 2, which means two edges need to be flipped so as to make the network balanced. After flipping the two positive edges into negative ones, a balanced network is obtained, shown in Fig. 4.36c.

It should be pointed out that, the ground truth of the SPP network is that it has been divided into two parts as shown in Fig. 4.37a.

It can be seen from Fig. 4.37a that nodes "SNS", "LDS", "ZS-ESS", "DS", and "ZLSD" belong to the same community. In the final population of the two-step approach, it is found the solution that corresponds to the ground truth partition of the SPP network, however, this solution is dominated by other solutions. From the perspective of MOEA, the solution is indeed dominated by other solutions. From the perspective of structural balance theory, although this solution corresponds to the same $H(s)$ value of 2 which means that flipping two edges can make the network balanced and the balanced structure is shown in Fig. 4.37b, it will need more effort to do so. It can be seen from Fig. 4.37 that it is needed to change the negative edge between nodes "SNS" and "DS" and the edge between nodes "SNS" and "ZS-ESS" into a positive edge so as to make the network balanced. However, from the view point of sociology, it is too hard for two enemies to become friends, but it is rather easy for two friends to become enemies. In other words, it is more difficult to flip a negative edge into a positive edge than the reverse way. Moreover, as for the topology of the SPP network, the node "SNS" is special. It has negative links with nodes "DS" and "ZS-ESS". The two-step approach simply detects this node as a separate cluster.

For the GGS network, the two-step approach discloses four communities. The corresponding energy function value is 4, which indicates four edges need to be changed. The detected community structure and the balanced network topology are exhibited in Fig. 4.38.

From Fig. 4.38, it is seen that the two-step approach has found four imbalanced edges, exhibited in Fig. 4.38b.

Figure 4.39 displays the ground truth of the GGS network and the corresponding balanced network structure. The GGS network is originally divided into three communities. However, the solution denoting the true partition is dominated by the other solutions. Although, by flipping the two edges connecting node "MASIL" with

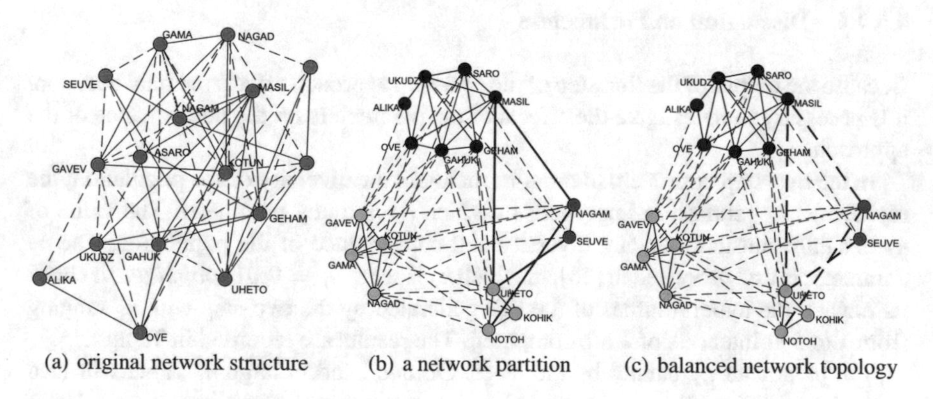

(a) original network structure (b) a network partition (c) balanced network topology

Fig. 4.38 Topology structures of the GGS network

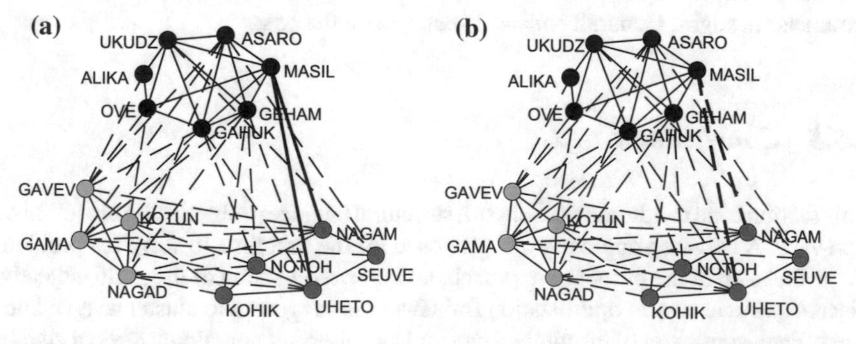

Fig. 4.39 Ground truth of the GGS network and the corresponding balanced network structure

nodes "NAGAM" and "UHETO", a balanced network structure is obtained as shown in Fig. 4.39b, it is thought that this kind of structure is not stable in real life. In Fig. 4.39 the four nodes "NAGAM", "SEUVE", "UHETO", and "NOTOH" form a quadrangle which is unstable in real life since a triangular relation is commonly recognized as the most stable status. The two-step approach separates the two members "NAGAM" and "SEUVE" from the original tribe to form a new subtribe.

Because the scales of the remaining six signed networks are large, it is hard to display their balanced network structures. However, from what are recorded in Table 4.14 it is noted that, after flipping the imbalanced edges discovered by the two-step approach, the $H_b(s)$ values became 0, which indicates that the changed networks have gotten structurally balanced. The WikiElec and Slashdot networks are large in size, however, the obtained $H(s)$ values for the two networks are comparable to those reported in [11].

The above experiments demonstrate that the two-step approach for addressing the network structural balance problem is effective. The two-step approach can simultaneously give a macroscopic (community structure) and a microscopic view (imbalanced edges) to analyze a signed social network.

4.5.4.4 Discussion on Parameters

Because the output of the first step of the two-step approach affects the final decision, it is necessary to investigate the effects of the parameters on the performance of the approach.

In the two-step approach, in order to maintain the diversity of the population, the replacement operation is improved, in which n_r is a key parameter. The value of n_r is significant in balancing diversity and convergence of the population. The n_r parameter is first proposed in [36], in which n_r is set to $n_r = 0.01 \cdot popsize$. To check its impact, the hypervolumes of the PFs generated by the two-step with n_r ranging from 1 to 5 at intervals of 1 are computed. The results are recorded in Table 4.15.

The parameter n_r cannot be too large. On one hand, a large n_r value will lead to the loss of diversity, on the other hand, a large n_r value will consume a lot of computational time. The results in Table 4.15 suggest that the influence of the n_r parameter is slight. Generally, $n_r = 2$ seems to be the best.

4.5.5 Conclusions

This section tried to address the network structural balance problem with an optimization view. A two-step approach was proposed for the first time to solve the problem. In the first step of the proposed approach, an improved version of the multi-objective discrete particle swarm optimization framework is suggested to cluster a signed network. Each single run of the method can yield a Pareto set containing a set of equally good solutions. Each solution represents a kind of network partition. In the second

Table 4.15 Hypervolumes of the PFs generated by the two-step with different settings of the parameter n_r. For each n_r, the two-step approach is run for 30 independent runs for the eight signed networks

Network	IH	n_r=1	n_r=2	n_r=3	n_r=4	n_r=5
SPP	Mean	1.4212	1.4210	1.4210	1.4212	1.4212
	Std	0.0020	0.0013	0.0013	0.0019	0.0021
GGS	Mean	1.3053	1.3146	1.3120	1.3060	1.2934
	Std	0.0147	0.0148	0.0176	0.0282	0.0381
EGFR	Mean	1.1215	1.1330	1.1377	1.1415	1.0604
	Std	0.0376	0.0318	0.0476	0.0430	0.1184
Macrophage	Mean	1.1919	1.2065	1.2111	1.2044	1.1718
	Std	0.0257	0.0294	0.0261	0.0392	0.0743
Yeast	Mean	1.1470	1.1695	1.1303	1.1394	1.1263
	Std	0.0301	0.0321	0.0323	0.0448	0.0391
Ecoli	Mean	1.0894	1.1257	1.1149	1.1141	1.0458
	Std	0.0414	0.0421	0.0358	0.0483	0.1064
WikiElec	Mean	0.0934	0.0980	0.1057	0.1305	0.1276
	Std	0.0251	0.0325	0.0282	0.0196	0.0147
Slashdot	Mean	0.5957	0.6121	0.6096	0.6073	0.6007
	Std	0.1738	0.1336	0.1124	0.1434	0.1401

step, a problem-specific model selection strategy is devised to select the best solution, then all the imbalanced edges based on the discovered network community structure are flipped. Extensive experiments on eight real-world signed networks have demonstrated the effectiveness of the two-step approach.

References

1. Antal, T., Krapivsky, P., Redner, S.: Dynamics of social balance on networks. Phys. Rev. E **72**(3), 036,121 (2005)
2. Antal, T., Krapivsky, P.L., Redner, S.: Social balance on networks: the dynamics of friendship and enmity. Phys. D: Nonlinear Phenom. **224**(1), 130–136 (2006)
3. Barahona, F.: On the computational complexity of ising spin glass models. J. Phys. A: Math. Gen. **15**(10), 3241 (1982)
4. Cai, Q., Gong, M., Ruan, S., Miao, Q., Du, H.: Network structural balance based on evolutionary multiobjective optimization: a two-step approach. IEEE Trans. Evol. Comput. **19**(6), 903–916 (2015)
5. Chen, X., Ong, Y.S., Lim, M.H., Tan, K.C.: A multi-facet survey on memetic computation. IEEE Trans. Evol. Comput. **15**(5), 591–607 (2011)
6. Deb, K., Pratap, A., Agarwal, S., Meyarivan, T.: A fast and elitist multiobjective genetic algorithm: NSGA-II. IEEE Trans. Evol. Comput. **6**(2), 182–197 (2002)
7. Doreian, P., Lloyd, P., Mrvar, A.: Partitioning large signed two-mode networks: problems and prospects. Soc. Netw. **35**(2), 178–203 (2013)

8. Doreian, P., Mrvar, A.: Partitioning signed social networks. Soc. Netw. **31**(1), 1–11 (2009)
9. Easley, D., Kleinberg, J.: Networks, crowds, and markets: reasoning about a highly connected world. Cambridge University Press (2010)
10. Esmailian, P., Abtahi, S.E., Jalili, M.: Mesoscopic analysis of online social networks: the role of negative ties. Phys. Rev. E **90**(4), 042,817 (2014)
11. Facchetti, G., Iacono, G., Altafini, C.: Computing global structural balance in large-scale signed social networks. Proc. Nat. Acad. Sci. **108**(52), 20953–20958 (2011)
12. Facchetti, G., Iacono, G., Altafini, C.: Exploring the low-energy landscape of large-scale signed social networks. Phys. Rev. E **86**(3), 036,116 (2012)
13. Gómez, S., Jensen, P., Arenas, A.: Analysis of community structure in networks of correlated data. Phys. Rev. E **80**, 016,114 (2009)
14. Gong, M., Cai, Q., Chen, X., Ma, L.: Complex network clustering by multiobjective discrete particle swarm optimization based on decomposition. IEEE Trans. Evol. Comput. **18**(1), 82–97 (2014)
15. Gong, M., Fu, B., Jiao, L., Du, H.: Memetic algorithm for community detection in networks. Phys. Rev. E **84**(5), 056,101 (2011)
16. Gong, M., Ma, L., Zhang, Q., Jiao, L.: Community detection in networks by using multiobjective evolutionary algorithm with decomposition. Phys. A: Stat. Mech. Appl. **391**(15), 4050–4060 (2012)
17. Heider, F.: Attitudes and cognitive organization. J. Psychol. **21**(1), 107–112 (1946)
18. Iacono, G., Ramezani, F., Soranzo, N., Altafini, C.: Determining the distance to monotonicity of a biological network: a graph-theoretical approach. IET Syst. Biol. **4**(3), 223–235 (2010)
19. Kułakowski, K., Gawroński, P., Gronek, P.: The heider balance: a continuous approach. Int. J. Mod. Phys. C **16**(05), 707–716 (2005)
20. Kunegis, J., Lommatzsch, A., Bauckhage, C.: The slashdot zoo: mining a social network with negative edges. In: Proceedings of the 18th International Conference on World Wide Web, pp. 741–750 (2009)
21. Leskovec, J., Huttenlocher, D., Kleinberg, J.: Signed networks in social media. In: Proceedings of the SIGCHI Conference on Human Factors in Computing Systems, pp. 1361–1370 (2010)
22. Ma, L., Gong, M., Du, H., Shen, B., Jiao, L.: A memetic algorithm for computing and transforming structural balance in signed networks. Knowl.-Based Syst. **85**, 196–209 (2015)
23. Ma, L., Gong, M., Liu, J., Cai, Q., Jiao, L.: Multi-level learning based memetic algorithm for community detection. Appl. Soft Comput. **19**, 121–133 (2014)
24. Marvel, S.A., Kleinberg, J., Kleinberg, R.D., Strogatz, S.H.: Continuous-time model of structural balance. Proc. Nat. Acad. Sci. **108**(5), 1771–1776 (2011)
25. Marvel, S.A., Strogatz, S.H., Kleinberg, J.M.: Energy landscape of social balance. Phys. Rev. Lett. **103**(19), 198,701 (2009)
26. Moscato, P., et al.: On evolution, search, optimization, genetic algorithms and martial arts: Towards memetic algorithms. Caltech concurrent computation program, C3P Report **826** (1989)
27. Newman, M.E.: Modularity and community structure in networks. Proc. Nat. Acad. Sci. **103**(23), 8577–8582 (2006)
28. Newman, M.E.J., Girvan, M.: Finding and evaluating community structure in networks. Phys. Rev. E **69**(2), 026,113 (2004)
29. Rosvall, M., Bergstrom, C.T.: Maps of random walks on complex networks reveal community structure. Proc. Nat. Acad. Sci. **105**(4), 1118–1123 (2008)
30. Sun, Y., Du, H., Gong, M., Ma, L., Wang, S.: Fast computing global structural balance in signed networks based on memetic algorithm. Phys. A: Stat. Mech. Appl. **415**, 261–272 (2014)
31. Szell, M., Lambiotte, R., Thurner, S.: Multirelational organization of large-scale social networks in an online world. Proc. Nat. Acad. Sci. **107**(31), 13636–13641 (2010)
32. Traag, V.A., Van Dooren, P., De Leenheer, P.: Dynamical models explaining social balance and evolution of cooperation. PloS one **8**(4), e60,063 (2013)
33. Wang, S., Gong, M., Du, H., Ma, L., Miao, Q., Du, W.: Optimizing dynamical changes of structural balance in signed network based on memetic algorithm. Soc. Netw. **44**, 64–73 (2016)

34. Yang, B., Cheung, W.K., Liu, J.: Community mining from signed social networks. IEEE Trans. Knowl. Data Eng. **19**(10), 1333–1348 (2007)
35. Zhang, Q., Li, H.: MOEA/D: A multiobjective evolutionary algorithm based on decomposition. IEEE Trans. Evol. Comput. **11**(6), 712–731 (2007)
36. Zhang, Q., Liu, W., Li, H.: The performance of a new version of MOEA/D on CEC09 unconstrained MOP test instances. In: Proceedings of IEEE Congress Evolutionary Computing, pp. 203–208 (2009)
37. Zitzler, E., Thiele, L.: Multiobjective optimization using evolutionary algorithms—a comparative case study. In: Parallel Problem Solving from Nature, pp. 292–301. Springer (1998)

Chapter 5
Network Robustness Analytics with Optimization

Abstract The community structure and the robustness are two important properties of networks for analyzing the functionality of complex systems. The community structure is crucial to understand the potential functionality of complex systems, while the robustness is indispensable to protect the functionality of complex systems from malicious attacks. When a network suffers from an unpredictable attack, its structural integrity would be damaged. It is essential to enhance community integrity of networks against multilevel targeted attacks. Coupled networks are extremely fragile because a node failure of a network would trigger a cascade of failures on the entire system. In reality, it is necessary to recover the damaged networks, and there are cascading failures in recovery processes. This chapter first introduces a greedy algorithm to enhance community integrity of networks against multilevel targeted attacks and then introduces a technique aiming at protecting several influential nodes for enhancing robustness of coupled networks under the recoveries.

5.1 Review on The State of the Art

In recent years, more and more researches have been focused on the robustness of networks under random failures or malicious attacks [2, 11, 18]. The robustness of networks is of great importance to guarantee the security of network systems, such as the airports and disease control networks. The structural integrity of networks would be damaged when unpredictable failures or attacks occur on them. This results in the loss of the functionalities of complex systems to some extent. For instance, in world airports networks, some airlines cannot work normally due to the terrible weather or the terrorist attacks. In power grids networks, the electricity cannot be transmitted due to the failures of generators.

The robustness of networks is usually measured by a criterion which considers the critical fraction of networks when they collapse completely [18]. This measure overlooks situations in which the networks suffer from a big damage but they are not completely collapsing [33]. Recently, Schneider et al. [33] proposed a measure, node robustness (R_n), to evaluate the robustness of networks under node attacks. When nodes are gradually damaged due to random failures or targeted attacks, a network

M. Gong et al., *Computational Intelligence for Network Structure Analytics*,
DOI 10.1007/978-981-10-4558-5_5

may be split into several unconnected parts. The node robustness (R_n) considers the size of the largest connected component during all possible node attacks, namely $R_n = \frac{1}{N}\sum_{q=1}^{N} s(q)$, where N is the number of nodes in the network and $s(q)$ is the integrity of nodes in the largest connected part after removing q nodes [33]. The normalization factor $1/N$ makes it possible to make a comparison of the node robustness between networks with different sizes. Generally, the larger the value of R_n, the more robust the network is. Schneider et al. [33] and Wu et al. [38] proposed some greedy techniques to optimize R_n. In their studies, they found that (1) the node robustness of networks can be greatly improved by modifying small parts of links without changing the total links and the degree of each node; (2) The optimal network for node robustness shares a common onion structure in which high degree nodes are hierarchically surrounded by rings of nodes with decreasing degree [17, 33]. Zeng and Liu extended their works and proposed a measure, link robustness R_l, for network robustness under malicious attacks on the links [39]. Similarly, the link robustness of networks can also be greatly improved by changing small parts of links.

Nowadays, many systems show a coupling property. The functionality of a complex system depends on not only itself, but also its coupled systems. The coupling property of networks is important to understand the controllability [22], robustness [30], synchronization [35], and cooperative evolution [20] of complex systems. Recent years, it has become a common focus for how to improve the robustness of coupled networks [4, 36]. For the systems to recover functionality soon, it is necessary to reconstruct the damaged systems [1, 3, 21]. However, there are also cascading failures in coupled systems during the recovery processes, which would make it difficult to recover the functionality. Majdandzic et al. [26] propose global recovery processes based on a general phenomenon. Damaged systems (e.g., human brain and the financial network) can be spontaneously recovered after an inactive period of time. Note that, in the real world, many complex systems (e.g., power system and airway system) have little ability to recover their functionality spontaneously. Moreover, it takes a long time and consumes much energy for the spontaneous recovery processes.

This chapter is organized into three sections. Section 5.2 introduces a heuristic algorithm to enhance the community integrity of networks against multilevel targeted attacks. Section 5.3 introduces a work, which aims to enhance robustness of coupled networks under targeted recoveries.

5.2 Enhancing Community Integrity Against Multilevel Targeted Attacks

The robustness of networks is an important property to guarantee the security of network systems. When a network suffers from an unpredictable attack, its structural integrity would be damaged. Earlier studies focused on the integrity of the node structure or the edge structure when a network suffers from a single-level malicious attack on the nodes or the edges.

When a network suffers from significant attacks, its communities may be damaged at different degrees. Correspondingly, the potential functionalities of real systems are destroyed at different degrees. The functional integrity of networks is crucial to judge whether the systems can work normally when they suffer from unpredictable damages. Moreover, previous works [17, 33, 39] are simply based on single-level targeted attacks on nodes or links. In reality, there exist more complicated attacks. In addition, under the original constraints keeping the total links and the degree of each node invariant for link changes, the optimized networks for network robustness may have changed their original community structures.

This section focuses on enhancing community integrity against multilevel targeted attacks. The advantages of this work are as follows. First, the malicious attack on the network is modelled as a two-level targeted one. The first level is a small-scale targeted node attack, and the second level is a large-scale targeted community attack. Second, a community robustness criterion R_c is proposed to measure the functional integrity of networks when they suffer from the modeled attacks. Third, to make the optimized networks for network robustness have the similar community structures with the original ones, new constraints are proposed for link changes. Finally, a greedy algorithm is proposed to improve the community robustness of networks by modifying small parts of links.

5.2.1 Model Malicious Attack on the Network as a Two-Level Targeted One

From the microscopic view, a network can be modeled as an undirected and unweighted graph $G = (V, E)$ with $|V| = N$ nodes and $|E| = M$ edges. The microscopic connections among nodes can be represented as an adjacency matrix A. If there is a connection between node v_i and v_j, the entry A_{ij} is 1 and 0 otherwise. From the macroscopic view, a network can also be modeled as $G = (S, E^{'})$, where $S = \{s_1, s_2, \ldots, s_k\}$ is the set of communities of networks and $E^{'}$ is the set of connections between different communities. The macroscopic connections among communities can also be represented as an adjacency matrix w. If there are connections between community s_i and s_j, the entry w_{ij} is larger than or equal to 1 and otherwise 0.

A schematic illustration of the microscopic and macroscopic representations of a network is given in Fig. 5.1 with a toy network. The toy network G_1 which is the largest component of the Santa Fe Institute Network consists of 118 nodes and 200 edges [15]. Nodes represent resident scientists coming from different fields at the Santa Fe Institute and their collaborators, and edges correspond to their collaborations in publishing at least one article during any part of calendar year 1999 or 2000. The toy network can be divided into 8 communities by the community detection algorithm BGLL. As shown in Fig. 5.1, the toy network G_1 is composed of 118 nodes and 200 connections in microscopic view or 8 communities and 19 connections in macroscopic view.

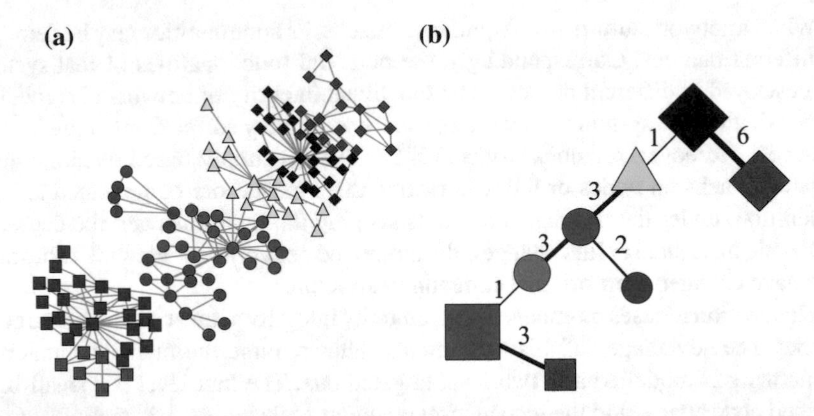

Fig. 5.1 The representation of the toy network G_1. **a** The toy network G_1 is composed of 118 nodes and 200 microscopic connections. **b** The toy network G_1 consists of 8 communities and 19 macroscopic connections. Nodes with different colors are in different communities. Each node in **b** represents one of communities in **a**, and each edge in **b** is the connection among these communities in **a**

The nodes or the edges of networks are possible to suffer from damages. There may be a situation in which a set of nodes which have similar properties are damaged at the same time. Namely, the communities of networks are also possible to suffer from damages. As shown in Fig. 5.1, the toy network will be divided into several unconnected parts when a few nodes, edges or communities are removed. Moreover, the damages caused by removing communities are greater than that by removing nodes or edges.

The network property having dense interconnections makes the real system resilient against random failures but vulnerable to targeted attacks [33]. Therefore, the studies on the robustness of networks under targeted attacks are useful to the security of real systems. In this work, the attack on the network is modelled as a two-level targeted one. The first level is the small-scale targeted attack which occurs on nodes with the largest degrees. The second level is the large-scale targeted attack under which the most influential communities which have the maximal number of inter-community links are removed gradually. A dynamic approach is used which recalculates the degrees of each node and the importance of each community during the attack. The dynamic way corresponds to a more harmful attack strategy [17].

In the real world, the attacks on nodes and communities are possible to happen. According to whether they can occur simultaneously, the attack strategies can be classified into two categories. The first one is the weighted strategy which simulates the situation in which the attacks on nodes and communities cannot happen simultaneously. In this case, the small-scale attacks on nodes and the large-scale attacks on communities are possible to occur but not simultaneously. In the absence of prior knowledge about which attacks will happen, it is necessary to consider the community robustness of networks under targeted attacks at both levels. By introducing a weighting parameter α, the community robustness of networks under both the small-scale and the large-scale attacks are considered. In the weighted strategy, the nodes with the largest degrees will be gradually removed when the network suffers from the small-scale attacks and the most important communities will be gradually cut when the network suffers from the large-scale attacks. The second one is the mixed strategy which simulates the situation in which the attacks on nodes and communities can happen simultaneously. In the mixed strategy, the nodes with the largest degrees will be removed with probability f and the most important communities will be cut with probability $1 - f$. The procedure ends when the largest connected component reaches 1 (in this case, the remaining network consists of a set of isolated nodes). In this section, the weighted strategy is used because it is more general in the real world. The mixed strategy is taken into consideration when considering the community robustness of a network.

A schematic illustration about the weighted and the mixed attack strategies on a toy network G_2 is shown in Fig. 5.2. The toy network which has 12 nodes and 15 links is divided into three communities (plotted with different colors). Assuming the network suffers from attacks three times, for the weighted strategy, the attacks are either the small-scale attacks on nodes or the large-scale attacks on communities. For the mixed strategy, the attacks on nodes and communities can happen simultaneously, e.g., attacks on nodes two times and the other one on communities, shown in Fig. 5.2b.

5.2.2 *Community Robustness of Networks*

The measures in [33, 39] for the robustness of networks consider the integrality of the node structure or the edge structure. However, these measures can hardly reflect the functional integrality of the network. Thus a measure, the community robustness R_c, is proposed to evaluate the community integrality of networks under malicious attacks. The community robustness of a network is defined as

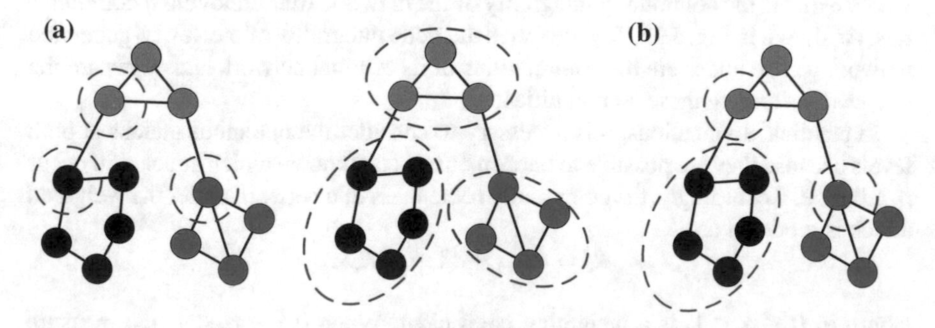

Fig. 5.2 The weighted and the mixed attack strategies on the toy network G_2. **a** The weighted attack strategy. **b** The mixed attack strategy. Nodes with different colors are in different communities. The *red dash circles* represent an attack

$$R_c = \frac{1}{m}\sum_{q=1}^{m}\Big[\frac{1}{k}\sum_{p=1}^{k}\frac{S_p(q)}{S_p}\Big], \tag{5.1}$$

where m is the number of the possible malicious attacks on the network, k is the number of communities, S_p is the number of nodes in the community p and $S_p(q)$ is the number of the remaining nodes in the community p when the qth attack happens. The normalization factors, $1/m$ and $1/k$, make it possible to make a comparison of the community robustness between networks with different sizes and different numbers of communities. A larger value of R_c usually indicates more robust community structure of the network.

For the weighted attack strategy, the community robustness of networks under the small-scale node attacks and the large-scale community attacks is necessary to be considered respectively. The community robustness of a network under the small-scale targeted attack can be written as

$$R_{c1} = \frac{1}{N}\sum_{q=1}^{N}\Big[\frac{1}{k}\sum_{p=1}^{k}\frac{S_p(q)}{S_p}\Big], \tag{5.2}$$

where N is the number of nodes in the network and $S_p(q)$ is the number of the remaining nodes in the community p when q nodes are removed. $S_1(q) = \frac{1}{k}\sum_{p=1}^{k}\frac{S_p(q)}{S_p}$ is the community integrality of the network after removing q nodes. When each community of the network has the same size, R_{c1} would equal to R_n.

The community robustness of a network under the large-scale targeted attack can be written as

$$R_{c2} = \frac{1}{k}\sum_{u=1}^{k} S_2(u), \tag{5.3}$$

where $S_2(u)$ is the community integrality of the network after removing u communities. As shown in Fig. 5.1b, R_{c2} measures the node integrality of the newly-generated network whose nodes are the communities of its original network and edges are the connections among these communities.

In practical applications, it is necessary to consider the malicious attacks at both levels because they are possible to happen but it is not known which attack will occur in advance. Accordingly, the community robustness of a network under the weighted attack is modeled as

$$R_c = \alpha R_{c1} + (1-\alpha) R_{c2}, \tag{5.4}$$

where α, $0 \le \alpha \le 1$, is a weighting coefficient. When $0.5 < \alpha \le 1$, the measure mainly focuses on the community robustness of a network under the small-scale targeted node attack. When $0 \le \alpha < 0.5$, the measure mainly considers the community robustness of a network under the large-scale targeted community attack.

For the mixed attack strategy, the community robustness of a network can be defined as follow:

$$MR_c = \frac{1}{L}\sum_{q=1}^{L}\Big[\frac{1}{k}\sum_{p=1}^{k}\frac{S_p(q)}{S_p}\Big] \tag{5.5}$$

where L is the total number of steps to reduce the size of the giant component to 1 [39]. MR_c evaluates the community robustness of a network when the network suffers from the small-scale and the large-scale attacks simultaneously. The probability with which the attack at each level occurs is controlled by a mixing parameter f ($0 \le f \le 1$). When $f = 0$, it means that the network is more likely to suffer from the large-scale attack. When $f = 1$, it indicates that the network is more likely to suffer from the small-scale attack.

5.2.3 Constraints for Improving Networks

For a given network, there are many ways to enhance the robustness of networks. A simple way is to add links without any constraints [33]. However, in practical applications, it is difficult to achieve due to the extra cost to create a link. Moreover, changing the degree of a node is more expensive than changing edges. In order to avoid producing additional cost as far as possible, Schneider et al. [33] propose two constraints which keep the number of links and the degree of each node invariant for link changes.

Note that, the above constraints are difficult to guarantee that the optimized networks for network robustness have the similar community structures with the original ones. Comparing the community structure in the optimized network for node robustness (in Fig. 5.3a) with that in the original one (in Fig. 5.1a), it is not easy to find the similarity between them. It means that the R_n-optimized network has changed the original community structure of the toy network G_1.

In order to identify that the optimized networks for network robustness and the original networks have the similar community structures, a constraint is added which keeps the number of intra-community links of each community invariant for link changes. As shown in Figs. 5.1a and 5.3b, under the new constraints the optimized network for node robustness has the similar community structure with the original one. The original and the optimized networks tend to be divided into 8 and 7 communities, respectively. The main difference between them is that the two communities (drawn in purple and orange) in Fig. 5.1a are merged into one community in Fig. 5.3b. Moreover, in some practical applications, adding an inter-community link is more expensive than creating an intra-community link. A simple example is that creating a highway between different countries is more expensive than that in the same country. Therefore, the constraints keeping the number of links, the degree of each node and the number of intra-community links of each community invariant for link changes are much closer to practical applications.

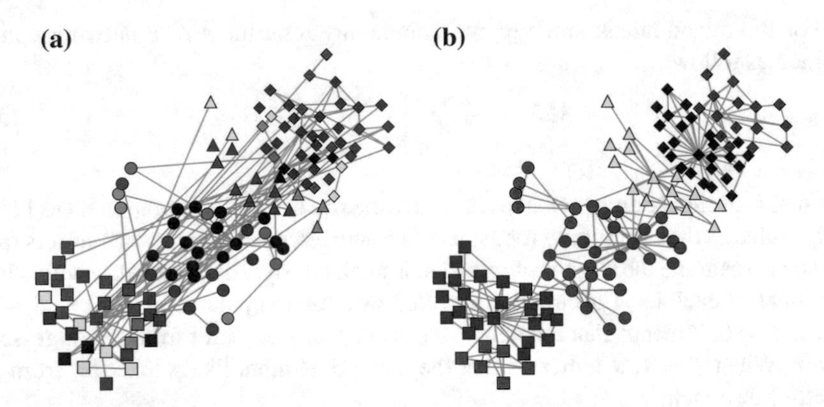

Fig. 5.3 The community structures of the optimized toy networks for node robustness under **a** the original constraints and **b** the new constraints for link changes. Nodes with different colors are in different communities. The optimized network under the original constraints for link changes has different community structure with the original one. Under the new constraints, the optimized network has the similar community structure with the original one

5.2.4 *Enhancing Community Robustness of Networks*

The framework to enhance the community robustness of networks under the above constraints is shown in Fig. 5.4. First, the network is divided into a set of communities, using any community detection algorithms. In this section, the community detection algorithm BGLL [6] is used. The algorithm BGLL is effective and efficient for uncovering the community structures of networks. More importantly, in the absence of prior knowledge of the number of communities, it can automatically detect the "right" number of communities. Then, a greedy algorithm is designed to optimize R_c under the proposed constraints. It works as follows: Starting from an original network G, two edges e_{ab} and e_{cd} are randomly selected. Swap the connections of e_{ab} and e_{cd} to e_{ad} and e_{bc} if the swap satisfies the constraints for link changes, and set the resulting network as G'. Update G with G' at a certain probability which is decided by the difference between R_c and R_c', where R_c and R_c' are the community robustness of G and G', respectively. As shown in Fig. 5.4, G is updated with G' at the probability $\exp^{-|R_c'-R_c|/T}$, where T is a parameter which controls the convergence speed of the algorithm to an optimal solution. The algorithm is easier to converge to an optimal solution when the value of T is small (here, it is set as 10^{-4}). The above operations will not stop until no further improvement can be achieved for a given large number of consecutive swapping trials t_{max} (here, it is set as 10^4). In this framework, for the toy network G_1, under the weighted attack, the improvement of the community robustness can reach 92.51%. Meanwhile, its node robustness is also increased by 94.15%. Under the mixed attack, the improvements of the community robustness and the node robustness can reach 162.3% and 8.04%, respectively.

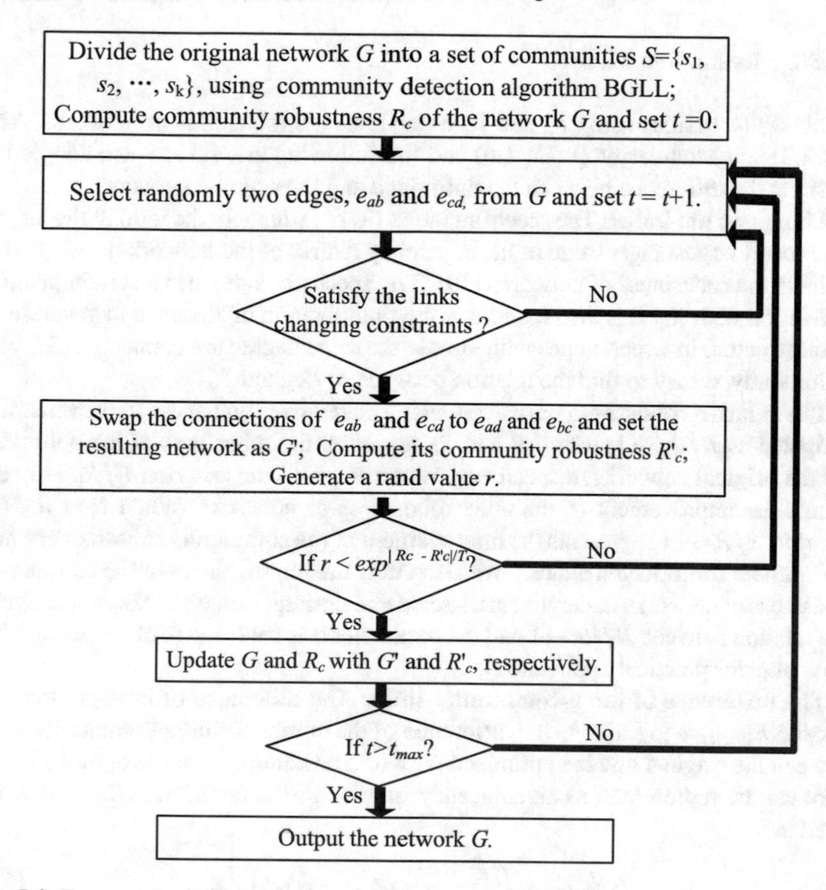

Fig. 5.4 Framework to enhance community robustness of networks

5.2.5 Experimental Results

In this section, first, the evaluation metrics is introduced. Then, the devised greedy algorithm is tested on three representative real-world networks: (i) The Electronic Circuits network [27]: It consists of 122 nodes and 189 edges. Nodes represent electrical elements and edges denote their interconnections. (ii) The USAir network: the network of US air transportation system. It contains 332 airports and 2126 airlines. (iii) The Road network [34]: a network of all roads included in the *International E-road Network*. It has 1177 nodes and 1417 edges (1469 edges for directed network). Nodes represent European cities and edges correspond to the road connections among them. All the networks considered here are undirected and unweighted. The results obtained by the algorithm [33] are also given for comparison.

5.2.5.1 Evaluation Metrics

In this study, besides using R_n and R_c to evaluate the robustness of networks, *NMI* (Eq. 2.3) and Modularity Q (Eq. 1.6) and the following criteria are also adopted to illustrate the difference between the optimized and the original networks.

The spectrum index: The spectrum index (λ_1/λ_2), namely the ratio of the largest and second largest eigenvalue of the adjacency matrix of the network, is adopted to evaluate the robustness of networks [19]. The spectrum index has a certain positive correlation with R_n. It is also found that the optimization of R_n tends to generate an onion structure in which nodes with almost the same degree are connected [33, 39]. In this study, we try to find the relation between λ_1/λ_2 and R_c.

The relative robustness improvement: The relative robustness improvement is computed as $R/R_0 - 1$, where R and R_0 represent the robustness of the optimized and the original networks, respectively. When $R = R_n$, the criterion $R/R_0 - 1$ represents the improvement of the node robustness of networks. When $R = R_c$, the criterion $R/R_0 - 1$ represents the improvement of the community robustness of networks under the modeled attacks. We also study the improvement of the community robustness of networks under the small-scale and the large-scale attacks, respectively. The relation between $R/R_0 - 1$ and the parameter α is analyzed to choose a suitable value of α for practical applications.

The difference of intra-community links: The difference of intra-community links (ΔE_{intra}) is to estimate the difference of the number of intra-community links between the original and the optimized networks. Assuming that the optimized network can be represented as an adjacency matrix A', the criterion ΔE_{intra} is computed as

$$\Delta E_{intra} = \Big(\sum_{i=1}^{k}\sum_{j,p\in s_i}(A'_{jp} - A_{jp})\Big)\Big/2. \tag{5.6}$$

When ΔE_{intra} is smaller than 0, it indicates that at least $|\Delta E_{intra}|$ intra-community links have been changed into inter-community links in the optimized networks. When ΔE_{intra} is larger than 0, it means that at least ΔE_{intra} inter-community links have been changed into intra-community links. When ΔE_{intra} is equal to 0, it indicates that the number of intra-community links in the original and the optimized networks has not changed.

5.2.5.2 Experiment on Real-World Networks Under the Weighted Attacks

In this section, the algorithm is tested on three real-world networks, the Electronic Circuits, the USAir, and the Road networks, to illustrate that the community robustness of networks can be largely enhanced when they suffer from the weighted attacks. The related results of the R_n-optimized and the R_c-optimized networks are recorded in Table 5.1. The R_n-optimized networks are generated by optimizing R_n under the original constraints for link changes. The R_c-optimized networks are generated by

Table 5.1 Results on different real-world networks: the node robustness criterion R_n, the community robustness index R_c (for $\alpha = 0.4$), normalize mutual information (*NMI*), the spectrum index λ_1/λ_2, the difference of intra-community links ΔE_{intra}, the ratio of removed links RE and the number of communities k. Results are averaged over 100 independent trials

Networks	Algorithms	R_n	R_c	NMI	λ_1/λ_2	ΔE_{intra}	RE	k
Electronic circuit	Original	0.1261	0.2272	1	1.131	0	0	13
	R_n-optimized	0.2017	0.3034	0.5845	1.171	−33.83	0.2431	15.71
	R_n'-optimized	0.1837	0.2472	0.7908	1.130	0	0.2200	14.72
	R_c-optimized	0.1549	0.3088	0.7958	1.135	0	0.3201	14.60
USAir	Original	0.1090	0.2436	1	2.382	0	0	9
	R_n-optimized	0.1568	0.2942	0.5937	2.463	−86.37	0.0557	8.570
	R_n'-optimized	0.1413	0.2475	0.7971	2.390	0	0.0412	8.030
	R_c-optimized	0.1388	0.2901	0.7913	2.392	0	0.0640	7.870
Road	Original	0.0548	0.0768	1	1.023	0	0	206
	R_n-optimized	0.0706	0.0971	0.9550	1.024	−23.85	0.0220	203.1
	R_n'-optimized	0.0630	0.0811	0.9896	1.023	0	0.0211	205.5
	R_c-optimized	0.0597	0.1035	0.9672	1.020	0	0.0342	204.5

optimizing R_c under the new constraints. As shown in Table 5.1, optimizing R_n can greatly improve the node robustness of networks. More specifically, R_n is increased by 59.95% in the Electronic Circuit network. In the USAir and the Road networks, the improvements of R_n can reach 43.85% and 28.83%, respectively. However, the R_n-optimized networks can hardly keep their original community structures. As shown in Table 5.1, in the R_n-optimized networks, the average values of ΔE_{intra} over 100 independent trials are −33.83, −86.37 and −23.85 for the Electronic Circuit, the USAir and the Road networks, respectively. This means that an average of 33.83, 86.37 and 23.85 intra-community links have been changed into inter-community links for the Electronic Circuit, the USAir and the Road networks, respectively. This would result in the change of community structures between the optimized and the original networks. The above opinion is confirmed by the criterion NMI. The similarities between the optimized and the original network partitions are only 58.45 and 59.37% for the Electronic Circuit and the USAir networks, respectively. Therefore, it is meaningless to make a comparison of the community robustness between the optimized and the original networks since the original community structures have changed.

In this study, one constraint keeping the total intra-community links of each community invariant is added. The related results of the $R_n^{'}$-optimized networks generated by optimizing R_n under the new constraints are also recorded in Table. 5.1. The results in Table. 5.1 clearly show that the $R_n^{'}$-optimized network partitions are more similar to the original ones. The similarities between the $R_n^{'}$-optimized and the original network partitions can reach 79.08, 79.71 and 98.96% for the Electronic Circuit, the USAir and the Road networks, respectively. Moreover, the improvement of node robustness can reach 45.68% for the Electronic Circuit network, 29.63% for the USAir network and 14.96% for the Road network. In addition, the changed links in the $R_n^{'}$-optimized networks are less than that in the R_n-optimized networks.

As it is seen from Table 5.1, under the new constraints, optimizing R_c can largely enhance both the node robustness and community robustness of networks. Optimizing R_n can largely enhance the node robustness of networks. However, optimizing R_n can hardly improve the community robustness of networks. The values of R_c in the $R_n^{'}$-optimized networks are only increased by 8.80% for the Electronic Circuit network, 1.60% for the USAir network and 5.60% for the Road network. The improvements of R_c in the R_c-optimized networks can reach even 35.92% for the Electronic Circuit network, 19.09% for the USAir network and 34.77% for the Road network. In order to further illustrate the above opinions, the community integrality of the tested networks under attacks at each level is plotted in Figs. 5.5, 5.6 and 5.7. By analyzing and comparing the results in Figs. 5.5, 5.6 and 5.7, it is noted that the R_c-optimized networks are the most robust. It means that the community robustness of networks can be enhanced by optimizing R_c. Moreover, the results also illustrate that the $R_n^{'}$-optimized networks are more robust than the original ones under the small-scale node attacks. Under the large-scale community attacks the $R_n^{'}$-optimized networks have the similar community integrity with the original ones. It indicates that optimizing R_n can hardly improve the community robustness of networks under the large-scale community attacks.

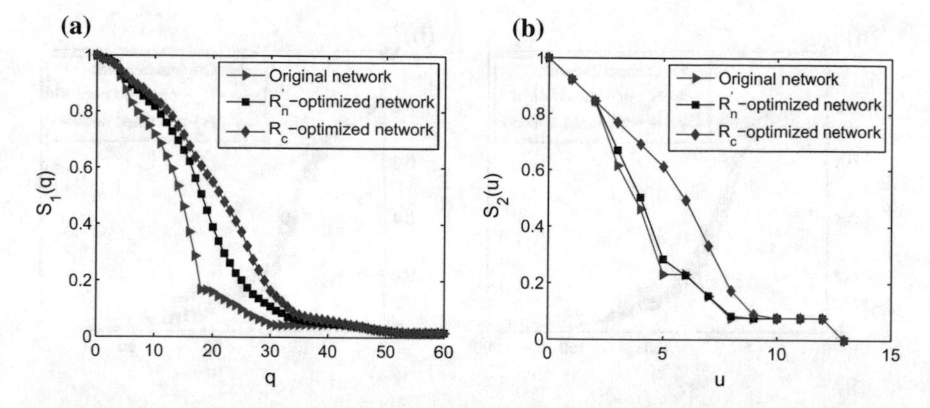

Fig. 5.5 The integrity of communities belonging to the largest components versus the number q of removed vertices or the number u of removed communities for the Electronic Circuit network. Three types of networks are compared: the original network, the $R_n^{'}$-optimized network and the R_c-optimized network. **a** The integrity of communities versus the number q of removed vertices. **b** The integrity of communities versus the number u of removed communities. Results are averaged over 100 independent trials

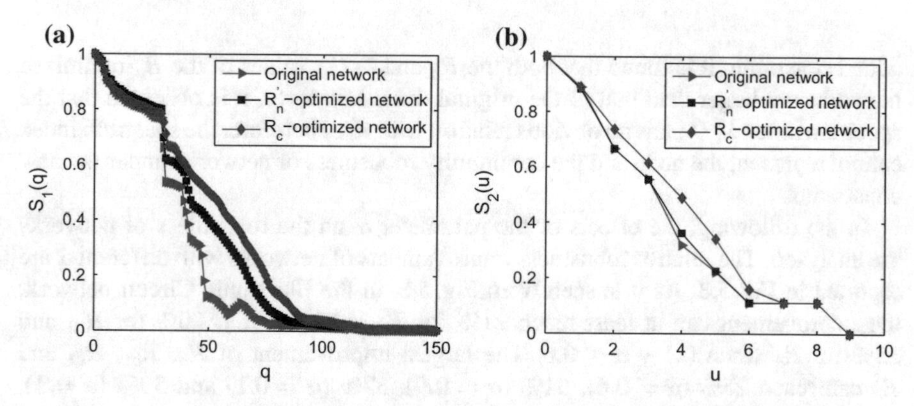

Fig. 5.6 The integrity of communities belonging to the largest components versus the number q of removed vertices or the number u of removed communities for the USAir network. Three types of networks are compared: the original network, the $R_n^{'}$-optimized network and the R_c-optimized network. **a** The integrity of communities versus the number q of removed vertices. **b** The integrity of communities versus the number u of removed communities. Results are averaged over 100 independent trials

The results in Table 5.1 show that the values of R_n and R_c in the $R_n^{'}$-optimized networks are larger and smaller than that in the R_c-optimized networks, respectively, which indicate that under the new constraints optimizing R_n cannot guarantee the improvement of R_c and vice versa.

The spectrum index λ_1/λ_2 has been used to evaluate the node robustness of networks. The studies in Ref. [17, 39] indicated that the spectrum index has a certain positive correlation with R_n. However, the results in Table 5.1 show that under the new constraints there is no obvious connection between R_n and λ_1/λ_2. Without the

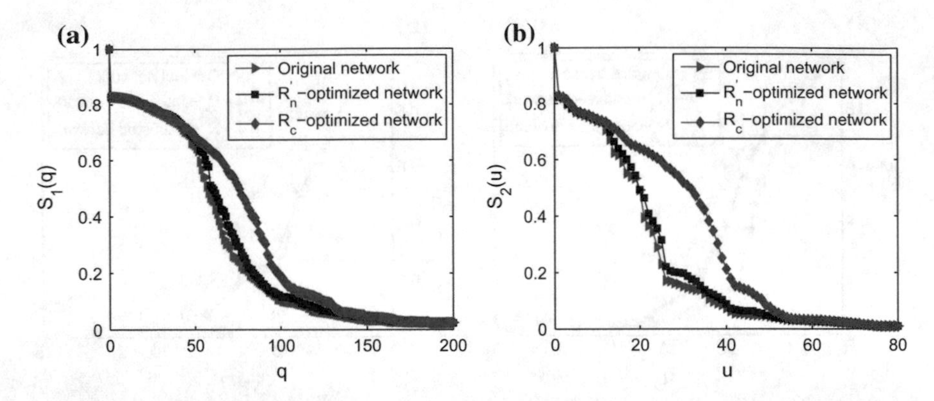

Fig. 5.7 The integrity of communities belonging to the largest components versus the number q of removed vertices or the number u of removed communities for the Road network. Three types of networks are compared: the original network, the $R_n^{'}$-optimized network and the R_c-optimized network. **a** The integrity of communities versus the number q of removed vertices. **b** The integrity of communities versus the number u of removed communities. Results are averaged over 100 independent trials

added constraint, it is found that both the R_n and λ_1/λ_2 values of the R_n-optimized networks are larger than that of the original ones. Moreover, it is observed that the spectrum index λ_1/λ_2 has no obvious relation with R_c. Therefore, the spectrum index cannot represent the node and the community robustness of networks under the new constraints.

In the following, the effects of the parameter α on the robustness of networks are analyzed. The relative robustness improvements of networks with different α are reported in Fig. 5.8. As it is seen from Fig. 5.8, in the Electronic Circuit network, the improvement can at least reach 21% for R_n, 40% for R_{c1}, 30% for R_{c2} and 35% for R_c when $0.1 \leq \alpha \leq 0.6$. The largest improvement of R_n, R_{c1}, R_{c2} and R_c can reach 25% ($\alpha = 0.6$), 51% ($\alpha = 0.9$), 37% ($\alpha = 0.1$) and 50% ($\alpha = 1$), respectively. In the USAir network, the improvement can at least reach 24% for R_n, 35% for R_{c1}, 13% for R_{c2} and 16% for R_c when $0.1 \leq \alpha \leq 0.6$. The largest improvement of R_n, R_{c1}, R_{c2} and R_c can reach 32% ($\alpha = 0.2$), 40% ($\alpha = 0.2$), 16% ($\alpha = 0.1$) and 38% ($\alpha = 1$), respectively. In the Road network, the improvement can at least reach 8% for R_n, 14% for R_{c1}, 36% for R_{c2} and 30% for R_c when $0.3 \leq \alpha \leq 0.6$. The largest improvement of R_n, R_{c1}, R_{c2} and R_c can reach 12% ($\alpha = 0.6$), 23% ($\alpha = 0.8$), 43% ($\alpha = 0.2$) and 41% ($\alpha = 0.0$), respectively. They indicate that when the network suffers from targeted attacks under various cases, both the node robustness and community robustness of the network can be improved by optimizing R_c.

The Q, *NMI* and λ_1/λ_2 values of the R_c-optimized networks with different α are also reported in Fig. 5.9. The results in Fig. 5.9 show that the community structures obtained by optimizing R_c are not sensitive to α. Generally, the values of Q are 0.653, 0.350 and 0.739, the values of *NMI* are 0.800, 0.790 and 0.965, and the

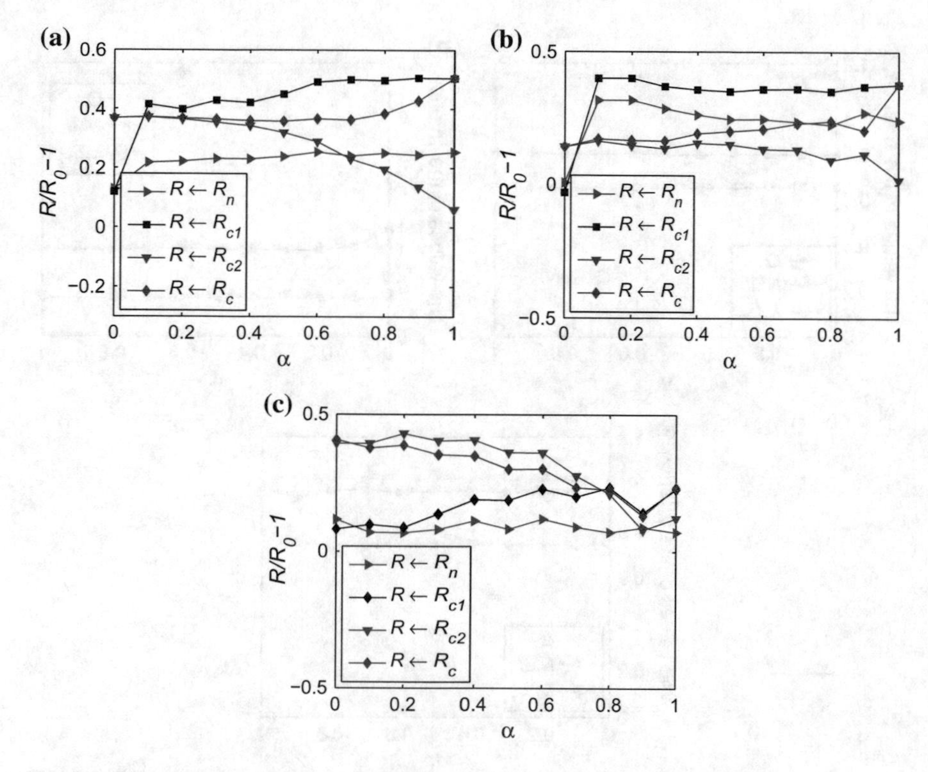

Fig. 5.8 The relative robustness improvement of the R_c-optimized networks when α changes from 0 to 1. Four kinds of robustness improvements are measured: node robustness improvement ($R \leftarrow R_n$), community robustness improvement under the small-scale attack on nodes ($R \leftarrow R_{c1}$), community robustness improvement under the large-scale attack on communities ($R \leftarrow R_{c2}$) and community robustness improvement under the two-level attack on both nodes and communities ($R \leftarrow R_c$). The experimental networks are **a** the Electronic Circuit network. **b** the USAir network and **c** the Road network. Results are averaged over 100 independent trials

values of λ_1/λ_2 are 1.13, 2.39 and 1.02, for the Electronic Circuit, the USAir and the Road networks, respectively. As both the node robustness and the community robustness of networks can be greatly improved and the community structures in the optimized networks basically remain unchanged compared with the original ones when $\alpha = 0.4$, we set the parameter α as 0.4 in this study.

5.2.5.3 Experiment on Real-World Networks Under the Mixed Attacks

In this section, the algorithm is tested on two real-world networks, the Electronic Circuits network and the USAir network, to illustrate that the community robustness of networks can be largely enhanced when they suffer from the mixed attacks. The MR_c values of the tested networks under different f are reported in Fig. 5.10. The

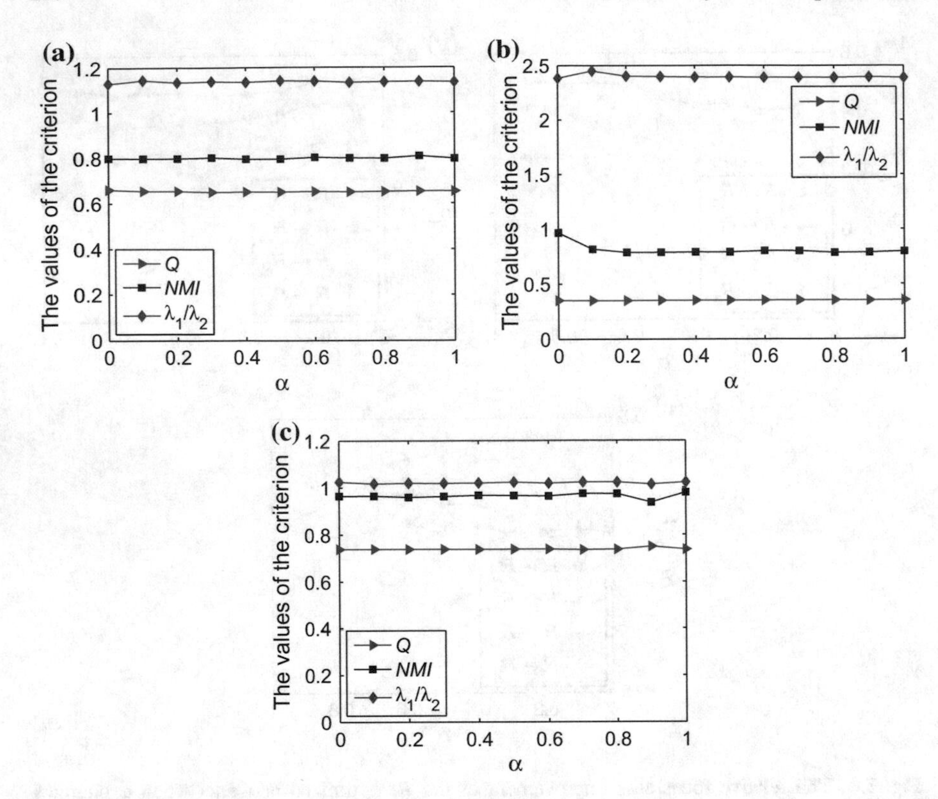

Fig. 5.9 The Q, NMI and λ_1/λ_2 values of the R_c-optimized networks when α changes from 0 to 1. The experimental networks are **a** the Electronic Circuit network. **b** the USAir network and **c** the Road network. Results are averaged over 100 independent trials

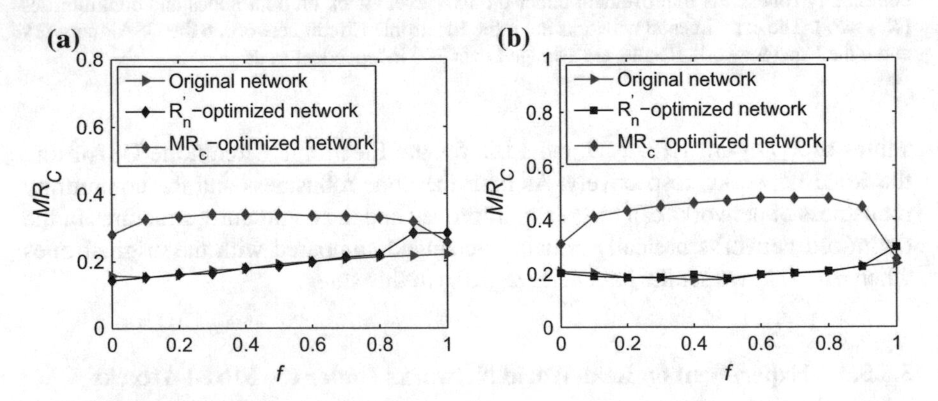

Fig. 5.10 The MR_c values of the tested networks when f changes from 0 to 1. Three types of networks are compared: the original network, the $R_n^{'}$-optimized network and the MR_c-optimized network. The experimental networks are **a** the Electronic Circuit network and **b** the USAir network. Results are averaged over 100 independent trials

results in Fig. 5.10 clearly show that optimizing R_n can improve the community robustness of networks when f is large. However, it can hardly improve the values of MR_c when f is small. More specifically, the MR_c values of the $R_n^{'}$-optimized networks are similar to that of the original networks when f is smaller than 0.8 and 0.9 for the Electronic Circuits and the USAir networks, respectively. These phenomena indicate that optimizing R_n can enhance the community robustness of networks under the small-scale node attacks. However, optimizing R_n can hardly improve the community robustness of networks under the large-scale community attacks. These conclusions are in accord with that described in Sect. 5.2.5.2. The results in Fig. 5.10 also show that the MR_c values of the MR_c-optimized networks are larger than that of the original and the $R_n^{'}$-optimized networks. In the Electronic and the USAir networks, the improvements of MR_c can reach 90 and 100%, respectively when $0.1 \leq f \leq 0.5$. It indicates that optimizing MR_c can indeed enhance the community robustness of networks under the mixed attacks.

5.2.6 *Conclusions*

As many real systems are fragile under random failures or malicious attacks, the robustness of networks has received an enormous amount of attentions in the last few years. In this section, a community robustness index is introduced to evaluate the integrality of the community structure when the network suffers from the modeled two-level targeted attacks. Moreover, new constraints for link changes are added. The optimized networks under the new constraints have the similar community structures with the original ones. Finally, a greedy algorithm is devised to improve the community robustness of networks. Experiments on three real-world networks show that both the node robustness and the community robustness of networks under the modeled attacks can be greatly improved by optimizing R_c (MR_c). Optimizing R_n can greatly improve the node robustness of networks under the small-scale node attacks. However, it can hardly enhance the community robustness of networks under the large-scale community attacks. The results also demonstrate that without the added constraint the optimized networks for network robustness cannot keep their original community structures. Under the new constraints, the optimized networks can effectively retain their original community structures.

5.3 Enhancing Robustness of Coupled Networks Under Targeted Recoveries

This section aims to enhance robustness of coupled networks under targeted recoveries. First, a damaged coupling model is proposed, and the cascading failures between coupled networks in the recovery processes is analyzed. Second, the work in [33] is extended, and an index R_{rc} is proposed to evaluate the robustness of coupled networks under the recoveries. Finally, a technique based on the protection of several influential nodes is presented to enhance the robustness of coupled networks under their recoveries. Thus, the influential nodes can work normally when they or their coupled nodes suffer from damages. Moreover, based on the network-specific knowledge, six strategies are adopted to find the influential nodes. It is shown that the recovery robustness can be greatly enhanced by protecting 5% influential nodes.

5.3.1 Algorithm for Enhancing Robustness of Coupled Networks Under Targeted Recoveries

5.3.1.1 Recovery Processes

In a system with two coupled subsystems, it is notable that both subsystems are likely to suffer from failures. In practical applications, the probability of damages on both subsystems is smaller than that on one of the subsystems. Moreover, the system cannot operate normally when one of its subsystems suffers from devastating damages. According to those phenomena, a damaged coupling network can be modeled as $G = (C, D, E_{CD})$, where E_{CD} denotes the coupled links between network C and D. In the model, each node in network C is randomly coupled with one node in network D, and network C suffers from devastating damages. In this case, the network D fails to work as well because its coupled network C is damaged.

In real-world applications, the damaged systems can be recovered by reconstructing the damaged entities gradually. Reconstructing a damaged system can be regarded as an inverse problem of attacking a system, and it can be modeled as a process in which the damaged nodes in networks are gradually recovered. In the recovery processes, there is an inter-propagation of recoveries in coupled networks. Nodes in network D can be triggered to work normally if their coupled nodes in network C have been recovered.

5.3.1.2 Cascading Failures in the Recovery Processes

There are cascading failures in coupled systems during the recovery processes. Assuming that a fraction of nodes p of network C has been reconstructed, first, the coupled nodes in network D are triggered to work normally. Then, cascading failures occur because the recovered nodes in network $C(D)$ may be scattered in a few unconnected clusters. The recovered nodes in network $C(D)$ which are not in the largest connected parts would lose their functionality, and the failures in network $C(D)$ will trigger the failures of the coupled nodes in network $D(C)$. The above failures recursively occur when there are no further failures in both networks C and D. A schematic illustration of a cascade of failures on a toy-coupling network with 4 recovered nodes of network C is given in Fig. 5.11.

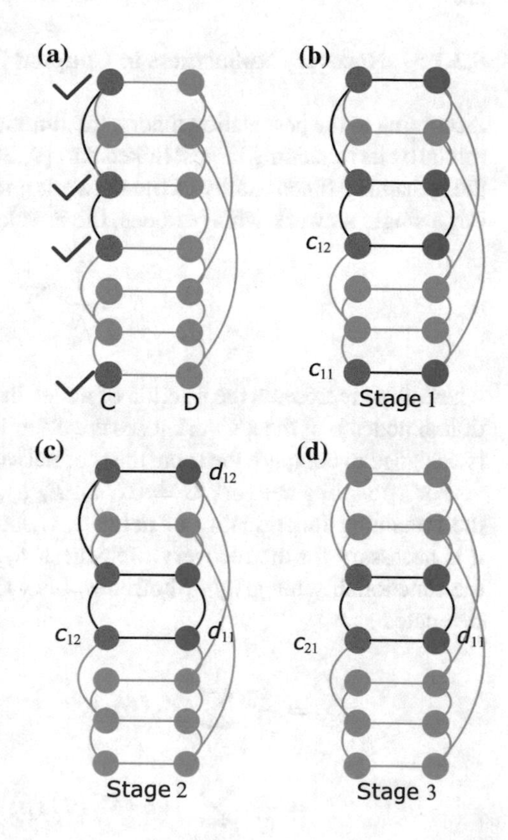

Fig. 5.11 Illustration of a cascade of failures on a toy coupling network with 4 recovered nodes. The *purple check marks* point out the recovered nodes. Nodes and links painted in gray cannot operate normally. **a** 4 nodes in network C are initially recovered. **b** Stage 1: the initial recoveries on network *C* trigger the nodes recoveries of network *D*. **c** Stage 2: the nodes of network *C* that are not in the largest connected part are failed, and the failures trigger the nodes failures of network *D*. **d** Stage 3: the nodes of network *D* that are not in the largest connected part are failed, and the failures, in turn, result in the nodes failures of network *C*

The cascading failures in the recovery process in which a fraction of nodes p of network C is recovered can be expressed as Eq. 5.7.

$$\begin{aligned}\psi_1^r(p) &= p,\\ \eta_n^r(p) &= q_{\eta,n}^r S_C^r(\psi_n^r(p))p,\\ \psi_n^r(p) &= q_{\psi,n}^r S_D^r(\psi_{n-1}^r(p))p\end{aligned} \tag{5.7}$$

where $\psi_n^r(p)$ $(\eta_n^r(p))$ is the fraction of surviving nodes in network $C(D)$ at the $(n-1)$-th coupled process when the fraction of nodes p in network C is initially recovered, $S_C^r\left(\psi_n^r(p)\right)\left(S_D^r\left(\eta_n^r(p)\right)\right)$ represents the ratio of nodes in the largest connected part of the recovered network $C(D)$, and $q_{\psi,n}^r$ $\left(q_{\eta,n}^r\right)$ is the fraction of nodes in network $C(D)$ which is coupled with the recovered nodes in network $D(C)$. When $\psi_{n+1}^r(p) = \psi_n^r(p)$ and $\eta_{n+1}^r(p) = \eta_n^r(p)$, the cascading failures end. In this case, we express $\psi_n^r(p)$ and $\eta_n^r(p)$ as $\psi^r(p)$ and $\eta^r(p)$, respectively.

5.3.1.3 Recovery Robustness in Coupled Networks

According to the percolation theory, the functionality of a damaged network is determined by its remaining largest linked part [9, 30], and the attack robustness considers the remaining functionality of the network under all possible damages [18, 25, 33]. For a single network with N nodes, the attack robustness can be calculated as [33].

$$R = \frac{1}{N} \sum_{p=1/N}^{1} S(p) \tag{5.8}$$

where $S(p)$ represents the fraction of nodes in the largest linked parts when the fraction of nodes p of the network loses the effectiveness. The normalization factor $1/N$ is designed to compare the robustness of networks that are with different scales [33].

For a coupling network $G = (C, D, E_{CD})$, its functionality depends on not only the remaining functionality of network C, but also that of network D. Therefore, it is necessary for the recovery robustness R_{rc} of the coupling network to consider the functionality integrity of both networks C and D during the recoveries. R_{rc} is computed as

$$\begin{aligned} R_{rc} &= \frac{1}{N} \sum_{p=1/N}^{1} f_{rc}(p) \\ &= \frac{1}{N} \sum_{p=1/N}^{1} \left[\left(p \cdot S_C^r\left(\psi^r(p)\right)\right)^{\lambda} \cdot \left(p \cdot S_D^r\left(\eta^r(p)\right)\right)^{1-\lambda} \right] \\ &= \frac{1}{N} \sum_{p=1/N}^{1} \left[p \cdot \left(S_C^r\left(\psi^r(p)\right)\right)^{\lambda} \cdot \left(S_D^r\left(\eta^r(p)\right)\right)^{1-\lambda} \right] \end{aligned} \tag{5.9}$$

where $f_{rc}(p)$ is the remaining functionality integrity of the coupled network after the fraction of nodes p in network C has been recovered, and $\psi^r(p)(\eta^r(p))$ is the fraction of nodes in the largest linked parts of network $C(D)$. The values of $\psi^r(p)$ and $\eta^r(p)$ can be computed by Eq. 5.7 when $\psi_1^r(p) = p$. λ is a mixing parameter ranging from 0 to 1. When $\lambda = 0$ or $\lambda = 1$, the robustness of the coupled network system is determined by that of D or C, respectively.

The recovery models are mainly divided into two categories: random and targeted. However, it is hard for a random model to recover the functionality of coupled networks with low cost. The degree-based targeted recovery model widely used in practical applications is adopted.

5.3.1.4 Enhancement of the Recovery Robustness of Coupled Networks by Protecting Influential Nodes

The functionality of many systems is controlled by a set of influential entities. These systems cannot operate normally once their influential entities are damaged. In real-world applications, complex systems have their own strategies for resisting unpredictable failures. For instance, in medical systems, all medical institutions have backup power generations for providing medical assistance for severe patients. In the coupled bank and computer systems, the influential transaction bank data are protected from hackers' targeted attacks. The common purposes of those strategies are that the influential entities are not to be damaged when they or their coupled entities suffer from damages. Based on the purposes above, a systematic technique aiming at protecting several influential nodes is introduced to reinforce the robustness of coupled networks under the recoveries.

The reasons why the proposed strategy can reinforce the robustness of coupled networks under the recoveries are as follows. First, damages on the protected nodes of one network will not lead to the nodes failures of its coupled network. It implies that the degree of coupling between two networks is reduced. According to the results found in [30, 33], the robustness of coupled networks can be enhanced with the degree of coupling decreasing. Second, the protected nodes would not lose their effectiveness when they are damaged. It means that the damages on the networks are decreased, which results in the improvement of the robustness of coupled networks. Moreover, the proposed strategy can reduce the influence caused by cascading failures during the recovery processes. Finally, extensive experiments demonstrate the effectiveness of the proposed strategy on the enhancement of the recovery robustness of coupled networks.

Many criteria based on the network-specific knowledge have been presented for evaluating the influence of nodes [31]. As many real networks have thousands of nodes and links, it is necessary to consider both the computational complexity and the efficiency of those criteria. In this study, the following well-known criteria are adopted to measure the influences of nodes.

Random (I_r). In Random centrality, the influence of a node $I_r(i)$ is computed as

$$I_r(i) = \text{rand}\,() \tag{5.10}$$

where rand () is the function for generating a random value in the range of 0 to 1.

Degree (I_d). In Degree centrality, the influence of a node $I_d(i)$ is highly related to its degree, and it is computed as [13]

$$I_d(i) = \sum_{j=1}^{N} a_{ij} \tag{5.11}$$

where a_{ij} represents the connection between nodes v_i and v_j. If there is an edge between nodes v_i and v_j, $a_{ij} = 1$, if not $a_{ij} = 0$.

Betweenness (I_b). In Betweenness centrality, the influence of a node $I_b(i)$ is evaluated by the number of the shortest paths that pass through the node [13].

$$I_b(i) = \sum_{j \neq i, q \neq i}^{N} \frac{\sigma_{jq}(i)}{\sigma_{jq}} \tag{5.12}$$

where σ_{jq} denotes the number of shortest paths from node v_j to node v_q, and $\sigma_{jq}(i)$ represents the number of shortest paths passing through node v_i from v_j to v_q.

PageRank (I_{p_r}). PageRank technique can be described by a random walk process on networks. In PageRank, the influence of a node $I_{p_r,t}(i)$ at time t in a network is calculated as [8, 31]

$$I_{p_r,t}(i) = \frac{1-\varepsilon}{N} + \varepsilon \sum_{j} \frac{a_{ij} I_{p_r,t-1}(i)}{k_{\text{out}}(j)} \tag{5.13}$$

where ε is a damping factor for a random walker to move along the links of the network, and it is usually set as a fixed value 0.85 [8]. 1ε is the probability for a random walker to jump to a randomly selected node. $k_{\text{out}}(j)$ represents the outdegree of node v_j. $I_{p_r,t}(i)$ will converge to a stationary value $I_{p_r}(i)$ with t increased, and the $I_{p_r}(i)$ value is the influence of node v_i in the network.

LeaderRank (I_{l_r}). In LeaderRank technique, all nodes, except for the ground node v_g, are initially assigned with one unit of resource. Then, at each state t, the resource of each node is evenly distributed to its neighborhoods. This process ends when the resource of each node keeps unchanged, and this state is recorded as t_c. Compared with PageRank, LeaderRank is parameter-free [23]. In LeaderRank centrality, the influence of a node $I_{l_r}(i)$ is determined by its resource value at the state t_c, and $I_{l_r}(i)$ is computed as [23]

$$I_{l_r}(i) = c_{t_c}(i) + \frac{c_{t_c}(g)}{N} \tag{5.14}$$

where $c_{t_c}(i)$ is the resource value of v_i at the state t_c, as computed in Eq. 5.15.

$$c_{t_c}(i) = \sum_{j=1}^{N+1} \frac{a_{ji} c_{t_c-1}(j)}{k_{\text{out}}(j)} \tag{5.15}$$

Local (I_l). In Local centrality, the influence of a node $I_l(i)$ is determined by its nearest and next nearest neighbors, and it is evaluated as [10].

$$I_l(i) = \sum_{v_j \in \Gamma_i} \sum_{v_u \in \Gamma_j} d(u) \tag{5.16}$$

where $d(u)$ is the number of nodes whose shortest paths from them to node v_u are no more than 2, and Γ_i (Γ_j) is the neighborhood of node v_i (v_j).

5.3.2 Experimental Results

To demonstrate the performance of the method in enhancing the robustness of networks under the recoveries, the method is tested on the following three damaged coupling networks.

(1) ER–ER-coupled system: Many traditional networks show a random connection property and they are widely modeled as Erdö-Rnyi (ER) random graphs. In ER random graphs, two nodes are linked with probability q, and the average degree $\overline{k}$ is computed as $N \cdot q$, where N is the number of nodes [12]. It is important to analyze the robustness of traditional network topology under attacks by studying on coupled ER–ER networks. The model is tested on a coupling system between a completely damaged ER network with $N = 10,000$ and $\overline{k} = 4$ and an ER network with $N = 10,000$ and $\overline{k} = 4$.

(2) ER–SF-coupled system: Many modern networks, such as Internet, scientific collaboration, telephone, power grid, and airline networks, can be approximated by scale-free (SF) networks with a power-law degree distribution $P(u) = k^{-u}$, where u is an exponential factor [32] and k represents the nodes degree of networks. Moreover, SF networks and ER networks have different statistic properties and topologies [5, 14]. Therefore, the analysis of the robustness of coupled ER–SF networks is also of great importance. The experimental ER–SF network is composed of a completely damaged ER network with $N = 10,000$ and $\overline{k} = 4$ and an SF network with $N = 10,000$ and $u = 2.5$.

(3) Power–SF-coupled system: ER networks have no modularity property with which a few nodes connect densely with each other but link sparsely with the remaining nodes of the network [33]. Besides, many systems show the modularity property [16, 25, 37]. Therefore, it is necessary to analyze the robustness of a coupling network with modularity property. A real US Power Grid network (power) with $N = 4,941$ nodes and $M = 6,954$ edges shows a high modularity property, and its coupled systems, e.g., communication networks, show the scale-free property with u ranging from 2 to 2.6 [7, 29]. Hence, a coupling system between a completely damaged power network and an SF network with $N = 4,941$ and $u = 2.2$ is analyzed.

All networks are coupled with each other using the model in [9, 30]. This model considers a pair of networks. Each node in one network is randomly coupled with one in the other network.

Table 5.2 records the robustness R_{rc} ($\lambda = 0.5$) of the tested networks under their recoveries. All experimental results are averaged over 50 independent trials, and all algorithms are simulated by MATLAB on a PC with Intel (R), Core (TM), 3 CPU with 3.2 GHZ, 3 GB memory.

The results in Table 5.2 show that the ER–ER network is robust while the ER–SF and Power–SF networks are fragile to cascading failures under the recoveries. Of the tested networks, the Power–SF network is the most fragile under the recoveries. This is due to two factors. First, the power network has the scale-free property [5, 29]. The scale-free degree distribution property results in the fact that the Power–SF network is more vulnerable than the ER–ER and the ER–SF networks. Second, the

Table 5.2 Comparisons of the recovery robustness $R_{rc}(\lambda = 0.5)$ obtained by protecting 5% influential nodes with different strategies

Networks	No	Random	Degree	Betweenness	LeaderRank	Local	PageRank
ER–ER	0.3652	0.4026	0.4319	0.4479	0.4328	0.4489	0.4302
ER–SF	0.2220	0.2521	0.4709	0.4755	0.4703	0.3242	0.4707
Power–SF	0.0317	0.0320	0.4561	0.5054	0.4562	0.4225	0.4560

power network has the modularity property [25] with which the nodes in the same modules are closely linked together. In this case, a node failure may trigger further failures on the nodes of the same module [24].

Six strategies are compared to select the 5% influential nodes: Random, Degree [13], Betweenness [13], PageRank [8], LeaderRank [23] and Loca choice [10], for the experimental networks, and the corresponding results are also recorded in Table 5.2. The results show that the recovery robustness of coupled networks can be greatly enhanced, especially for the Betweenness protection strategy. More specifically, the improvement of recovery robustness can reach 22.65% for the ER–ER network, 114.2% for the ER–SF network and 1,494% for the Power–SF network. Moreover, compared with the random protection, the targeted protections make the ER–ER network more robust to cascading failures during the recoveries. The results also show that the Degree, the PageRank, and the LeaderRank protection strategies can get similar results since for the ER and SF networks these three criteria are highly correlated [28, 33]. It is notable that the Betweenness and the Local protection strategies have similar results in the ER–ER network. However, the Local strategy obtains much smaller R_{rc} values than the Betweenness strategy in the ER–SF and Power–SF networks. This is because the Betweenness criterion reflects the global nodes information of networks while the Local criterion evaluates the local nodes information. It is in the ER networks rather than the SF and Power networks that the global connections can be reflected by the local information of nodes.

In order to further analyze the fragility of experimental networks during the recoveries, the variations of the remaining fraction of nodes $f_{rc}(p)$ in the largest connected parts with the fraction of recovered nodes p are analyzed in Fig. 5.12. It shows that without protection strategies, the experimental networks undergo a discontinuous percolation transition where the $f_{rc}(p)$ value abruptly changes from zero to a finite value. More specifically, the ER–ER network, the ER–SF network and the Power–SF network begin to recover their functionality when close to 43%, 63%, and 97% nodes are recovered, respectively. With the protection strategies, especially for the Betweenness protection strategy, the experimental networks begin to recover their functionality when few nodes are recovered. The results in Fig. 5.12 further demonstrate the effectiveness of the proposed protection strategies on the enhancement of the recovery robustness of coupled networks.

Figure 5.13 shows the influences of the mixing parameter λ and the coupling degree on the recovery robustness of the Power–SF networks. As shown in Fig. 5.13, R_{rc} has negative relation with the coupling degree and the parameter λ. For a strong

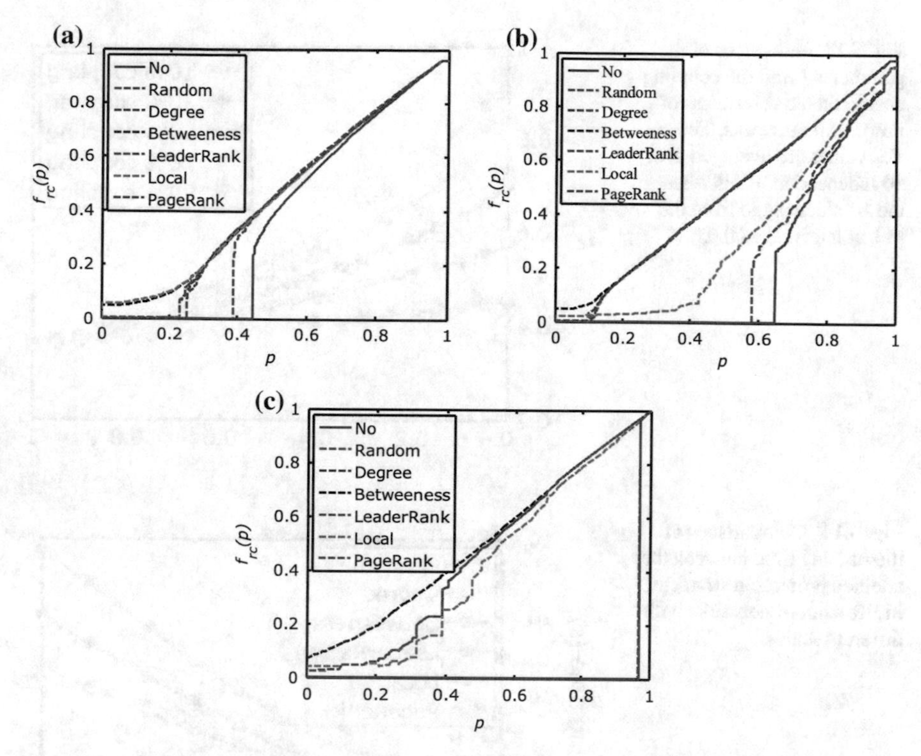

Fig. 5.12 Variations of the fraction of remaining nodes $f_{rc}(p)$ in the largest connected parts with the fraction of recovered nodes p for **a** the ER–ER network. **b** the ER–SF network and **c** the Power–SF network. It can be seen that the recovery robustness can be greatly enhanced by protecting 5% influential nodes. We choose the 5% influential nodes in six different ways: Random (*blue dotted line*), high Degree (*brown dotted line*), high Betweenness (*black dotted line*), high LeaderRank (*green dotted line*), high Local (*light green dotted line*) and high PageRank (*purple dotted line*). Results are averaged over 50 independent trials

coupling network, say 90% coupling, the R_{rc} value is less sensitive to the parameter λ. This is because the fraction of surviving nodes in one network is close to that of the ones in its coupled network. However, for a weak coupling network, say 10% coupling, R_{rc} has negative relation with λ. This is because the cascading failures in a weak coupling network are decreased and the damages on the SF network are smaller than those on the power network during the recoveries.

In order to compare the computational performances of the six protection strategies, they are employed on the ER networks with different scales and record the consumed time in Fig. 5.14. When the scale of the ER network is small, the six protection strategies can find the influential nodes quickly. When the scale of the ER network is large, it is hard for the LeaderRank and the Betweenness protection strategies to evaluate the importance of nodes in a short period of time. Many systems (e.g., Internet and communication systems) have millions of nodes and links. The LeaderRank and the Betweenness protection techniques cannot tackle it well.

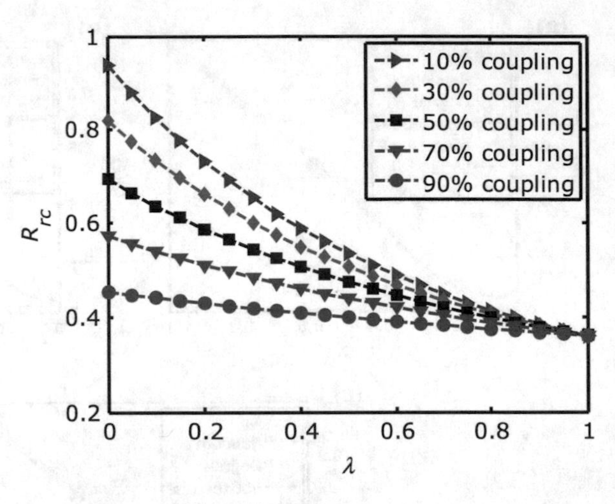

Fig. 5.13 Influences of the parameter λ and the coupling degree on the robustness of Power–SF networks. The R_{rc} values are averaged over 50 independent trials when the λ values range from 0.0 to 1 at intervals of 0.05

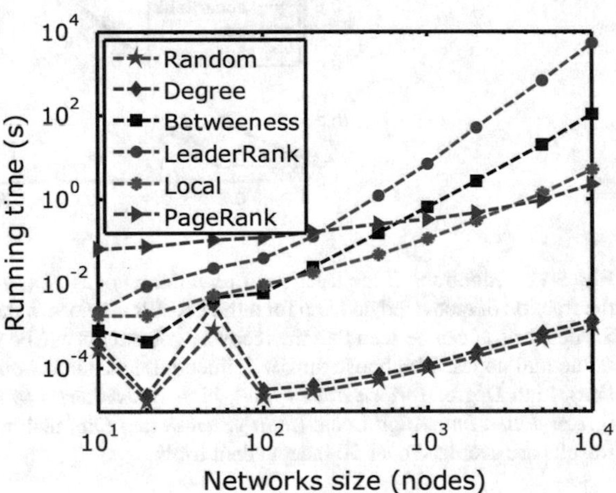

Fig. 5.14 Comparisons of the running time between the adopted protection strategies in ER random networks with different scales

Moreover, the Random and Local protection strategies can evaluate the influences of nodes in a short time. However, it is hard for them to improve the recovery robustness of networks by protecting 5 influential nodes, as shown in Table 5.2. The results in Fig. 5.14 and Table 5.2 also show that both the Degree and the PageRank protection strategies can greatly enhance the recovery robustness of large-scale networks in a reasonable time.

5.3.3 Conclusions

The coupling property makes complex systems fragile to cascading failures during the recoveries. How to enhance the robustness of coupled systems with low cost has received many attentions in recent years. In this section, the cascading failures between coupled networks in the recovery processes are represented by a mathematical model, and the resilience of coupled systems to cascading failures under the recoveries is evaluated by the proposed recovery robustness index. And a protection method for protecting several influential nodes is proposed to enhance the recovery robustness of coupled systems. Experimental results have shown that the recovery robustness of coupled systems is greatly enhanced by the proposed method. The experiments also demonstrate that the coupled networks with modularity property are more fragile to failures than those with no modularity property. The Degree and the PageRank protection strategies have low computational complexity, and can be used to enhance the robustness of large-scale coupled networks with millions of nodes.

References

1. Akkaya, K., Senel, F., Thimmapuram, A., Uludag, S.: Distributed recovery from network partitioning in movable sensor/actor networks via controlled mobility. IEEE Trans. Comput. **59**(2), 258–271 (2010)
2. Albert, R., Albert, I., Nakarado, G.L.: Structural vulnerability of the North American power grid. Phys. Rev. E **69**(2), 025,103 (2004)
3. Ammann, P., Jajodia, S., Liu, P.: Recovery from malicious transactions. IEEE Trans. Knowl. Data Eng. **14**(5), 1167–1185 (2002)
4. Babaei, M., Ghassemieh, H., Jalili, M.: Cascading failure tolerance of modular small-world networks. IEEE Trans. Circuits Syst. II: Express Briefs **58**(8), 527–531 (2011)
5. Barabási, A.L., Albert, R.: Emergence of scaling in random networks. Science **286**(5439), 509–512 (1999)
6. Blondel, V.D., Guillaume, J.L., Lambiotte, R., Lefebvre, E.: Fast unfolding of communities in large networks. J. Stat. Mech.: Theor. Exp. **2008**(10), P10,008 (2008)
7. Brandes, U., Erlebach, T.: Network analysis: methodological foundations, vol. 3418. Springer Science & Business Media (2005)
8. Brin, S., Page, L.: The anatomy of a large-scale hypertextual web search engine. Comput. Netw. ISDN Syst. **30**(1), 107–117 (1998)
9. Buldyrev, S.V., Parshani, R., Paul, G., Stanley, H.E., Havlin, S.: Catastrophic cascade of failures in interdependent networks. Nature **464**(7291), 1025–1028 (2010)
10. Chen, D., Lü, L., Shang, M.S., Zhang, Y.C., Zhou, T.: Identifying influential nodes in complex networks. Phys. A: Stat. Mech. Appl. **391**(4), 1777–1787 (2012)
11. Cohen, R., Havlin, S., Ben-Avraham, D.: Efficient immunization strategies for computer networks and populations. Phys. Rev. Lett. **91**(24), 247,901 (2003)
12. Erdos, P., Rényi, A.: On the evolution of random graphs. Publ. Math. Inst. Hung. Acad. Sci **5**(1), 17–60 (1960)
13. Freeman, L.C.: Centrality in social networks conceptual clarification. Soc. Netw. **1**(3), 215–239 (1978)
14. Gao, J., Buldyrev, S.V., Stanley, H.E., Havlin, S.: Networks formed from interdependent networks. Nat. Phys. **8**(1), 40–48 (2012)
15. Girvan, M., Newman, M.E.: Community structure in social and biological networks. Proc. Nat. Acad Sci. **99**(12), 7821–7826 (2002)
16. Gong, M., Cai, Q., Chen, X., Ma, L.: Complex network clustering by multiobjective discrete particle swarm optimization based on decomposition. IEEE Trans. Evol. Comput. **18**(1), 82–97 (2014)

17. Herrmann, H.J., Schneider, C.M., Moreira, A.A., Andrade Jr, J.S., Havlin, S.: Onion-like network topology enhances robustness against malicious attacks. J. Stat. Mech.: Theor. Exp. **2011**(01), P01,027 (2011)
18. Holme, P., Kim, B.J., Yoon, C.N., Han, S.K.: Attack vulnerability of complex networks. Phys. Rev. E **65**(5), 056,109 (2002)
19. Hoory, S., Linial, N., Wigderson, A.: Expander graphs and their applications. Bull. Am. Math. Soc. **43**(4), 439–561 (2006)
20. Jiang, L.L., Perc, M.: Spreading of cooperative behaviour across interdependent groups. arXiv preprint arXiv:1310.4166 (2013)
21. Kvalbein, A., Hansen, A.F., Čičic, T., Gjessing, S., Lysne, O.: Multiple routing configurations for fast ip network recovery. IEEE/ACM Trans. Netw. (TON) **17**(2), 473–486 (2009)
22. Liu, Y.Y., Slotine, J.J., Barabási, A.L.: Controllability of complex networks. Nature **473**(7346), 167–173 (2011)
23. Lü, L., Zhang, Y.C., Yeung, C.H., Zhou, T.: Leaders in social networks, the delicious case. PloS one **6**(6), e21,202 (2011)
24. Ma, L., Gong, M., Cai, Q., Jiao, L.: Enhancing community integrity of networks against multilevel targeted attacks. Phys. Rev. E **88**(2), 022,810 (2013)
25. Ma, L., Gong, M., Liu, J., Cai, Q., Jiao, L.: Multi-level learning based memetic algorithm for community detection. Appl. Soft Comput. **19**, 121–133 (2014)
26. Majdandzic, A., Podobnik, B., Buldyrev, S.V., Kenett, D.Y., Havlin, S., Stanley, H.E.: Spontaneous recovery in dynamical networks. Nat. Phys. **10**(1), 34–38 (2014)
27. Milo, R., Itzkovitz, S., Kashtan, N., Levitt, R., Shen-Orr, S., Ayzenshtat, I., Sheffer, M., Alon, U.: Superfamilies of evolved and designed networks. Science **303**(5663), 1538–1542 (2004)
28. Newman, A.: Introduction. Camden Third Series **94**, vii–xiv (1963)
29. Nguyen, D.T., Shen, Y., Thai, M.T.: Detecting critical nodes in interdependent power networks for vulnerability assessment. IEEE Trans. Smart Grid **4**(1), 151–159 (2013)
30. Parshani, R., Buldyrev, S.V., Havlin, S.: Interdependent networks: reducing the coupling strength leads to a change from a first to second order percolation transition. Phys. Rev. Lett. **105**(4), 048,701 (2010)
31. Pei, S., Makse, H.A.: Spreading dynamics in complex networks. J. Stat. Mech.: Theor. Exp. **2013**(12), P12,002 (2013)
32. Scheffer, M., van Nes, E.H.: Self-organized similarity, the evolutionary emergence of groups of similar species. Proc. Nat. Acad Sci. **103**(16), 6230–6235 (2006)
33. Schneider, C.M., Moreira, A.A., Andrade, J.S., Havlin, S., Herrmann, H.J.: Mitigation of malicious attacks on networks. Proc. Nat. Acad Sci. **108**(10), 3838–3841 (2011)
34. Šubelj, L., Bajec, M.: Robust network community detection using balanced propagation. Eur. Phys. J. B-Condens. Matter Complex Syst. **81**(3), 353–362 (2011)
35. Um, J., Minnhagen, P., Kim, B.J.: Synchronization in interdependent networks. Chaos: Interdisc. J. Nonlinear Sci. **21**(2), 025,106 (2011)
36. Wang, J.: Robustness of complex networks with the local protection strategy against cascading failures. Saf. Sci. **53**, 219–225 (2013)
37. Watts, D.J., Strogatz, S.H.: Collective dynamics of small-world-networks. Nature **393**(6684), 440–442 (1998)
38. Wu, Z.X., Holme, P.: Onion structure and network robustness. Phys. Rev. E **84**(2), 026,106 (2011)
39. Zeng, A., Liu, W.: Enhancing network robustness against malicious attacks. Phys. Rev. E **85**(6), 066,130 (2012)

Chapter 6
Real-World Cases of Network Structure Analytics

Abstract In complex systems, except for the issues discussed in previous chapters, the issues, including recommender system, network alignment and influence maximization etc. are also NP-hard problems, and they can be modeled as optimization problems. Computational intelligence algorithms, especially evolutionary algorithms, have been successfully employed to these network structure analytics topics. In this chapter, we will present how to use computational intelligence techniques to tackle the recommendation system, the network alignment, and the influence maximization problem in complex networks. First, an evolutionary multiobjective algorithm is used for recommendation. And then, a memetic algorithm for influence maximization is introduced. Finally, a memetic algorithm for global biological network alignment is presented.

6.1 Review on the State of the Art

Previous sections give a brief introduction of computational intelligence algorithms and their applications to some network structure analytics, including community discovery, structure balance, and network robustness. In Chap. 1, we have listed 12 critical issues that concern network analytics. Other issues, such as recommendation systems, influence maximization, and biological network alignment et al. have also received great attentions in recent years. Recent researches have shown that these issues also can be modeled as optimization problems and computational intelligence algorithms are effective tools for solving these problems. In this section, we will give a brief introduction of computational intelligence algorithms on network structure analytics, especially on recommendation systems, influence maximization, and biological network alignment.

Recommender systems (RSs) [52], which use statistical and knowledge discovery techniques to provide recommendations automatically, are considered to be the most promising tools to alleviate the overload of information. Ever since their advent, RSs have attracted considerable attention in both theoretical research and practical applications [9].

M. Gong et al., *Computational Intelligence for Network Structure Analytics*,
DOI 10.1007/978-981-10-4558-5_6

A pressing challenge for RSs is how to develop personalized recommendation techniques that can generate recommendations with both high accuracy and diversity. To achieve a proper balance between accuracy and diversity, a variety of recommendation techniques have been developed. Zhang et al. [63] modeled the trade-off between accuracy and diversity as a quadratic programming problem and developed several strategies to solve this optimization problem. A control parameter should be used to determine the importance of diversification in the recommendation lists. Zhou et al. [64] proposed a hybrid recommendation algorithm, which combines Heat-spreading (HeatS) algorithm specifically to address the challenge of diversity and probabilistic spreading (ProbS) algorithm to focus on accuracy. Note that the hybrid algorithm is produced by using a basic weighted linear aggregation method. As a result, the weight parameter should be appropriately tuned so as to make accurate and diverse recommendations. Adomavicius et al. [1], developed a number of item ranking techniques to generate diverse recommendations while maintaining comparable levels of recommendation accuracy.

Influence maximization is to extract a small set of nodes (or seeds) from a social network which can generate the propagation maximally under a cascade model. This problem is first studied by Domingos and Richardson [23]. They view a market as a social network and propose a probabilistic method with modeling the influence between network users as a Markov random field. Then the problem is formalized as a discrete optimization problem which is proved to be NP-hard under the independent cascade model (IC) and the linear threshold model (LT) by Kempe et al. [33].

Many methods have been proposed for influence maximization [14, 16, 28, 43, 51, 59]. Leskovec et al. [43] present a method called CELF which is reported to be 700 times faster than the greedy algorithm. Goyal et al. [28] propose an improvement on [43] named as CELF++ which further improves the efficiency. Chen et al. [14] put forward two new greedy algorithms. One is the NewGreedy. It attempts to reduce computations by generating a new smaller graph with all edges not participating in the propagation removed. The other one is the MixedGreedy. It runs the NewGreedy in the first round and runs the CELF [43] in the later rounds. Although these improvements outperform the greedy algorithm in efficiency, they still cannot be scalable to large-scale networks. Recently, some authors use the communities of the networks to improve the efficiency of algorithms. Wang et al. [59] introduce a community-based greedy algorithm called CGA which narrows down the search space of the influential nodes from the whole graph to the communities. They exploit dynamic programming to select the community which has the largest increase of influence spread and adopts MixGreedy algorithm [14] to find the most influential node as the seed in the chosen community. Chen et al. [16] propose a community-based algorithm under the Heat Diffusion Model, termed as CIM. They select seeds from the candidate nodes by comparing scores of them. Rahimkhani et al. [51] present a fast algorithm, called ComPath, combining with the community character and introduce an influence spread estimation under the LT model.

High-throughput experimental screening techniques, such as yeast two-hybrid [24] and co-immunoprecipitation [3], have produced large amount of *protein-protein interactions* (PPI) data. Network alignment, which matches the nodes in different networks to gain a maximum similarity, can be used to identify the conserved subgraphs in different networks. And the conserved subgraphs can be used to predict protein functions and understand evolutionary trajectories across species [7, 10, 13, 25, 29, 38, 60]. Additionally, the knowledge acquired may give us a deeper insight into the mechanisms of various diseases [55–57]].

Network alignment is usually divided into two categories, local network alignment and global network alignment. *Local network alignment* (LNA) mainly aims to find highly similar subgraphs. These subgraphs are independent of each other and usually represent conserved protein complexes and metabolic pathways [31, 32, 54]. *Global network alignment* (GNA) mainly aims to generate a comprehensive mapping in the whole network graph, which often implies functional orthologs [40, 55]. These two types network alignment have different emphases. LNA pays more attention to find local optimal subgraphs and allows mutually inconsistent mapping (e.g., one protein can be mapped to different proteins in other networks). Contrary to LNA, GNA aims to find a best single consistent mapping that covers all of the input nodes. It can compare PPI networks in a species level and is crucial to understand cross-species variations.

Over the past few years, network alignment has become a hot research topic and many state-of-the-art algorithms have been put forward. These network alignment algorithms can be divided into two classes: two-step methods and the objective function-based search methods. Two-step approaches first calculate the similarity between any two nodes from different input networks to form a similarity matrix. Then they transform network alignment into a maximum weight bipartite matching problem [19]. The first step has high efficiency by utilizing priori knowledge and the second step determines the alignment quality by using different extracting strategies. Elements of the similarity matrix are regarded as the weights of bipartite graph [19]. Most existing algorithms such as IsoRank [55], MI-GRAAL [40], NETAL [48], Mat3 [36] and HubAlign [29] belong to the first classes. The objective function-based search methods are proposed recently. They first define an objective function and then adopt different kinds of heuristic strategies to optimize it. Some algorithms like Piswap [17], MAGNA [53], MAGNA++ [58], and Optnetalign [20] belong to this class. There are advantages and limits in these two kinds of approaches. Two-step methods are relatively mature, but it is usually hard to make a balance between topology and sequence similarities. The objective function-based search methods are likely to make a good balance after sufficient search. Nevertheless, it is difficult for them to incorporate the priori knowledge into the algorithms.

This chapter is organized into four sections. Section 6.2 introduces an evolutionary multiobjective optimization for community-based personalized recommendation. Section 6.3 introduces a memetic algorithm for influence maximization in social networks. Section 6.4 introduces a memetic algorithm for global biological network alignment.

6.2 Community-Based Personalized Recommendation with Evolutionary Multiobjective Optimization

This section introduces a general multiobjective recommendation model, termed as MOEA-ProbS, to address the challenge of striking a balance between accuracy and diversity. The model considers two conflicting objectives. The first one is the measurement of recommendation accuracy, which can be estimated by an accuracy-based recommendation technique, and the other one is a diversity metric. The task of personalized recommendation is thus modeled as a multiobjective optimization problem (MOP). A multiobjective evolutionary algorithm (MOEA) is then performed to evolve the population to maximize these two objectives. Finally, a set of different recommendations can be provided for users.

The advantages of MOEA-ProbS are as follows. A general multiobjective recommendation model is proposed to balance recommendation accuracy and diversity. Different from traditional recommendation techniques, MOEA-ProbS can simultaneously provide multiple recommendations for multiple users in only one run. The proposed algorithm can provide a set of diverse and accurate recommendations. Especially, the coverage of recommendations is greatly improved, which is a promising property for RSs. A clustering technique is employed to improve the computational efficiency.

6.2.1 MOEA-Based Recommendation Algorithm

In order to balance recommendation accuracy and diversity, the recommendation problem is modeled as an MOP. In this algorithm, NSGA-II [22] is adopted as the MOEA to solve the modeled MOP, due to its robustness and effectiveness. In the following, the proposed MOEA-based recommendation algorithm is described in detail.

6.2.2 User Clustering

To reduce the computational complexity, a clustering technique is used to split a large number of users into several clusters. Since the users from different clusters have different habits and preferences, diverse recommendations can be easily provided for these users by RSs. In contrast, the users belonging to the same cluster are similar, and therefore similar items tend to be recommended to these users. The aim of the proposed algorithm is to improve the diversity of recommendations to these similar users. In particular, the quality of recommendations can be improved by using a clustering technique sometimes.

As discussed in [61], there exist many clustering techniques, such as k-means, fuzzy c-means (FCM), and hierarchical clustering. In addition, community-based methods [27], which are able to mine potential relationship between different users, can be also used for clustering. In MOEA-ProbS, the k-means clustering method is employed. The performance of clustering will be influenced by the used similarity strategy. Here the cosine index [2] is used to measure the similarity s_{ij} between two users i and j, which is defined as

$$s_{ij} = \frac{r_i \cdot r_j}{|r_i||r_j|} \tag{6.1}$$

where r_i and r_j are rating vectors given by i and j on items, respectively.

6.2.3 Problem Formation

In MOEA-ProbS, two conflicting objectives are considered. The first one measures the accuracy of recommendations. In fact, it is impossible to compute the true preferences of users in the training stage. Therefore, the estimated ratings of items are used. For a user i and an item α, the predicted rating of α given by i is $pr_{i\alpha}$. For all the users in one cluster S, the predicted rating is defined as

$$PR = \frac{\sum_{i \in S} \sum_{\alpha=1}^{L} pr_{i\alpha}}{|S| \times L} \tag{6.2}$$

where $|S|$ is the number of users in S, and L is the length of the recommendation list.

The second one is to measure the diversity of recommendations. There are several diversity metrics [44], such as inter-user diversity, intra-user diversity, and coverage. Due to its simplicity, coverage is used in this paper. The specific definition is given as follows:

$$CV = \frac{N_{dif}}{N} \tag{6.3}$$

where N_{dif} is the number of different items in the recommendation lists for the users in the same cluster, and N is the total number of items. Obviously, within a certain level of accuracy, a higher value of coverage indicates a better recommendation.

6.2.4 Representation

Directly, items recommended to a user are encoded by a vector of integer values, each of which represents the corresponding item number. Since MOEA-ProbS aims to provide recommendations for all the users in the same cluster, the chromosome is encoded by a matrix. Assuming that L items will be recommended to each user and

Table 6.1 Illustration of chromosome encoding

	Item 1	Item 2	...	Item L
User 1	5	3		13
User 2	16	27		8
...			...	
User K	19	5		7

there are K users in the cluster, the scale of the matrix is thus $K \times L$. An illustration of the encoding method is given in Table 6.1, where rows represent users and columns represent items. Usually, RSs will not recommend one item to one user twice. This means that duplicate alleles are not allowed in the same row. In addition, for a given user, there is no need to suggest items rated by the given user in the past. However, different users often rate different items, leading to different search spaces of decision variables. Hence, the modeled optimization problem can be considered as a complex discrete MOP.

6.2.5 Genetic Operators

The genetic operators used in MOEA-ProbS include crossover and mutation, which are performed to produce new solutions. In MOEA-ProbS, the uniform crossover is adopted. Since one item is not allowed to be suggested to one user twice, an additional operation should be executed to avoid generating invalid solutions. The procedure of the crossover operator can be described as follows. First, the same items from two parents are identified and propagated to the child. Then, the remaining alleles perform crossover. A random number in [0, 1] is produced for each remaining position in the child. If the number is larger than 0.5, the child receives the corresponding allele from the first parent. Otherwise, it receives the allele from the second parent. An illustration of the crossover operator is shown in Fig. 6.1. Note that the crossover operator is performed row by row and then the two parent matrices complete crossover.

The mutation operator is applied to a single individual. If one allele in the parent matrix is to be mutated, another available item is randomly selected from the item set to replace the initial one. An item available means that the item does not exist in the parent. In this way, the mutation operator can always generate feasible solutions.

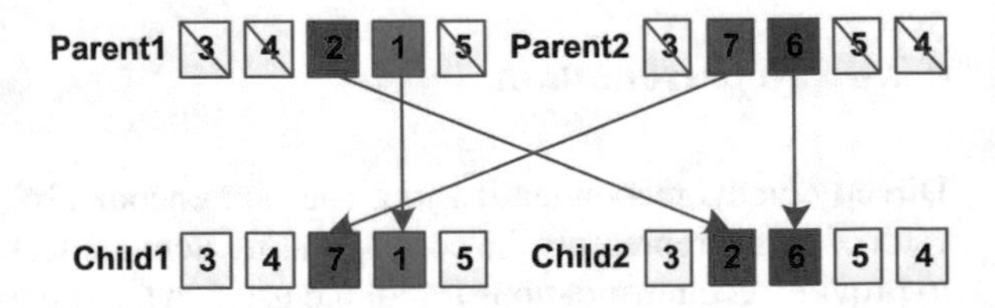

Fig. 6.1 Illustration of the crossover operator. Only the positions without slash perform crossover. Two generated random numbers are 0.1 and 0.8, respectively

6.2.6 *Experimental Results*

To evaluate the performance of the proposed algorithm, a classical benchmark data set, Movielens is used. This data set contains 943 users and 1682 movies. Since Movielens uses a rating system (ratings 1–5), the data set is preprocessed with the same method applied in [65]. An item is considered to be liked by a user, if the user rated this item at least 3. Then randomly select 80% of the data as the training set, and the remaining data constitutes the probe set. The training set is treated as known information for generating recommendations, while the probe set is used to evaluate the performance of RSs. In order to accelerate the search process, the users are divided into several clusters. In the experiments, the users are divided into four clusters, thus four different relatively small data sets are generated. The properties of these data sets are presented in Table 6.2, where the sparseness of each data set is defined as the number of links divided by the total number of user-object pairs [64]. A sparse data set indicates that only a few items are rated by users.

As is known to all, some common parameters in the MOEA need to be predetermined. The specific values of the parameters used in the computational experiments are listed in Table 6.3. To obtain statistical results, 30 independent runs are performed for each data set.

6.2.6.1 Performance Metrics

Precision is widely used to measure the accuracy of recommendations [44]. For a given user i, precision $P_i(L)$ is defined as

$$P_i(L) = \frac{d_i(L)}{L} \tag{6.4}$$

Table 6.2 Properties of the test data sets

Data set	Users	Items	Sparsity
Movielens 1	200	1682	1.39×10^{-2}
Movielens 2	258	1682	5.17×10^{-2}
Movielens 3	227	1682	2.38×10^{-2}
Movielens 4	258	1682	6.89×10^{-2}

Table 6.3 Parameter settings of the algorithm

Parameter	Meaning	Value
L	The length of the recommendation list	10
S_{pop}	The size of population	100
P_c	The crossover probability	0.8
P_m	The mutation probability	$1/L$
G_{max}	The number of generations	3000

where $d_i(L)$ is the number of relevant items, which are in the recommendation list and also preferred by user i in the probe set. L is the length of the recommendation list. The obtained mean precision of all users can reveal recommendation accuracy of RSs generally.

Coverage denotes the ability of a recommendation algorithm to recommend diverse items. It is considered as one of the two conflicting objectives in our proposed algorithm. Here it is also adopted as a performance metric to measure the diversity of recommendations.

Novelty is used to measure how well RSs recommend unknown items to users. To measure the unexpectedness of the recommended items, self-information is used. Given an item α, the probability to collect it by a random-selected user is k_α/M, where M is the total number of users, and k_α is the degree of item α (i.e., the popularity of item α) [44]. The self-information of item α is thus:

$$N_\alpha = \log_2\left(\frac{M}{k_\alpha}\right). \tag{6.5}$$

A user-relative novelty is obtained by calculating the average self-information of items in the target user's recommendation list. Then the mean novelty $N(L)$ over all users can be obtained according to:

$$N(L) = \frac{1}{ML}\sum_{i=1}^{M}\sum_{\alpha\in O_L^i} N_\alpha \tag{6.6}$$

where O_L^i is the recommendation list of user i and L is the length of the recommendation list.

6.2.6.2 Effectiveness of MOEA-ProbS

In this subsection, the experimental results of MOEA-ProbS on the four data sets are given. To show the effectiveness of MOEA-ProbS, the final non-dominated solutions with the highest hypervolume [66] for each data set are displayed. Since each solution represents recommendations to all the users in the same cluster, the accuracy and coverage of these recommendations can be calculated to examine the quality of the solution. The final solutions of MOEA-ProbS in the accuracy-coverage space are then plotted. In addition, to observe the convergence trend of the MOEA, the hypervolume values of the non-dominated solutions among 30 independent runs with different generations are recorded. Since the two objectives are non-negative obviously, the reference point for computing hypervolume is set to the origin.

From Figs. 6.2, 6.3, 6.4 and 6.5, it can be concluded that there exists a trade-off between recommendation accuracy and diversity. After a certain number of generations, the proposed MOEA-ProbS can generate a set of recommendations. Note that the accuracy and coverage of different recommendations determined by a set

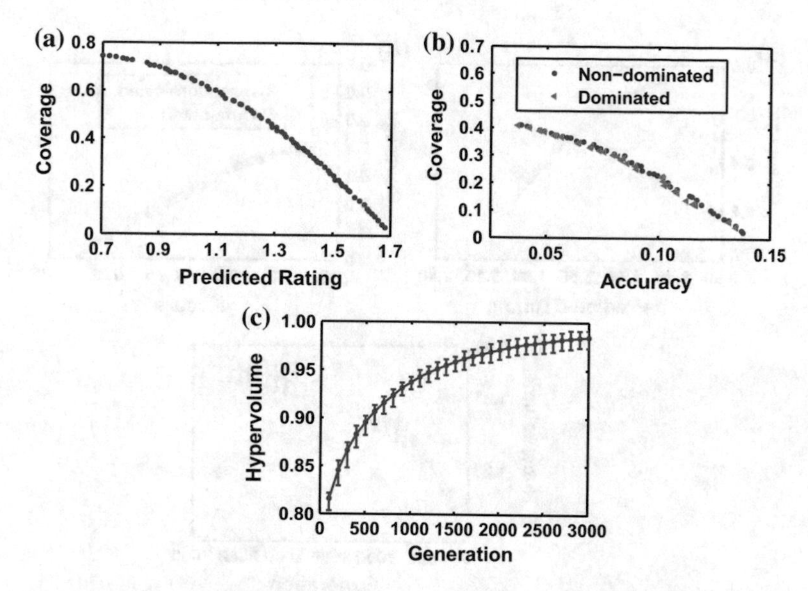

Fig. 6.2 Results of MOEA-ProbS on Movielens 1. **a** Plots of final non-dominated solutions with the highest hypervolume. **b** Plots of final solutions in the accuracy-coverage space (**c**) The error-bar of hypervolume metric of population among 30 independent runs with different generations

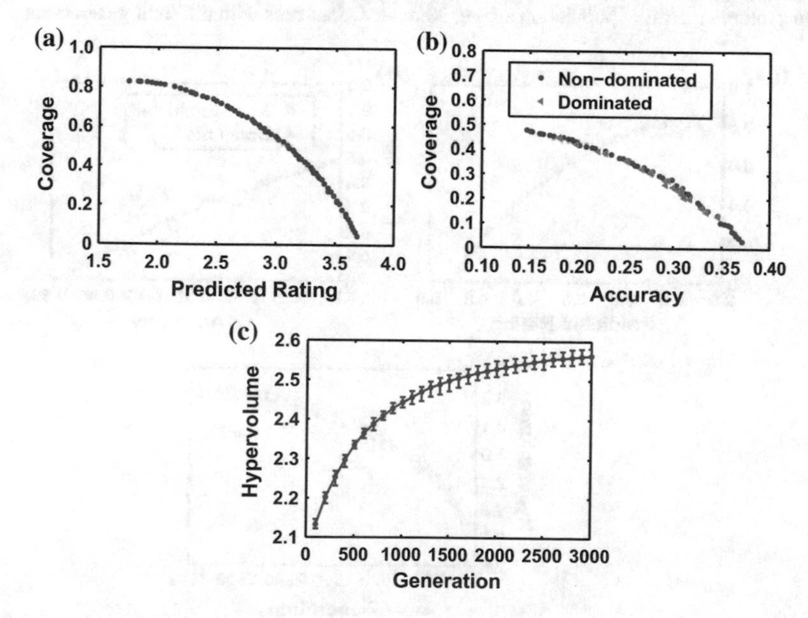

Fig. 6.3 Results of MOEA-ProbS on Movielens 2. **a** Plots of final non-dominated solutions with the highest hypervolume. **b** Plots of final solutions in the accuracy-coverage space (**c**) The error-bar of hypervolume metric of population among 30 independent runs with different generations

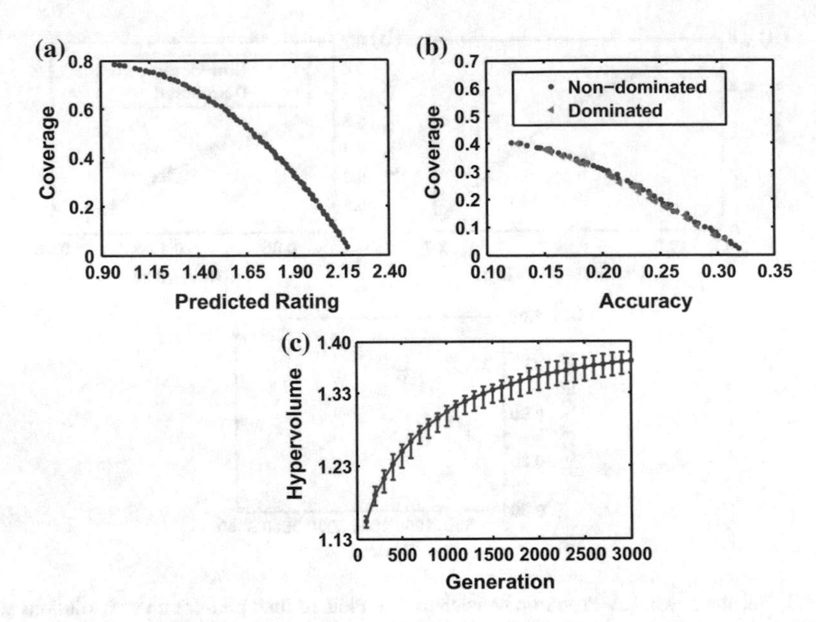

Fig. 6.4 Results of MOEA-ProbS on Movielens 3. **a** Plots of final non-dominated solutions with the highest hypervolume. **b** Plots of final solutions in the accuracy-coverage space (**c**) The error-bar of hypervolume metric of population among 30 independent runs with different generations

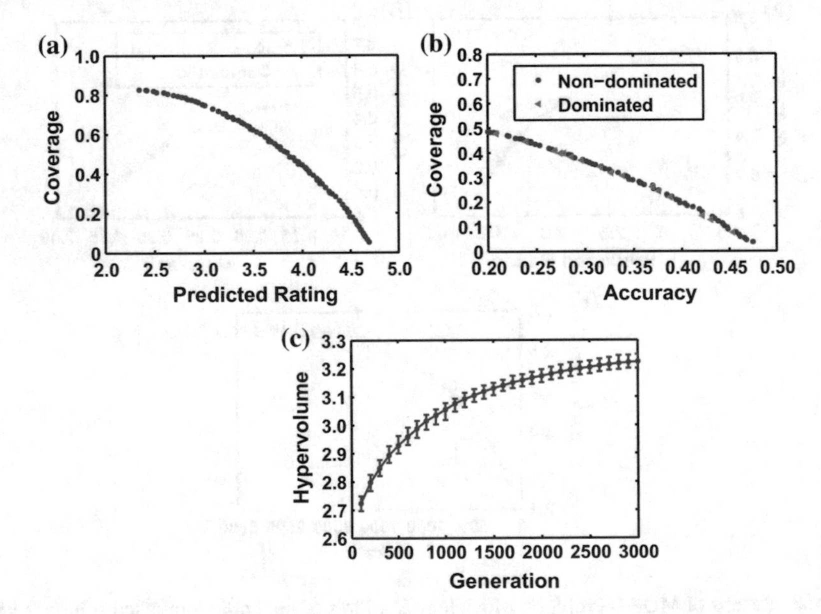

Fig. 6.5 Results of MOEA-ProbS on Movielens 4. **a** Plots of final non-dominated solutions with the highest hypervolume. **b** Plots of final solutions in the accuracy-coverage space (**c**) The error-bar of hypervolume metric of population among 30 independent runs with different generations

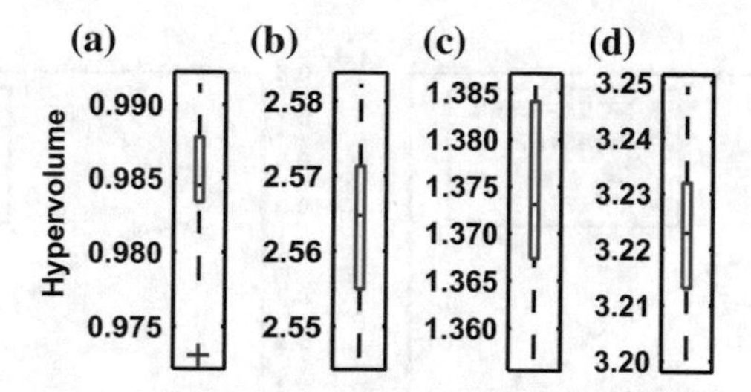

Fig. 6.6 Statistical values of hypervolume for four data sets. **a** Movielens 1. **b** Movielens 2. **c** Movielens 3. **d** Movielens 4. On each box, the central mark is the median and the edges of the box mean the 25th and 75th percentiles. The whiskers extending to the most extreme data points are not outliers. The outliers are denoted by symbol "+"

of non-dominated solutions can also form a non-dominated front. However, there may exist some dominated points in the accuracy-coverage space. The reason is that the predicted rating is not exactly equal to the true accuracy of recommendations. For example in Fig. 6.2 (b), the points denoted by green left triangle represent the dominated solutions. According to the error bars of hypervolume metric in Figs. 6.2, 6.3, 6.4 and 6.5, the proposed MOEA-ProbS is robust and effective, which can be proved further by the box plots in Fig. 6.6.

To show the advantages of the proposed MOEA-ProbS, MOEA-ProbS is compared with several well-known recommendation techniques, including CF [30] and matrix factorization method (MF) [37], which can produce only one solution. In addition, a hybrid recommendation algorithm [64] is selected as a comparative algorithm, which combines an accuracy-based method (ProbS) and a diversity-focused method (HeatS). For convenience, the hybrid algorithm is denoted by ProbS+HeatS. A weight parameter $\lambda \in [0, 1]$ is used to incorporate these two algorithms with completely different features. Different from searching for one optimal λ through extensive experiments in [64], a number of λ evenly sampled in $[0, 1]$ is generated. Then the experiments with different λ are conducted to get a set of recommendations. A non-dominated front can be obtained by eliminating the dominated points in the accuracy-diversity or accuracy-novelty space. For fair comparison, the number of different λ is equal to the size of population used in MOEA.

In the experiments, three performance metrics are considered, namely, accuracy, coverage and novelty. As displayed in Figs. 6.7 and 6.8, the solutions of CF and MF are dominated by those of MOEA-ProbS on all the data sets, which demonstrates the effectiveness of our algorithm. Figure 6.7 shows that MOEA-ProbS is able to generate multiple recommendations with higher coverage and similar accuracy compared to ProbS+HeatS. This is a promising property, especially for online business. Diverse items can be discovered to stimulate the purchase desire of customers. However, MOEA-ProbS is beaten by ProbS+HeatS according to the accuracy metric. The reason is twofold. First, the performance of MOEA-ProbS is mainly influenced by the introduced accuracy-based recommendation technique. Hybrid recommendation methods, which have been proved to provide more accurate recommendations [12],

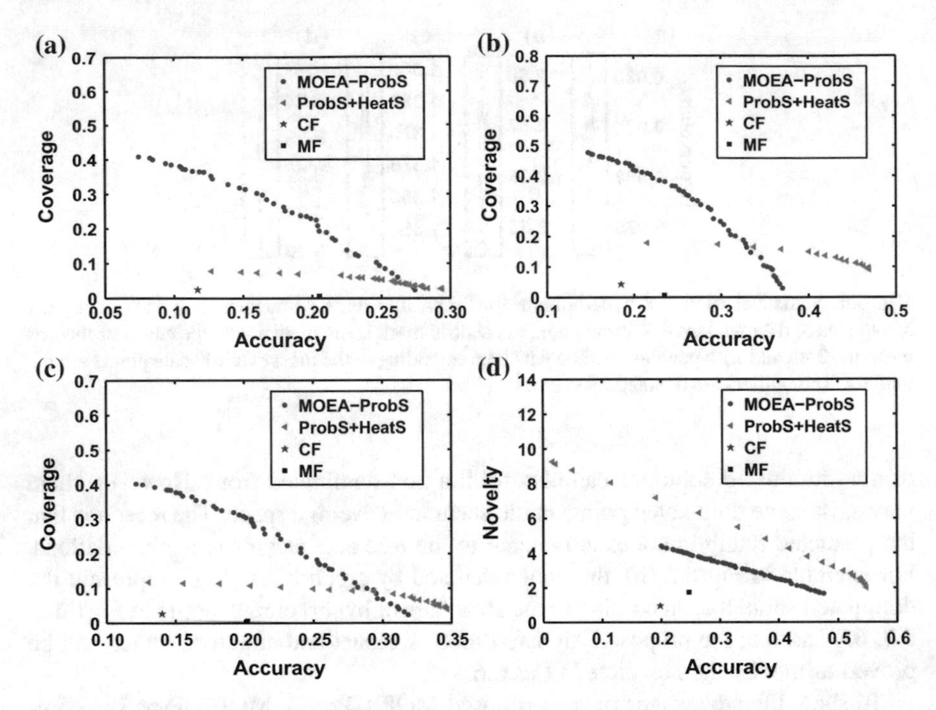

Fig. 6.7 Final non-dominated solutions of CF, MF, MOEA-ProbS, and ProbS+HeatS in the accuracy-coverage space. **a** Movielens 1. **b** Movielens 2. **c** Movielens 3. **d** Movielens 4

can be employed in this model. Second, the large-scale search space may cause difficulty. In order to improve the search ability, MOEAs should be elaborately designed and some local search methods can be taken into account. According to the values of hypervolume reported in Table 6.4, the proposed MOEA-ProbS gains a slight superiority over ProbS+HeatS in terms of accuracy and coverage.

However, Fig. 6.8 shows the deficiency in generating novel recommendations compared to ProbS+HeatS. This is due to the use of HeatS, which is inclined to suggest less popular items. In fact, high novelty can be obtained by recommending items as less popular as possible to users. Particularly, the value of novelty reaches the maximum, when only HeatS works ($\lambda = 0$). Nevertheless, it will result in a low accuracy, which can be observed by the extreme point close to y-axis in Fig. 6.8. In contrast, the accuracy of recommendations obtained by MOEA-ProbS is well maintained. Note that the performance of ProbS+HeatS is determined by the parameter λ, which varies with the data sets. However, it is difficult and computationally expensive to choose a suitable parameter λ in the training stage. Moreover, several runs need to be performed to get multiple recommendations. Different from ProbS+HeatS, the proposed MOEA-ProbS can make multiple recommendations in one run without additional parameters. To improve the novelty of recommendations obtained by MOEA-ProbS, a good way is to introduce the novelty as the third objective to be optimized, which will be discussed in the following subsection.

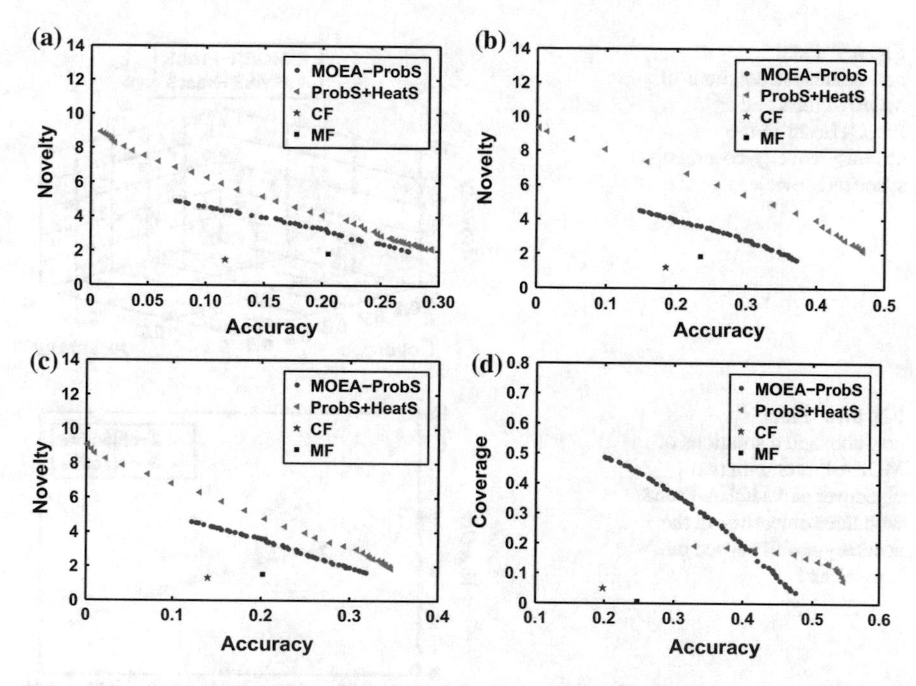

Fig. 6.8 Final non-dominated solutions of CF, MF, MOEA-ProbS and ProbS+HeatS in the accuracy-novelty space (**a**) Movielens 1. **b** Movielens 2. **c** Movielens 3. **d** Movielens 4

Table 6.4 Results of CF, MF, ProbS+HeatS and ProbS-MOEA on four data sets. HV denotes hypervolume in the accuracy-coverage space

	Movielens 1		Movielens 2		Movielens 3		Movielens 4	
Method	HV	Time(s)	HV	Time(s)	HV	Time(s)	HV	Time(s)
CF	–	7.36	–	10.95	–	8.30	–	10.43
MF	–	3.51	–	6.27	–	4.78	–	5.62
ProbS+HeatS	0.0198	927.51	0.0772	1538.42	0.0361	1156.27	0.0958	1573.05
MOEA-ProbS	0.0794	3636.76	0.1384	4233.58	0.0958	3869.20	0.1763	4198.32

In addition, the computational time in seconds (Time) required by each algorithm is given in Table 6.4, where the execution time of ProbS+HeatS is the total time used by multiple runs to get multiple recommendations. It is evident that ProbS+HeatS and MOEA-ProbS are much more time-consuming than CF and MF. For ProbS+HeatS, the combination of two algorithms increases the computational burden, and the implementation of multiple runs brings about expensive time cost further. Due to the complexity of the modeled MOP, some computational cost is needed for our algorithm to search for a set of optimal solutions. Note that the computational time of ProbS+HeatS for selecting suitable parameter λ is not involved. The efficiency of the proposed MOEA-ProbS is thus comparable to that of ProbS+HeatS.

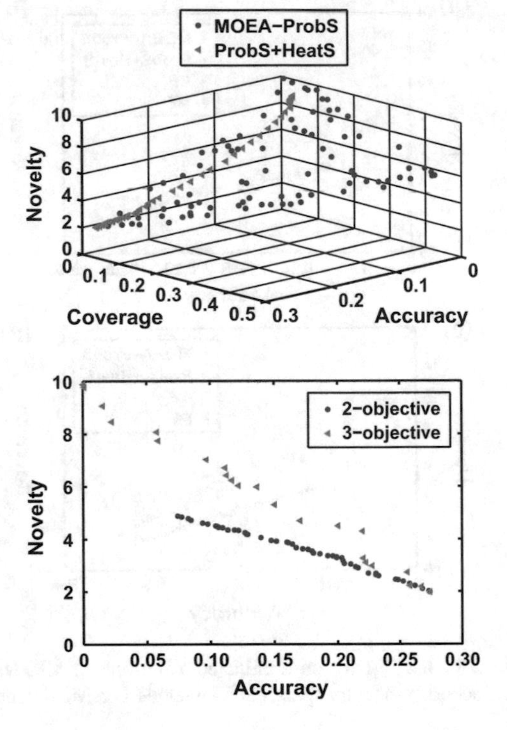

Fig. 6.9 Final non-dominated solutions of MOEA-ProbS and ProbS+HeatS in the accuracy-novelty-coverage space on Movielens 1

Fig. 6.10 Final non-dominated solutions of MOEA-ProbS with two objectives and MOEA-ProbS with three objectives in the accuracy-novelty space on Movielens 1

6.2.6.3 Discussions of Objective Functions

In this subsection, the extended three-objective model is considered, by introducing the novelty as the third objective. The maximum number of generation is set to be 6000, and other parameters remain the same as in Table 6.3. The results of MOEA-ProbS and ProbS+HeatS on Movielens 1 are displayed in Fig. 6.9. It can be observed that the solutions of ProbS+HeatS in the three-dimensional space form a curve, while those of MOEA-ProbS form a two-dimensional surface. This reveals the effectiveness of the extended three-objective model. Similar results are also observed for other data sets. In addition, to test the performance of recommendation novelty, the results of bi-objective and three-objective models in the accuracy-novelty space are presented in Fig. 6.10. It is obvious that the novelty of recommendations is greatly improved by introducing the third objective. However, little effect is obtained for the solutions with a relatively high precision. Moreover, it will inevitably increase the computational cost in the three-objective model.

6.2.6.4 Impact of the Clustering Method

To investigate the impact of the clustering method, several experiments are carried out in this subsection. First, CF and ProbS are used to test the rationality of using the clustering method. More specifically, the results obtained by CF and ProbS are

Table 6.5 Results of CF_C, CF, ProbS_C, and ProbS on Movielens

	Movielens 1		Movielens 2		Movielens 3		Movielens 4		Movielens
Method	Accuracy	Time(s)	Accuracy	Time(s)	Accuracy	Time(s)	Accuracy	Time(s)	Time(s)
CF_C	0.115	7.36	0.185	10.95	0.140	8.30	0.198	10.43	37.04
CF	0.070	–	0.164	–	0.128	–	0.192	–	104.72
ProbS_C	0.270	12.48	0.416	15.24	0.317	12.98	0.490	15.75	56.45
ProbS	0.274	–	0.371	–	0.318	–	0.475	–	82.19

compared with those by their variants using the clustering method, denoted as CF_C and ProbS_C, respectively. The k-means clustering method is also applied. The detail results, including the accuracy and the computational time required by the related algorithms, are reported in Table 6.5. Note that the last column in Table 6.5 displays the computational time of the related algorithms run on the complete Movielens data set. For CF_C and ProbS_C, the time presented in the last column is the sum of time used on the four data sets. From Table 6.5, it can be concluded that the clustering method can reduce the computational time greatly. Actually, the computational time required by CF is more than the sum of time used by CF_C on the four data sets. Similar performance is also observed for ProbS. In particular, the accuracy of CF is improved for all the four data sets by using the clustering method. Since CF is based on the similarity of users, more effective information from similar users in the same cluster are available to produce better results. For ProbS, by using the clustering method, the accuracy is improved on Movielens 2 and Movielens 4 while slightly decreased on other data sets.

Next, the MOEA-ProbS is compared with its variant without clustering (MOEA-ProbS-noC). MOEA-ProbS-noC is tested on Movielens 3 with a computational time limit, which is set to be twice the sum of time required by MOEA-ProbS on the four data sets. As shown in Fig. 6.11, MOEA-ProbS-noC performs slightly better than MOEA-ProbS in terms of coverage metric. It is easy to understand that as more users participate in rating items, more items will be found and recommended to users, leading to higher coverage. Nevertheless, the accuracy of recommendations becomes worse by using the clustering method.

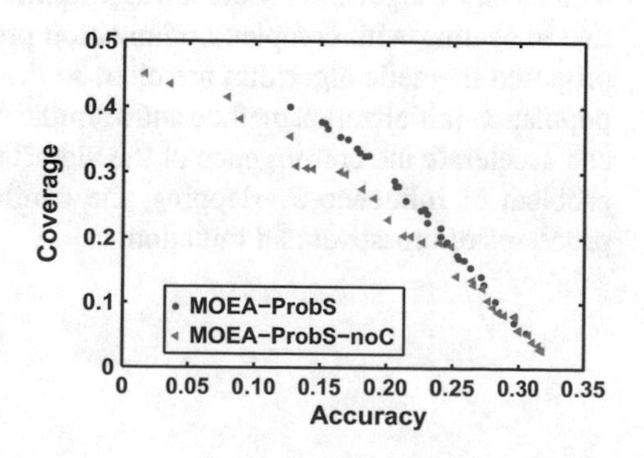

Fig. 6.11 Final non-dominated solutions of MOEA-ProbS and MOEA-ProbS-noC in the accuracy-coverage space on Movielens 3

6.2.7 Conclusions

This section introduces a general multiobjective recommendation model to simultaneously optimize recommendation accuracy and diversity. The accuracy was predicted by the ProbS method while the diversity was measured by recommendation coverage. NSGA-II was adopted to solve the modeled MOP for personalized recommendation. To reduce the computational cost, the k-means clustering technique is used to split the users into several relatively small clusters. The proposed MOEA-based recommendation algorithm can make multiple recommendations for multiple users in only one run. The experimental results show that MOEA-ProbS can provide a set of diverse and accurate recommendations for users. In addition, the experiments for the clustering method indicate that it can improve the algorithmic efficiency.

6.3 Influence Maximization in Social Networks with Evolutionary Optimization

This section introduces a novel memetic algorithm for influence maximization in social networks, termed as CMA-IM. Figure 6.12 provides the framework of CMA-IM, which comprises three steps: (1) Network clustering; (2) Candidate selection; and (3) Seed generation. In the first step, a fast two-phase heuristic algorithm BGLL [8] is used to detect communities in networks. In the second step, select the communities that are significant and then choose a few nodes from each significant community to form the candidate pool. In the final step, the influence maximization problem is modeled as the optimization of a 2-hop influence spread [41] and a problem-specific memetic algorithm is proposed to find the ultimate seeds. CMA-IM combines a genetic algorithm as the global search method and a similarity-based strategy as the local search procedure.

The advantages of CMA-IM are as follows: A novel memetic algorithm for the community-based influence maximization problem is proposed. To the best of the knowledge, CMA-IM is the first attempt for dealing with the influence maximization by a memetic algorithm. Memetic algorithms have already been proved to be effective in dealing with complex optimization problems, so solutions generated by the proposed memetic algorithm are close to the optimal solution. A problem-specific population initialization method and a similarity-based local search procedure, which can accelerate the convergence of the algorithm, are designed. In order to solve the problem of influence overlapping, the similarity-based strategy is applied to the processes of crossover and mutation.

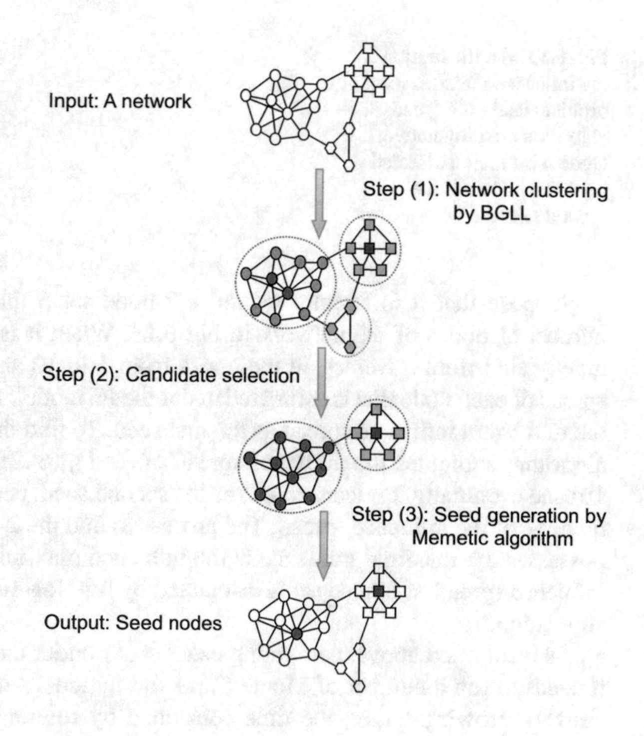

Fig. 6.12 The framework of CMA-IM. In Step (1), the input network is divided into communities by network clustering algorithm BGLL. In Step (2), candidates are selected from each significant community. And in Step (3), the ultimate seed set is determined by the proposed memetic algorithm

6.3.1 *Memetic Algorithm for Influence Maximization in Social Networks*

A social network is modeled as an undirected network denoted as $G = (V, E)$ where V represents the node set denoting users in the social network and E represents the edge set denoting the relationships between users [14]. n and m are the number of nodes and edges, respectively.

Given a node set S that includes k nodes, the influence spread produced by S which is denoted as $\sigma(S)$ is the number of nodes that S can influence in the network G under a certain cascade model. In CMA-IM, the IC model is selected as the cascade model, which is widely used in the previous works [14, 28, 33, 41, 43]. In the IC model, the state of a node has only two types, either active or inactive. The inactive nodes can be changed into the active nodes, but not vice versa. Each edge is associated with a propagation probability p and $p(u, v)$ represents the probability of an inactive node v to be influenced by its active neighbor u. Under the IC model, $\sigma(S)$ works as follows. Let S_t be the set of nodes that are active in the step t, and $S_0 = S$ which is initialized by a k-node set. At step t, each node $v \in S_{t-1}$ has only one chance to independently activate every inactive neighbors with the probability p. This influence diffusion process stops at the step t when $S_t = \emptyset$ and $\sigma(S)$ is the union of S_t obtained at each step. Thus, the influence maximization problem is to find the k-node set S that can make $\sigma(S)$ maximal under the IC model. Let us illustrate this problem using the greedy algorithm as an example on a toy network in Fig. 6.13.

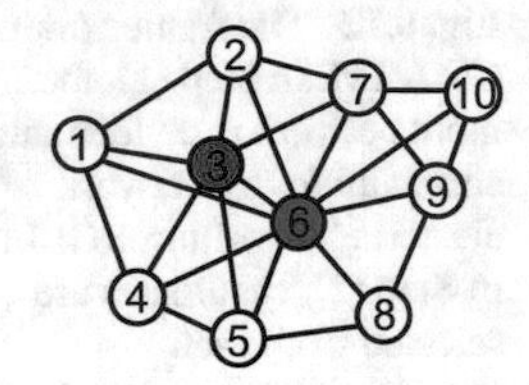

Fig. 6.13 An illustration for the influence maximization problem using the greedy algorithm on a toy network. Node 6 is the first selected seed and node 3 is the second selected node

Suppose that it is attempt to find a 2-node set S that can influence the most number of nodes of the network in Fig. 6.13. When it is to find the first seed, the greedy algorithm traverses all the nodes from 1 to 10 and calculates the influence spread of each node that is estimated under the IC model. It selects node 6 which has the maximum influence spread as the first seed. To find the second seed, the greedy algorithm computes the influence spread of (6, 1), (6, 2), ... , (6, 5), (6, 7), ... , (6, 10) and eventually it selects node 3 as the second seed, because it results in the most increase of the influence spread. The process to find the 2-node set $S = \{6, 3\}$ which possesses the maximal influence is the influence maximization problem. Here, the influence spread of a node set is calculated by running 10,000 times a Monte Carlo simulation.

As mentioned above, computing exact $\sigma\ (S)$ under the IC model is #P-hard and it needs to run a number of Monte Carlo simulations. And with the size of a social network growing larger, the time consumed by running Monte Carlo simulation becomes not negligible.

In [41], Lee et al. propose a fast approximation for influence maximization, they consider the influence spread on the nodes within 2-hops away from seed node set instead of all nodes in the network. The 2-hop influence spread of a node set $\hat{\sigma}\ (S)$ is calculated as Eq. 6.7.

$$\hat{\sigma}\ (S) = \sum_{s \in S} \hat{\sigma}\{s\} - \left(\sum_{s \in S} \sum_{c \in C_s \cap S} p\ (s, c) \left(\sigma_c^1 - p\ (c, s) \right) \right) - \chi, \tag{6.7}$$

where C_s denotes the 1-hop nodes cover of node s, i.e., the neighbors of node s, p is the propagation probability in the IC model, $\chi = \sum_{s \in S} \sum_{c \in C_s \cap S} \sum_{d \in C_c \cap S \setminus \{s\}} p\ (s, c)\ p\ (c, d)$ and $\sigma_c^1 = 1 + \sum_{c \in C_u} p\ (u, c)$. The term σ_c^1 is the 1-hop influence spread of node c. In Eq. 6.7, the first term is the sum of the 2-hop influence spread of each seed in S, the second term considers the redundant situation that a seed is a neighbor of another seed and the third term considers the redundant situation that a seed is 2-hops away from another seed.

The authors show that the 2-hop influence spread is sufficiently valid and efficient to estimate the influence spread of a node set. Therefore, the 2-hop influence spread is used in CMA-IM. The important variables mentioned above are listed in Table 6.6.

As illustrated earlier in Fig. 6.12, CMA-IM consists of three steps: (1) Network clustering; (2) Candidate selection; (3) Seed generation. A detailed description of each step will be given in the following.

Table 6.6 the variables used in CMA-IM

Variables	Descriptions
$G(V,E)$	An undirected network with node set V and edge set E
N	The number of nodes in G
M	The number of edges in G
k	The number of seeds ($1 \leq k \leq N$)
S	The seed set with k-node
p	Propagation probability

6.3.2 Network Clustering

BGLL proposed by Blondel et al. [8] is a fast heuristic method based on modularity optimization, which consists of two phases. At the first phase, each node of the network is considered as a community. Then they remove a node from its original community to its neighbor's community which has the maximal positive gain in modularity. This phase is applied repeatedly for all nodes until no further improvement can be achieved. The first phase is then complete [8]. The second phase considers the communities obtained in the first phase as nodes such that a new network can be built. Then BGLL runs these two phases iteratively until achieving an unchanged result and obtaining the maximal modularity.

BGLL algorithm can discover natural structures of networks because it needs no prior knowledge about the community number. So communities obtained by the BGLL get closer to the inherent communities in networks. Meanwhile BGLL only needs a few iterations to obtain a maximal modularity, which makes the BGLL algorithm have better performance in terms of efficiency when applied to large-scale networks.

6.3.3 Candidate Selection

The candidate selection step aims to determine a set of candidate nodes according to the information about communities obtained in the first step. Because social networks in realistic settings are usually extremely huge, the search space for selecting seeds is also huge. Therefore, there is a need to effectively reduce the number of candidate nodes.

By analyzing the structures of communities, it is found that not all communities are significant enough to accommodate seed nodes. For example, in Fig. 6.14, although the network is divided into three communities, community 3 may be insignificant compared with community 1 and 2 due to its smaller community size. Suppose a seed node from community 3 is chosen, it may only activate three nodes initially. The community 1 and 2 are chosen as the significant communities because their sizes are large and there may be more influential nodes in them. Here, significant communities

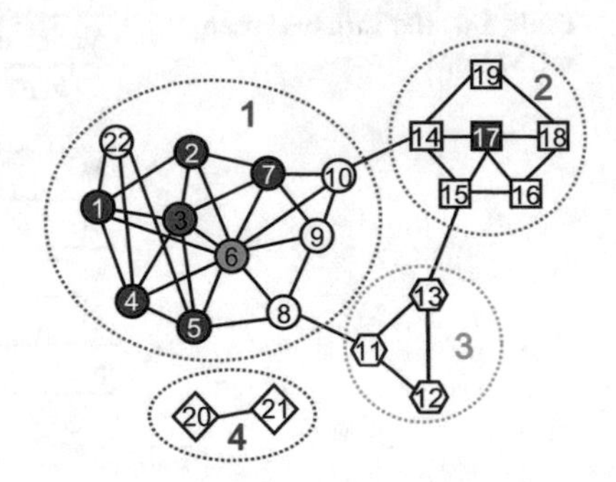

Fig. 6.14 A network with three communities. The communities are numbered as 1, 2, and 3, respectively. The *red nodes* are common neighbors of the node 3 and the node 6

are defined as the first n large communities, where n is varied with networks. However, the nodes in community 1 have many common neighbors. For example, node 3 and node 6 have 5 common neighbor nodes, which is not beneficial to influence spread. In order to solve the problem of overlapping influence, a similarity-based high degree method called SHD is used.

Let *Candidate* be a candidate node pool. Next, the task is to choose a number of potential nodes from each significant community to fill in the *Candidate*. If considering the influence ability of a node, degree centrality may be the most intuitive way to estimate the ability. The higher degree of a node is, the more neighbors the node has, which means that node with higher degree can influence more nodes with the same propagation probability. Therefore, potential nodes are chosen from each significant community based on the degree centrality. Here, the way of [51] is chosen, which is shown in Eq. 6.8, to decide the number of candidate nodes selected from each significant community.

$$\left(\frac{C_i - MinC}{MaxC - MinC}\right) * \beta + \alpha \tag{6.8}$$

where C_i is the size of the i-th significant community, $MaxC$ is the size of the largest significant community and $MinC$ is the size of the smallest significant community. The term $\left(\frac{C_i - MinC}{MaxC - MinC}\right)$ is the ratio of the i-th community among all selected communities and its value is confined to [0, 1]. β is the amplification term and α is the constant term that guarantees the least selection in each community. After selecting the *Candidate*, the last step of CMA-IM is to generate the ultimate seeds.

6.3.4 Seed Generation

After the two aforementioned steps, the search space has been reduced. In the next step, the proposed problem-specific memetic algorithm is employed, named as Meme-IM, to generate the ultimate seeds by optimizing the 2-hop influence spread shown as Eq. 6.7. The whole framework of Meme-IM is shown as Algorithm 26.

Algorithm 26 Framework of Meme-IM

Input: Maximum generation: $maxgen$, population size: pop, mating pool size: $pool$, tournament size: $tour$, crossover probability: pc, mutation probability: pm, spread probability: p, seed size: k, the candidate nodes pool: $Candidate$ and the connection matrix: A.
Output: The most influential k-node set.
1: **Step 1) Initialization**
2: **Step 1.1)** Population initialization:
$\boldsymbol{P} = \{\boldsymbol{x}_1, \boldsymbol{x}_2, ..., \boldsymbol{x}_{pop}\}^T$;
3: **Step 1.2)** Best individual initialization: $\boldsymbol{P}_{best} = \boldsymbol{x}_i$;
4: **Step 2) Set** $t = 0$; // the number of generations
5: **Step 3) Repeat**
6: **Step 3.1)** Select parental chromosomes for mating;
$\boldsymbol{P}_{parent} \leftarrow$ Selection($\boldsymbol{P}$, $pool$, $tour$);
7: **Step 3.2)** Perform genetic operators:
$\boldsymbol{P}_{child} \leftarrow$ GeneticOperation($\boldsymbol{P}_{parent}$, pc, pm);
8: **Step 3.3)** Perform local search:
$\boldsymbol{P}_{new} \leftarrow$ LocalSearch($\boldsymbol{P}_{child}$);
9: **Step 3.4)** Update population:
$\boldsymbol{P} \leftarrow$ UpdatePopulation($\boldsymbol{P}$, $\boldsymbol{P}_{new}$);
10: **Step 3.5)** Update the best individual $\boldsymbol{P}_{best}$;
11: **Step 4) Stopping criterion**: If $t < maxgen$, then
$t = t + 1$ and go to **Step 3)**, otherwise, stop the algorithm and output.

In **Step 1)**, Meme-IM mainly completes the population initialization task. First, it creates the initial population of solutions $\boldsymbol{P} = \{\boldsymbol{x}_1, \boldsymbol{x}_2, ..., \boldsymbol{x}_{pop}\}^T$ according to a problem-specific strategy. And then it selects the individual with the maximum fitness as $\boldsymbol{P}_{best}$. **Step 3)** is the evolution procedure. In **Step 3.1)**, Meme-IM first uses the deterministic tournament selection method to select parental individuals $\boldsymbol{P}_{parent}$ for mating in genetic algorithm. Then in **Step 3.2)**, Meme-IM reproduces the chosen parental individuals $\boldsymbol{P}_{parent}$, i.e., performs crossover and mutation operation on $\boldsymbol{P}_{parent}$. **Step 3.3)** is an individual reinforcement procedure. **Step 3.4)** is to refresh the current population by taking the best pop individuals from $\boldsymbol{P} \cup \boldsymbol{P}_{new}$. And in **Step 4)**, when the algorithm terminates on convergence, Meme-IM stops and outputs the ultimate k-node set.

6.3.4.1 Representation and Initialization

In Meme-IM, each chromosome (individual) $\boldsymbol{x}_a$ ($1 \leq a \leq pop$) in the population represents a k-node set, which is encoded as an integer string

$$x_a = \{x_a^1, x_a^2, ..., x_a^k\},$$

where k is the number of seeds, each gene x_a^i of the chromosome corresponds to a node selected from $Candidate$. An illustration of this representation is shown as Fig. 6.15. It is noticed that there is no repeated node in $\boldsymbol{x}_a$. Considering that the

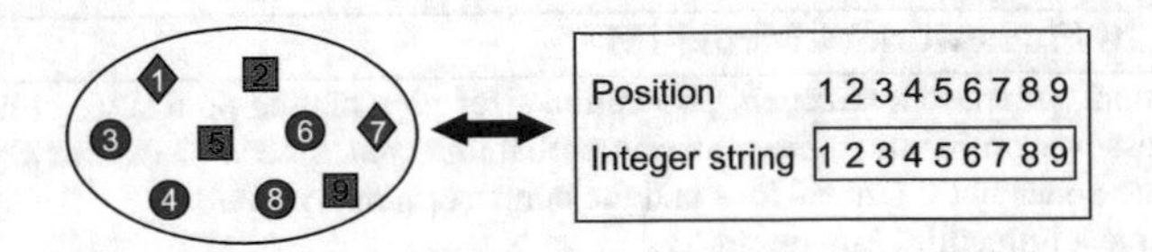

Fig. 6.15 Illustration of the representation. *Left* a 9-node set selected from *Candidate*. *Right* the individual encoding of the 9-node set

solution produced by selecting k nodes randomly from *Candidate* is of low quality and may result in a long time to converge, we attempt to initialize a higher quality population to speed up the convergence. Here, a similarity-based high degree method called SHD is proposed. The SHD and the random mechanism can guarantee the convergence and diversity of the individuals. The population initialization procedure is shown as Algorithm 27.

Algorithm 27 Population Initialization

Input: Population size: *pop*
Output: Population $\boldsymbol{P}$
1: Generate a half of population based on SHD, see **Algorithm 28** for more information;
2: **for** i from 1 to $(pop/2)$ **do**
3: **for** j from 1 to k **do**
4: **if** $rand\,(1) > 0.5$ **then**
5: select a random node different from each node in $\boldsymbol{x}_i$ from the *Candidate* to replace x_i^j;
6: **end if**
7: **end for**
8: **end for**
9: **for** i from $(pop/2 + 1)$ to pop **do**
10: select k different nodes from the *Candidate* to initialize $\boldsymbol{x}_i$ based on SHD;
11: **end for**

High degree centrality is a standard method for influence maximization on social and other networks [6]. But high degree centrality may produce overlapping influence spread between nodes. To solve this problem, a similarity-based high degree method (SHD) is proposed. SHD starts with choosing the node with the highest degree in *Candidate*. After choosing a node, it excludes the neighbor nodes that are similar with the existing nodes. Then SHD chooses the next node with the highest degree in the left candidate nodes. This procedure iteratively operates until k nodes are chosen. Here, the degree of similarity between nodes is measured by the structure similarity defined as Eq. 6.9. The process is described as Algorithm 28.

Algorithm 28 SHD algorithm

1: Start with $\boldsymbol{x}_a = \emptyset$;
2: $TempCandidate = Candidate$;
3: **for** i from 1 to k **do**
4: choose a node $v_i \in TempCandidate$ with the highest degree;
5: $\boldsymbol{x}_a \leftarrow \boldsymbol{x}_a \cup \{v_i\}$;
6: $SimNeighbor \leftarrow \{u \in N(v) \,|Similarity(u, v) \geq sim\}$;
7: $TempCandidate \leftarrow \{v|v \in TempCandidate, v \notin SimNeighbor, v \neq v_i\}$;
8: **if** $TempCandidate = \emptyset$ **do**
9: $\boldsymbol{x}_a \leftarrow \boldsymbol{x}_a \cup \{v_{i+1}, v_{i+2}, \cdots, v_k\}, v_{i+1}, v_{i+2}, \cdots, v_k$ are selected from *Candidate* randomly;
10: **break**;
11: **end if**
12: **end for**
13: return $\boldsymbol{x}_a$

In Algorithm 28, $N(v) = \{\exists u \in V, uv \in E\}$ represents the neighbors of node v and the structural similarity between nodes u and v are defined as Eq. 6.9.

$$Similarity(u, v) = \frac{|NB(u) \cap NB(v)|}{|NB(u)| + |NB(v)|}, \tag{6.9}$$

where $NB(v) = \{v|v \cup N(v)\}$ includes the node v and its neighbors. In Algorithm 28, *sim* is a threshold confined to [0, 1] which is set according to different data sets. When the structure similarity between two nodes is larger than *sim*, the nodes are similar. The similarity between nodes is a significant criterion to present from overlapping influence spread.

6.3.4.2 Genetic Operators

Crossover. In Meme-IM, the one-point crossover is employed because it is simple. The one-point crossover works as follows. Given two parent chromosomes $\boldsymbol{x}_a$ and $\boldsymbol{x}_b$, first randomly select a crossing over position i $(1 \leq i \leq k)$, then exchange each node j after the position i between the two parents, i.e., $\left(\boldsymbol{x}_a^j \leftrightarrow \boldsymbol{x}_b^j, \forall j \in \{j|i \leq j \leq k\}\right)$, and then two new offspring chromosomes $\boldsymbol{x}_c$ and $\boldsymbol{x}_d$ return. It needs to guarantee the validity of $\boldsymbol{x}_c$ and $\boldsymbol{x}_d$, i.e., there are no same nodes in $\boldsymbol{x}_c$ and $\boldsymbol{x}_d$, respectively. Specifically, when the j-th node in $\boldsymbol{x}_b$ $(\boldsymbol{x}_a)$ is not similar to the nodes in $\boldsymbol{x}_a$ $(\boldsymbol{x}_b)$ except for x_a^j (x_b^j), then we exchange x_b^j and x_a^j.

Mutation. Here, the similarity-based mutation is employed on the generated population after crossover. For each gene in a chromosome, if the generated random value $r \in [0, 1]$ is smaller than the mutation probability pm, then the gene is mutated to another gene in *Candidate*. The other gene is selected randomly from the genes which are dissimilar with the genes in the chromosome. The similarity is evaluated by Eq. 6.9. However, when the value of r is larger than pm, there is also a situation we mutate the gene. When the gene is similar with the most influential gene in the chromosome, the gene is mutated to another gene in *Candidate* which is dissimilar with the most influential gene and its similar neighbors.

6.3.4.3 Local Search Procedure

For a chromosome, when a node is changed to another node in the *Candidate* which is different from any node in the chromosome, a neighbor of the chromosome is obtained. Here, a similarity-based strategy is employed as the local search procedure. The procedure is performed on an arbitrary individual in the population, then it attempts to find a better individual from the neighborhoods of the individual. The fitness is calculated by Eq. 6.7 and the neighbor individual is better if its fitness is larger than that of the original individual. If the change can produce a better individual, this change is accepted. The procedure repeats until no further improvement can be made. Here, find the fittest chromosome in $\boldsymbol{P}_{child}$ which is obtained after the genetic operators and apply the local search procedure on it. The implementation of the local search procedure is shown in Algorithm 29.

Algorithm 29 The Local Search Procedure

Input: $\boldsymbol{P}_{child}$
Output: $\boldsymbol{P}_{child}$
1: $N_{current} \leftarrow FindBest\ (\boldsymbol{P}_{child})$;
2: $islocal \leftarrow FALSE$;
3: **repeat**
4: $N_{next} \leftarrow FindBestNeighbor\ (N_{current})$;
5: **if** $Eval\ (N_{next}) > Eval\ (N_{current})$;
6: $N_{current} \leftarrow N_{next}$;
7: **else**
8: $islocal \leftarrow TRUE$;
9: **end if**
10: **until** *islocal* is *TRUE*

In Algorithm 29, first perform FindBest() function to select the individual with the maximum fitness in the input individuals. Then apply the local search procedure on it. The Eval() function is to compute the fitness of a solution based on Eq. 6.7. The FindBestNeighbor() function is to find the best neighbor chromosome with the largest fitness value.

6.3.5 Experimental Results

In this section, the effectiveness and the efficiency of CMA-IM are tested and the influence spread and the running time are compared with other six algorithms on three real-world social networks, Dolphin network [45], NetGRQC network [42] and NetHEPT network [14].

The basic characteristics of the real-world networks described above are given in Table 6.7.

Six comparison algorithms are CGA-IM, MA-IM, CELF [43], CMA-HClustering [16], CMA-SLPA [51], Degree centrality [33], PageRank [11], Random [33]. CGA-IM is the variant version of CMA-IM by removing the local search procedure and MA-IM is the variant of CMA-IM by removing the network clustering step. CMA-IM with CGA-IM and MA-IM are compared on two real-world networks to demonstrate the effectiveness of the network clustering and the local search procedure in CMA-IM.

All experiments are implemented under the IC model. In order to compare the accuracy of different algorithms, the influence spread of the ultimate k-node set of each algorithm is computed by running Monte Carlo simulation for 10,000 times and the average influence spread is taken. All algorithms are independently run 30 times on each network. All the experiments are conducted on a PC with 1.70 GHz Inter Core i5 and 8.00 GB Memory. The experimental parameters of our algorithm are listed in Table 6.8.

Table 6.7 Statistics of the three real-world networks

Network	Nodes	Edges	Average degree
Dolphin	62	159	5.129
NetGRQC	5242	14496	5.5261
NetHEPT	15233	58891	3.8635

Table 6.8 The parameters of the algorithm

Parameter	Meaning	Value
α	The constant term in (2)	4
β	The amplification term in (2)	10
G_{max}	The number of iterations	50
S_{pop}	Population size	200
S_{pool}	Size of the mating pool	100
S_{tour}	Tournament size	2
P_c	Crossover probability	0.8
P_m	Mutation probability	0.2

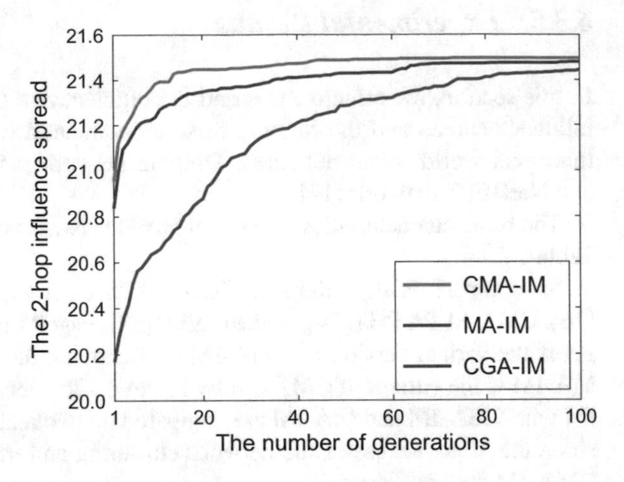

Fig. 6.16 Comparisons between CMA-IM and its variants CGA-IM and MA-IM in terms of convergence on the Dolphin social network, respectively

6.3.5.1 Experiments on Real-World Networks

In order to compare the convergence of CMA-IM with its two variants CGA-IM and MA-IM, the parameter *maxgen* is set larger than the original setting. However, the parameters in each pair of comparisons remain the same.

First, the comparison between CMA-IM and its variant CGA-IM is made to show the effectiveness of the local search procedure. CMA-IM and CGA-IM are tested on two real-world networks, the Dolphin social network and the NetHEPT network. In these experiments, *maxgen* is set as 100 for the Dolphin network and 1,000 for the NetHEPT network. The results are shown as the green and blue lines in Figs. 6.16 and 6.17.

The green and blue lines in Fig. 6.16 show the 2-hop influence spread obtained by CMA-IM and CGA-IM with generation increasing from 1 to 100 on the Dolphin social network. And the green and blue lines in Fig. 6.17 show the 2-hop influence spread obtained by CMA-IM and CGA-IM with generation increasing from 1 to 1,000 on the NetHEPT network. For the small Dolphin social network, when the generation is up to 100, the 2-hop influence spread produced by CMA-IM and CGA-IM does not differ a lot from each other. Both of them can reach an optimal solution. CMA-IM with local search can converge within 50 generations while CGA-IM without local search needs more generations. However, for the large NetHEPT network, CGA-IM cannot find the optimal solution within 1,000 generations while CMA-IM can evolve to a better solution efficiently within 50 generations. These results demonstrate that local search can speed up the convergence and produce a higher quality solution, especially when the search space is large.

Next, the comparison between CMA-IM and its variant MA-IM is made to illustrate the effectiveness of the network clustering step. We test CMA-IM and MA-IM on two real-world networks, the Dolphin network and the NetHEPT network. The parameter *maxgen* is set as 100 for the Dolphin network and 1,000 for the NetHEPT network. The results are shown as the green and red lines in Figs. 6.16 and 6.17.

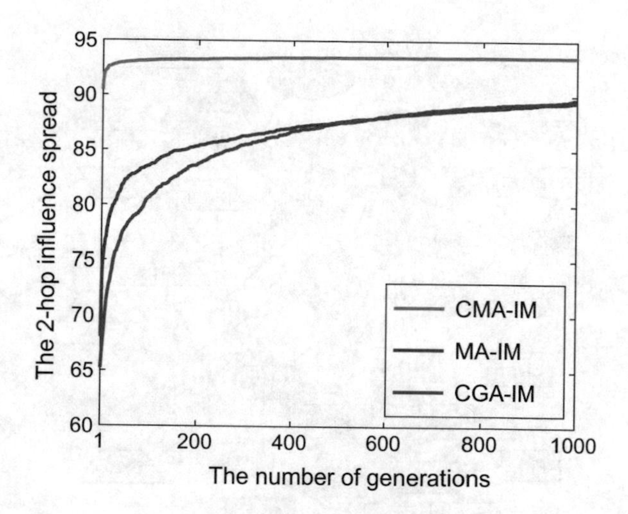

Fig. 6.17 Comparisons between CMA-IM and its variants CGA-IM and MA-IM in terms of convergence on the NetHEPT network, respectively

Table 6.9 The information about communities and candidates of the three networks

Network	Dolphin	NetGRQC	NetHEPT
Nodes	62	5242	15233
Seeds	10	30	30
Communities	5	392	1820
Significant communities	5	35	50
Candidates	47	227	315

The green and red lines in Fig. 6.16 show the 2-hop influence spread obtained by CMA-IM and MA-IM with generation increasing from 1 to 100 on the Dolphin social network. The green and red lines in Fig. 6.17 show the 2-hop influence spread obtained by CMA-IM and MA-IM with generation increasing from 1 to 1,000 on the NetHEPT network. In CMA-IM, network clustering step is used to narrow down the search space of the seed nodes. Table 6.9 gives the number of candidates resulted in reducing the search space. For the Dolphin social network, all the five communities are taken as significant communities. The search space is not reduced obviously, so MA-IM achieves a similar result as that of the CMA-IM. However, CMA-IM can converge within 50 generations while MA-IM needs more than 60 generations. For the NetHEPT network, the influence spread of CMA-IM is much better than that of MA-IM because the search space is reduced nearly 50 times. When the generation increases up to 1,000, MA-IM still cannot achieve the optimal solution. However, CMA-IM only needs less than 50 generations. Therefore, the combination of the network clustering and the memetic algorithm is effective that can improve the performance of CMA-IM apparently (Fig. 6.18).

Finally, the comparisons between CMA-IM and other six state-of-the-art algorithms are made. The influence spread and the running time of each algorithm with

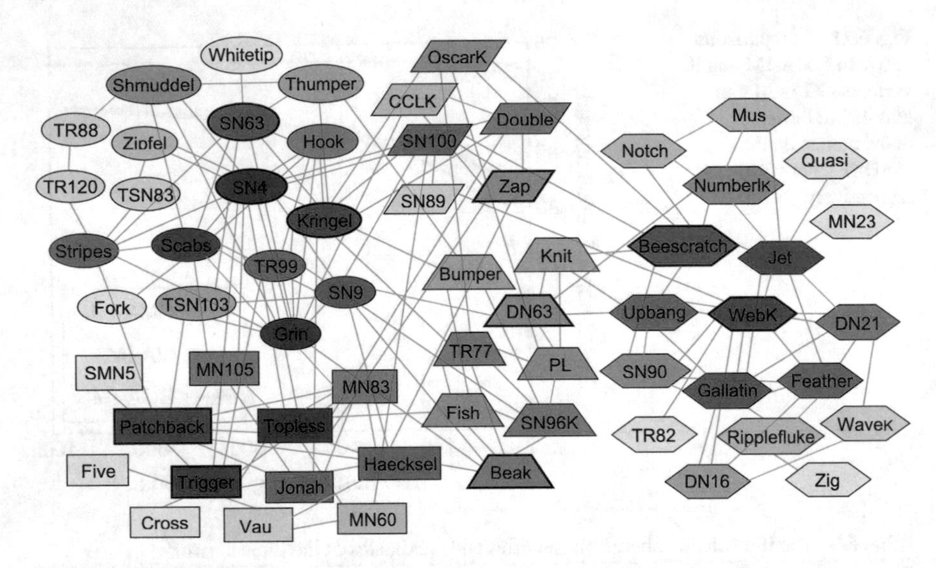

Fig. 6.18 An illustration for the community structure and the ultimate seeds of the Dolphin network. The Dolphin network clusters into five communities, the nodes of the same shape belong to the same community. The color of nodes from *dark to light* corresponds to the degree of nodes from large to small. The ten bold nodes are the seeds generated by CMA-IM

the number of seeds increasing on the three real-world networks are compared. In these experiments, the parameter *maxgen* is set as 50. Figures 6.19, 6.20, and 6.21 show the influence spread of seven algorithms on the three networks under the IC model whose x-axis represents the seed set size and y-axis represents the influence spread estimated by running Monte Carlo simulation for 10,000 times. And Fig. 6.22 shows the running time of seven algorithms whose x-axis represents seven algorithms on the three networks and y-axis represents the running time of each algorithm in log scale. The percentages below about influence spread are computed when the seed set size is 30 (The seed set size is 10 for the case of the Dolphin network).

Figure 6.18 is an illustration for the community structure and the ultimate seeds of the Dolphin network. From Fig. 6.18, it is seen that the network is partitioned into five communities by BGLL algorithm and the nodes of the same shape belong to the same community. The color of nodes from dark to light red corresponds to the degree of nodes from large to small. The bold nodes are the ultimate seeds generated by CMA-IM. The network is low scale and connected sparsely, and set the seed set size ranging from 1 to 10 and set the propagation probability p as 0.1.

Figure 6.19 shows the influence spread on the Dolphin network. From Fig. 6.19, it is seen that the differences in the influence spread of seven algorithms are not obvious. The result of CELF greedy algorithm is the best and CMA-IM essentially matches CELF. The results of CMA-IM, CMA-HClustering and CMA-SLPA are extremely close. Compared with other heuristics, CMA-IM is 2.3, 4.6, and 21.4% better than Degree centrality, PageRank and the baseline method Random, respectively. In Fig. 6.18, it is seen that the generated seeds belong to five communities

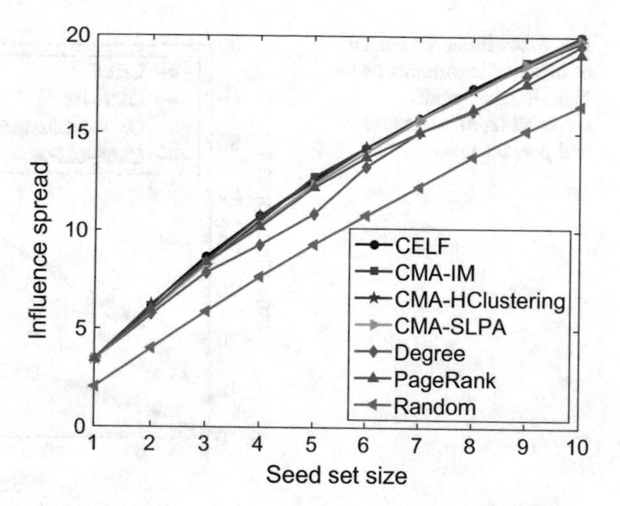

Fig. 6.19 Influence spread of different algorithms on the Dolphin network ($N = 62$, $M = 159$, and $p = 0.1$)

while the nodes with top 10 degree belong to three larger communities. It shows that selecting many top degree nodes in large communities may not influence more nodes. Clustering network into communities and selecting a number of top degree nodes in each significant community can get efficient candidate nodes. For the running time, Fig. 6.22 shows that the CELF is quite slow, CMA-IM is one order of magnitude better than CELF. CMA-IM, CMA-HClustering, and CMA-SLPA have close running time. The heuristic algorithms outperform CMA-IM in running time while their performances are poor.

Figure 6.20 shows the influence spread on the NetGRQC network. The result in Fig. 6.20 indicates that CELF produces the largest influence spread and the result of CMA-IM is close to that of CELF with 1.72% lower. CMA-IM outperforms CMA-HClustering and CMA-SLPA slightly. And CMA-IM outperforms the three heuristics, Degree, PageRank and Random with 32.3, 10.9 and 100.8% higher, respectively. It is found that the nodes with top 30 degrees are only within two communities while the seeds discovered by CMA-IM cover twelve communities. As a result that Degree centrality produces overlapping influence and limits the influence spread. It can be seen from Fig. 6.20, the influence spread increases more and more gently with selecting more and more top degree nodes. This shows CMA-IM can reduce the overlapping influence spread and find more influential nodes. For the running time, when the network grows larger, the low effectiveness of CELF becomes apparent. It takes hours to find 30 seeds while CMA-IM only needs minutes.

Figure 6.21 shows the influence spread on the NetHEPT network. From Fig. 6.21, it is seen that the influence spread produced by CMA-IM almost matches that of CELF with 1.5% lower. The results of the three community-based algorithms are still close to each other. CMA-IM is 12.5, 13.2 and 173.5% higher than Degree, PageRank, and Random, respectively. When looking at the running time, CELF takes two orders of magnitude longer time than CMA-IM. From the comparison between the running

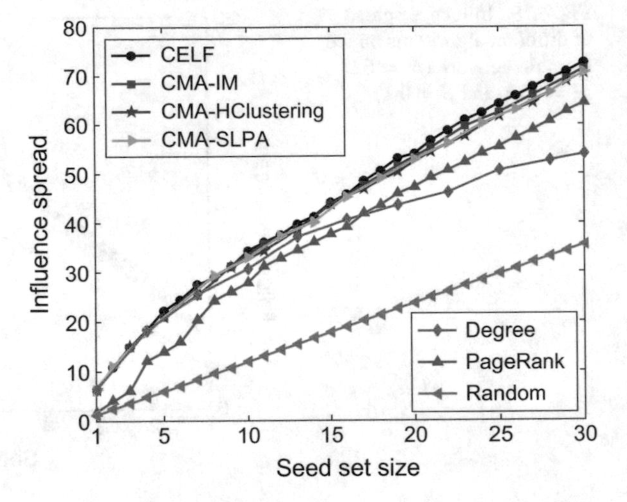

Fig. 6.20 Influence spread of different algorithms on the NetGRQC network ($N = 5242$, $M = 14496$, and $p = 0.01$)

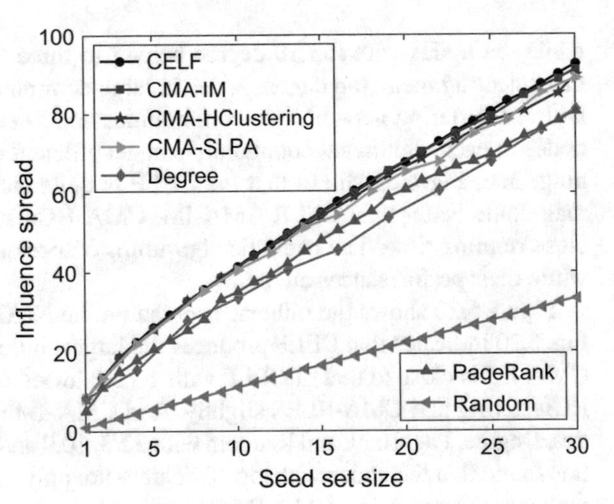

Fig. 6.21 Influence spread of different algorithms on the NetHEPT network ($N = 15233$, $M = 58891$, and $p = 0.01$)

time of CMA-IM on NetGRQC and NetHEPT, it is seen that the running time on NetHEPT is close to NetGRQC due to the network clustering step which reduces the search space of candidate nodes efficiently.

As summarized from the experimental comparisons, the memetic algorithm plays an important role in speeding up the convergence and finding the promising solutions in a low running time. For the running time, the Random has the best performance. The Degree centrality and PageRank also perform better than CMA-IM. However, the Degree centrality, PageRank and Random cannot provide a seed set with good quality. Although the CELF method can provide a reliable seed set for influence maximization, it is not scalable for large-scale networks. The CMA-IM algorithm can solve the problem of the influence maximization both effectively and efficiently on social networks with different sizes.

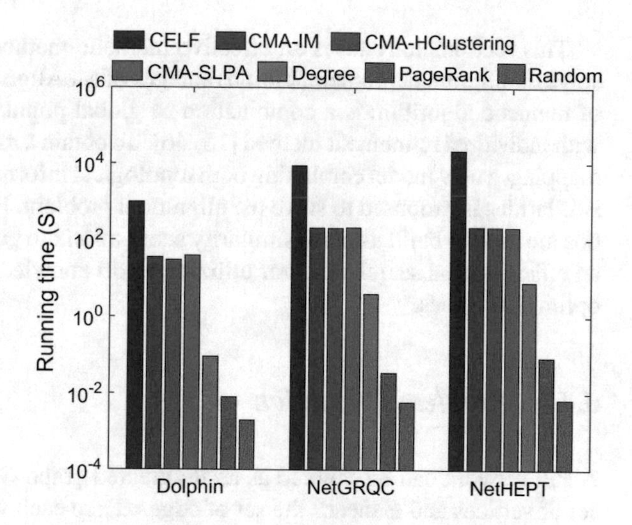

Fig. 6.22 The running time of different algorithms for three real-world networks

6.3.6 Conclusions

This section introduces an efficient memetic algorithm for information maximization. The community property has been incorporated to reduce the search space of seed nodes effectively. Then a problem-specific memetic algorithm has been proposed to optimize the 2-hop influence spread which can estimate the influence spread of a node set effectively. In the memetic algorithm, a problem-specific population initialization method and a similarity-based local search procedure are proposed, which can accelerate the convergence of the algorithm. The similarity between nodes is taken into consideration to solve the problem of influence overlapping. The experiments on three real-world networks illustrate that the proposed CMA-IM algorithm has a good performance in terms of the effectiveness and efficiency on social networks with different sizes.

6.4 Global Biological Network Alignment with Evolutionary Optimization

Global biological network alignment (GNA) can be used to identify conserved subgraphs and understand evolutionary trajectories across species. In biological network alignment, it is difficult to consider information of both network structure and nodes properties simultaneously. In addition, GNA resembles subgraph isomorphism problem which is proved to be NP-complete [21]. That is to say, there are no efficient methods to solve it in polynomial time. What's more, the search space of GNA problem is $n!$ where n is the number of nodes of the large input network. When the size of the network becomes larger, the growth of search space gets even faster than that of an exponent.

This sections introduces an effective memetic method, denoted as MeAlign, to solve the alignment problem. The advantages of MeAlign are as follows. The essence of memetic algorithm is a combination of global population-based search method with individual refinement method [15, 46]. To obtain a more biologically significant mapping, a new model combining both topological information and protein sequence similarities is proposed to solve the alignment problem. Besides, the two-step methods are used to build a coarse similarity score matrix to guide the initialization. Then an efficient local search operator utilizing priori knowledge is designed to find local optimal solutions.

6.4.1 Problem Formation

A PPI network can be denoted as an undirected graph $G(V, E)$ where V denotes the set of vertices and E means the set of edges. Here each vertex in $G(V, E)$ represents a protein and each edge is an interaction. Let $|V| = n$, the topological structure of $G(V, E)$ can be represented by an adjacency matrix $A_{n \times n}$. A_{ij} is 1 if there is an edge between node x_i and node x_j in $G(V, E)$, otherwise it equals to 0 [26].

When given two input graphs $G_1 = (V_1, E_1)$ (with $|V_1| = n_1$)and $G_2 = (V_2, E_2)$ (with $|V_2| = n_2$), the global network alignment is defined as follows. The nodes sequence similarities are given by BLAST scores and these similarity scores are normalized into a $n_1 \times n_2$ matrix H. In Fig. 6.23, the weight w_{ij} is the sequence similarity score between node x_i in G_1 and node y_j in G_2. We define the global network alignment as an injective function $f : V_1 \to V_2$. Suppose that $n_1 \leq n_2$, the goal is to find a good alignment that each node in V_1 is mapped to a node in V_2 [39, 53, 55, 62]. It is easy to find that a good alignment should consider both topological structure and nodes sequence properties and make a trade-off between them. Figure 6.23 shows the illustration of global network alignment.

6.4.2 Optimization Model for Biological Network Alignment

The optimization model, which considers both the topological similarity and nodes sequence similarities, is given as

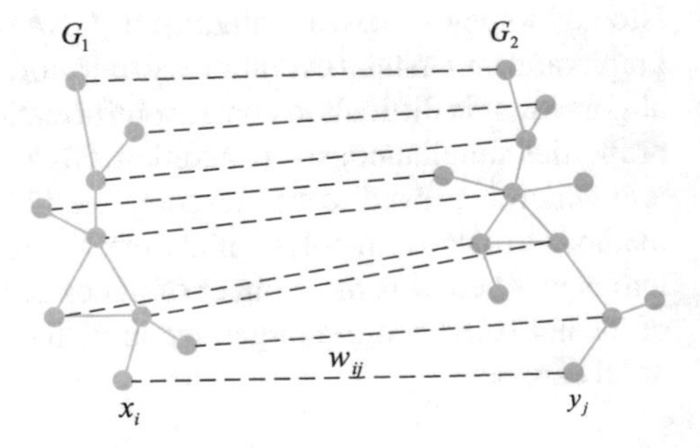

Fig. 6.23 Illustration of global network alignment, where w_{ij} is the sequence similarity score between node x_i in G_1 and node y_j in G_2

$$f = \alpha T_s + (1 - \alpha) N_s$$

where T_s is the topological similarity of input networks and N_s is the nodes sequence similarities. T_s can be chosen as edge correctness (EC), induced conserved structure (ICS) and symmetric substructure score (S^3). The detail definition of EC, ICS and S^3 will be given in the next chapter. EC, ICS and S^3 are also chosen as the evaluation metrics to measure the topological similarity of an alignment. N_s is the nodes sequence similarities and is calculated as

$$N_s = \frac{\sum\limits_{(i,j)\in A} H_{i,j}}{\sum \max(H)}$$

where A is an alignment and H is a normalized nodes sequence similarity score matrix. $\sum \max(H)$ is the sum of the max sequence similarity score in each row of matrix H. α is a weight parameter to control the relative importance of topological similarity and nodes sequence similarities and its value is confined to [0,1].

6.4.3 Memetic Algorithm for Network Alignment

In the memetic algorithm, first design a coarse similarity matrix to guide our initialization. Then an objective function, consisting of both topological measure and node sequences measure, is defined to illustrate the structural similarity and biological similarity of an alignment. After that, the crossover operator [53] is used to produce offsprings and a designed local search algorithm is adopted to optimize them. At last, all parents and the optimized offsprings are put together and they are sorted by the fitness values. Then choose the top p of them to form the new population. Here p is the population size. The framework of MeAlign is outlined in Algorithm 30.

Algorithm 30 Framework of MeAlign

1: **Input**: Networks $G_1(V_1, E_1)$ and $G_2(V_2, E_2)$, Maximum number of iteration: N; Population size: S_{pop}; Weight of relative importance of topology and sequence similarities: α; Size of mating pool: S_{pool}.
2: Building the coarse similarity matrix S;
3: $P \leftarrow$ Initialization(S_{pop}, S);
4: **repeat**
5: $\quad P_{parent} \leftarrow$ Roulette(P, S_{pool})
6: $\quad P_{offsping} \leftarrow$ Crossover(P_{parent})
7: $\quad P_{new} \leftarrow$ LocalSearch($P_{offsping}$)
8: $\quad P \leftarrow$ Update(P, P_{new})
9: **until** Termination(N)
10: **Output**: a mapping ma $\rightarrow$ mb

First, the functions used in Algorithm 30 are explained. First of all, The Initialization() function generates the initial population on the basis of the coarse similarity matrix S. The Roulette() procedure is the selection operator. Here the widely used roulette wheel selection is chosen to select parental population for genetic operators. The Crossover() function is the crossover operator that is used to generate offsprings. The LocalSearch() function is a local search operator used to optimize offsprings generated by the crossover operator. The Update() function uses updated strategy to select the next population from the current population and the offspring population. The Termination() procedure judges when to stop the program. Here a maximum iteration value is set in advance. In what follows, there is detailed description of each function.

6.4.4 Representation and Initialization

Given two graphs $G_1 = (V_1, E_1)$ and $G_2 = (V_2, E_2)$, let $n_1 = |V_1|$ and $n_2 = |V_2|$. Supposing $V_1 = \{x_1, x_2, ..., x_{n_1}\}$ and $V_2 = \{y_1, y_2, .., y_{n_2}\}$, each node in set V_1 is allocated a digital label which is equal to its subscript. In other words, a seriation $\sigma_1 = (1, 2, \ldots, n_1)$ is used to represent the node set V_1. And for V_2, a random permutation σ_2=$(a_1, a_2, ..., a_{n_2})$ is used to stand for it. A rule is made in advance that the final mapping is the permutation σ_3=$(a_1, a_2, ..., a_{n_1})$ which consists of the first n_1 element of σ_2. Seriation σ_1 is fixed and permutation σ_2 is changed, so each different permutation σ_2 represents a different alignment. At last, the network alignment problem is converted to find a permutation that has a high fitness value of objective function in n_2 dimensional space. Figure 6.25a shows the representation. Permutations σ_1 and σ_2 represent two different networks and we map σ_1 to the first n_1 elements of σ_2 to form an alignment.

As is well-known, a good initialization is important to evolutionary computation. So a coarse similarity matrix is constructed to initialize the population instead of initializing the population randomly. This similarity matrix makes use of both topological information and sequence similarities information of input networks. Relative degree difference and relative clustering coefficient difference [47] are used to represent topological knowledge. A normalized sequence similarity score matrix is taken to express sequence information.

The relative degree difference between node a and b is calculated as

$$DD(a, b) = \frac{|degree(a) - degree(b)|}{\max\{degree(a), degree(b)\}}$$

In Fig. 6.24a, $DD(a, b) = 0$; $DD(a, c) = 0.6$. It can clearly be shown that node a is more similar with node b than node c in terms of topology.

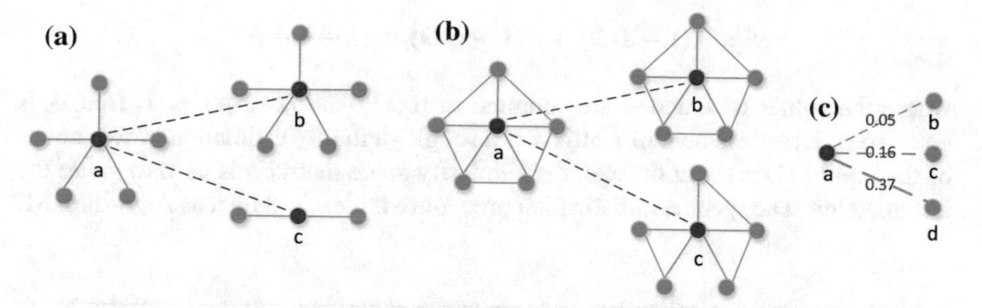

Fig. 6.24 Illustration of initialization. **a-c** show usefulness of relative degree difference, relative clustering coefficient difference and sequence similarity scores, respectively

Before computing the relative clustering coefficient difference, first give the definition of the clustering coefficient of node a as follows:

$$coefficient(a) = \frac{2 \times E_a}{degree(a) \times (degree(a) - 1)}$$

where E_a is the number of edges between neighborhoods of node a. The relative clustering coefficient difference between node a and b is calculated as

$$CD(a, b) = \frac{|coefficient(a) - coefficient(b)|}{\max\{coefficient(a), coefficient(b)\}}$$

In Fig. 6.24b, $CD(a, b) = 0$; $CD(a, c) = 0.4$. In can be concluded that node a is more similar with node b than node c in terms of topology. Obviously, if the relative degree difference and relative clustering coefficient difference between node a and node b is small, then node a and b are similar.

The original sequence similarity matrix of the input networks H can be normalized as follows:

$$\mathrm{H} = \frac{\mathrm{H}}{max(\max(H))}$$

where $max(\max(H))$ is the biggest value of matrix H. Figure 6.24c shows that the normalized similarity scores between node a and nodes b, c, d. Because $H(a, d) > H(a, c) > H(a, b)$, It is said that node a is more similar with node d than node b and node c.

The main purpose of the initialization is to find seed protein pairs. In other words, the initialization purpose is to find protein pairs that have high similarity scores. Because the sequence similarity matrix is very sparse, it is needed to consider topological similarity to compensate for it, which will generate more accurate results.

After all the above preparation, the final similarity score matrix of two input networks is computed as

$$S = \lambda(1 - DD) + \mu(1 - CD) + (1 - \lambda - \mu)H$$

where the values of λ and μ are confined to [0,1]. And $(\lambda + \mu) \leq 1$. Here λ is equal to μ. Every element of matrix S is a rough similarity estimation of two nodes of the input networks. In the end, the similarity score matrix S is used to guide the initialization. The specific initialization procedure is clearly shown as Algorithm 31.

Algorithm 31 Population initialization process

1: **Input**: Population size: S_{pop}; Similarity score matrix: S; Size of the input networks: $n_1, n_2(n_1 < n_2)$.
2: **repeat**
3: Generate a random permutation σ, with length n_1;
4: **for** $i \leftarrow 1$ to n_1 **do**
5: choose the maximum score in row $\sigma(i)$ of S and map the corresponding node in G_2 to node $\sigma(i)$ in G_1;
6: **end for**
7: add the unmatched nodes in G_2 to the end of the mapping to form a permutation;
8: **until** the number of individuals generated is equal to S_{pop}.

6.4.5 Genetic Operators

In this memetic algorithm, the crossover operator proposed by Saraph in 2014 [53] is used. Utilizing Knuths canonical decomposition [35] and cycle decomposition algorithm, the crossover operator can make sure two parent permutations generate one child permutation and the child permutation can inherit almost half characters of its parents.

The crossover operator is shown as Fig. 6.25b. Two parents permutations σ_{21} and σ_{22} generate an offspring permutation σ_{23}. The offspring σ_{23} inherits node 5, node 3, node 2 from parent σ_{21} and node 4, node 6 from parent σ_{22}. The result also shows that this crossover operator can make sure offspring conserve almost half characters of its parents.

6.4.6 The Local Search Procedure

The neighborhood concept is important to design a local search strategy. First, the definition of the neighborhood is used and then a novel neighborhood-based heuristic method is used to search a local optimum solution.

A permutation can represent an alignment. Given a permutation, its neighborhood is defined as permutations that keep the position of "contributors" unchanged. Here "contributors" are nodes that have made contributions to the objective function which consists of both topology and node sequence similarities. More specifically, these points are either nodes of conserved edges or nodes with nonzero value of sequence similarities. In Fig. 6.25c, permutation σ_1 and permutation σ_2 represent two different networks. The red balls are "contributors. For example, if node 2 in σ_1 and node 4 in σ_2 have a nonzero sequence similarity score, then node 4 is reserved. Suppose that there is an edge between node 4 and node 5 in σ_1 and there is also an edge between node 6 and node 3 in σ_2. When mapping node 4 and node 5 in σ_1 to node 6 and node 3 in σ_2, respectively, there will have a conserved edge. So node 6 and node 3 in σ_2 are also reserved.

The neighborhood of a permutation σ is a set of permutations that keep all "contributors" unchanged. In most instances, there are many neighborhoods for a permutation σ. The local search method belongs to very large-scale neighborhood search (VLSN) algorithm [4]. In order to efficiently solve this computationally intractable problem, a heuristic particular neighborhood-based local search (*short for PNLS*) algorithm is used to search for the best solution. *PNLS* algorithm searches the particular neighborhood space of a permutation randomly. Here the particular space is the set of permutations that consist of nodes making no contribution to the fitness function. This searching strategy can promise a better alignment each time at least not worse. Algorithm 32 shows more details of the local search strategy.

Algorithm 32 Local Search procedure

1: **Input**: An alignment: σ; Local search count: c_0. ($c = c_0$);
2: **while** count c is zero **do**
3: find out "contributors" of σ to determine the space of its neighborhood;
4: $\sigma_{new} \leftarrow PNLS(\sigma)$;
5: **if** Fitness(σ_{new}) > Fitness(σ) **then**
6: σ=σ_{new};
7: $c = c_0$;
8: **else**
9: $c \leftarrow c - 1$;
10: **end if**
11: **end while**
12: **Output**: a better alignment σ.

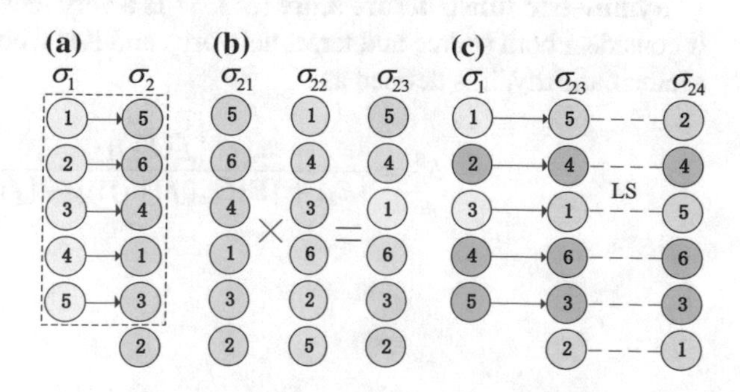

Fig. 6.25 **a-c** show the illustration of Representation, Crossover operator and Local Search operator, respectively

6.4.7 *Experiments Results*

6.4.7.1 Alignment Quality Evaluation

MeAlign is compared with several state-of-the-art methods. Before making comparisons, first give definitions of the used evaluation indexes: Edge correctness (EC) [39], Induced conserved structure(ICS) [50], Symmetric substructure score(S^3) [53] and Functional coherence (FC). EC, ICS and S^3 are metrics to measure topological similarity and FC is to measure biological similarity of proteins.

Given two networks $G_1(V_1, E_1)$ and $G_2(V_2, E_2)$, $f : V_1 \rightarrow V_2$ represents a matching. The conserved edges are expressed as $f(E_1) = \{(f(u), f(v)) \in E_2 : (u, v) \in E_1\}$. Let $f(V_1) = \{f(v) \in V_2 : v \in V_1\}$ be the set of nodes in network G_2 that mapped to G_1. Suppose V_s is the subset nodes of network G_2. Let $G_2(V_s)$ represents the induced subgraph of $G_2(V, E)$ with nodes set V_s. In like manner, $E(G_2(V_s))$ means the edges set of $G_2(V_s)$ which is a subgraph of network $G_2(V, E)$.

Edge correctness (EC). EC is the percentage of conserved edges relative to the source network. Conserved edges are edges in the source network which are aligned with edges in the target network [39]. It can be viewed as the normalization of conserved edges relative to the source network. It is defined as

$$\mathrm{EC} = \frac{|f(E_1)|}{|E_1|}$$

Induced conserved structure (ICS). ICS is an improvement of EC proposed by Patrol in 2012. It is the percentage of conserved edges relative to edges between a set of nodes in the target network that maps to nodes in the source network [50]. It is defined as

$$\mathrm{ICS} = \frac{|f(E_1)|}{|E(G_2(f(V_1)))|}$$

ICS is proposed to penalize alignment that maps sparser regions to denser ones. The drawback of ICS emerges when the size of the two networks is not in the order of magnitude. Imaging when align a small network to a big one, it tries to give a decentralized mapping which is not our original intention.

Symmetric substructure score (S^3). S^3 is a very recent topological metric [53]. It considers both source and target networks and it is a combination of EC and ICS. Mathematically, it is defined as

$$S^3 = \frac{|f(E_1)|}{|E_1| + |E(G_2(f(V_1)))| - |f(E_1)|}$$

Functional coherence (FC). To tell the quality of an alignment, a more important evaluation index is whether the alignment results can reflect biological relevance. In the biological field, GO (gene ontology) terms are used to mark biological properties of proteins. We use GO to calculate functional coherence score that reflects the level of similarity of proteins. This method can also be referred in [5, 17, 55]. GO terms are organized in a hierarchical structure. First collect all GO terms of each pair of aligned proteins and then extract GO terms at level five in GO hierarchy. Choosing level five can avoid comparing GO terms at different levels. What's more, GO terms that are at a closer level from the root of the GO tree denote a more abstract function. Choosing level five also helps remove GO annotations that are not specific enough. Finally, the functional coherence score of aligned proteins is calculated as

$$F(u, v) = \frac{|S_u \cap S_v|}{|S_u \cup S_v|}$$

where S_u and S_v are the set of GO terms of protein u and protein v at level five in GO hierarchy, respectively. That proteins both have GO terms at level five is the requirement to calculate functional coherence score. All FC scores over aligned protein pairs are used to measure biological similarity of an alignment. Additionally, the percentage of "exact" aligned proteins is also given. Two proteins can be regarded as "exact" only when they are annotated by identical GO terms in level five.

6.4.7.2 Data Sets

MeAlign is tested on five eukaryotic species which are Homo sapiens (human), Drosophila melanogaster (fly), Saccharomyces cerevisiae (yeast), Caenorhabditis elegans (worm), and Mus musculus (mouse).

The source data can be obtained from Isobase which is a database of functionally related orthologs [49]. Besides, the source data of these species can also be obtained from database IntAct [34]. The nodes and edges information of these five species are shown in Table 6.10.

Table 6.10 Nodes and edges information of species of human, fly, yeast, worm, and mouse

Species	Abbreviation	Nodes	Edges
H. sapiens (human)	hs	9633	34327
D. melanogaster (fly)	dm	7518	25635
S. cerevisiae (yeast)	sc	5499	31261
C. elegans (worm)	ce	2805	4495
M. musculus (mouse)	mm	290	242

Table 6.11 Parameter values used in MeAlign and the comparing algorithms

Algorithm	Parameter	Meaning	Value
0MI-GRAAL	p	Choose measure types	3
NETAL	a	Weight of similarity and interaction score	0.01
	i	Iterations	2
MAGNA++	m	Optimizing measure	EC
	p	Population size	300
	n	Iterations	300
IsoRank	α	Weight of network and sequence data	0.8
	K	Iterations	30
	thresh	Threshold for L1 norm of the change in R	1e-4
	maxvecln	Number of nonzero entries in eigenvector	1000k
MeAlign	α	Weight of network and sequence data	0.1
	p	Population size	300
	n	Iterations	300
	S_{pool}	Mating pool size	30
	c_0	Local search count	30

6.4.7.3 Experimental Description

MeAlign is compared with five algorithms which are IsoRank [55], MI-GRAAL [40], NETAL [48], MAGNA++ [58] and MeAlign's variant without local search operator.

For MAGNA++ and MeAlign, the population size and the maximum number of generations are set as 300. Besides, α, a parameter to control the relative importance of topology and sequence similarities, is set as 0.1 for both MAGNA++ and MeAlign. MI-GRAAL uses graphlet degree signature distance and relative degree difference to construct the confidence scores. Parameters not mentioned are default values that are recommended by the presented references. The detailed information of parameters used in comparing algorithms is shown in Table 6.11.

MeAlign is applied to optimize a fitness function composed of edge correctness and nodes sequence similarities. MeAlign is compared with MI-GRAAL, NETAL, MAGNA++, IsoRank and Non-LS over five eukaryotic species which are Homo sapiens (human), Drosophila melanogaster (fly), Saccharomyces cerevisiae (yeast), Caenorhabditis elegans (worm), and Mus musculus (mouse), respectively. The difficulty of PPI network alignment problem is to obtain both high topological and biological properties simultaneously. In order to evaluate MeAlign and the other comparing algorithms, EC, ICS, and S^3 are used to measure topological similarity and FC to evaluate biological similarities.

Figure 6.26a-c show the values of EC, ICS and S^3 of six methods over five species, respectively. As a whole, the EC, ICS and S^3 values of MeAlign over ten species pairs are lower than those of MI-GRAAL and NETAL. But it is more effective than

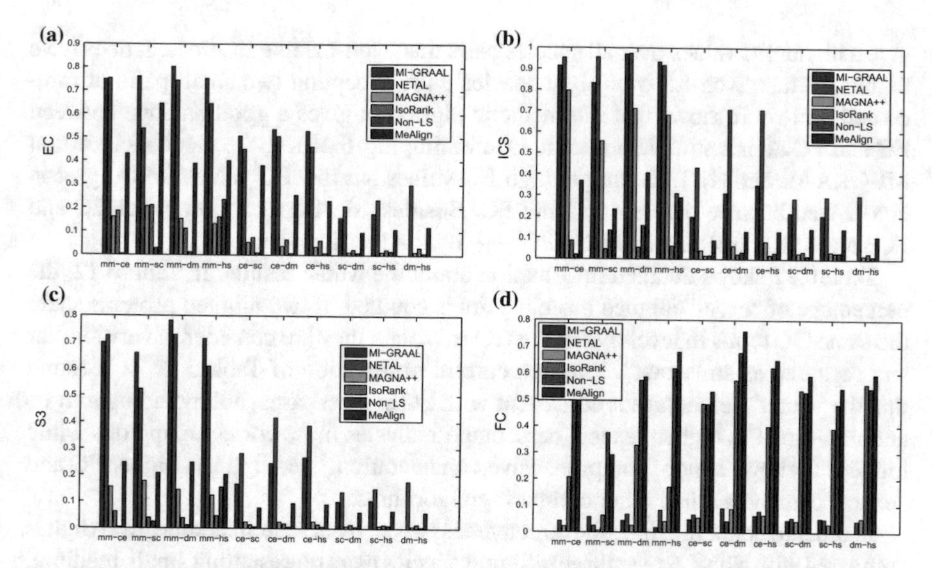

Fig. 6.26 Values of topological metrics EC, ICS, S^3 and biological metric FC of algorithms MI-GRAAL, NETAL, MAGNA++, IsoRank, Non-LS and MeAlign over five species human (hs), fly (dm), yeast (sc), worm (ce), and mouse (mm). Here MAGNA++ represents a genetic algorithm

that of IsoRank, MAGNA++ and Non-LS. As to topological metrics, MI-GRAAL has a little advantage over NETAL when the size of the network is relatively small. But NETAL shows advantages in large-scale network pairs, e.g., sc-dm, sc-hs, and dm-hs. Additionally, MI-GRAAL crashes over the biggest species pair dm-hs (fly-human). NETAL can get good topological values because it only considers topology and leaves sequence similarities out of consideration. With the same population and the same iterations, MeAlign has higher EC and S^3 values than that of MAGNA++ and Non-LS. These results show the effectiveness of the local search operators. In Fig. 6.26b, MAGNA++ has higher ICS values than MeAlign over four species pairs that include mm. As is explained in preceding part of the text, ICS shows its disadvantages when the sizes of the two networks differ greatly. ICS tends to map nodes to sparse areas. When the sizes of the input networks have the same order of magnitude, such as ce-sc, ce-dm, ce-hs, sc-dm, sc-hs, and dm-hs, our algorithm shows advantages over MAGNA++ regardless of EC,ICS and S^3. From Fig. 6.26a, it is also seen that Non-LS has lower EC values than MAGNA++ over all species pairs. This may be caused by two reasons. One is that the designed initialization tends to find more biological similar pairs and it limits a higher topological similarity to some extent. The other reason is the smaller size of the crossover mating pool. The size of mating pool of MAGNA++ and Non-LS is 300 and 30, respectively.

Figure 6.26d shows FC of MeAlign and the comparing algorithms over different species. In this term, MeAlign wins. The results clearly show that MeAlign gets higher FC values than MI-GRAAL, NETAL, MAGNA++ and IsoRank over all species pairs regardless of big or small network size. In expectations, MeAlign will give a higher FC values over all species pairs than Non-LS like EC does. It does have higher FC than Non-LS over eight species pairs excepting two small pairs of mm-ce and mm-sc. It shows that the memetic algorithm gives a good balance between EC and FC after a sufficient search. Combining Fig. 6.26a, d, we can conclude that MI-GRAAL and NETAL can get high EC values but low FC values. MeAlign can give a well balance between EC and FC. Besides, MeAlign can get better EC and FC values than IsoRank and MAGNA++ over all testing species.

Table 6.12 shows detailed information about the whole results. In Table 6.12, the percentage of "exact" aligned protein pairs is counted. If two aligned proteins share the same GO terms in level five of the GO tree, then they are considered very similar and regarded as an "exact". Through careful observation of Table 6.12, it is found that the "exact" percentage is consistent with FC in most cases, not including mm-ce and mm-dm. The higher "exact" percentage indicates more correct mappings while higher FC shows a more comprehensive consideration. MeAlign has higher FC and "exact" percentage than other comparing algorithms.

In order to better illustrate the effectiveness of the local search module, MeAlign is compared with Non-LS over three different species pairs representing small, medium, and large size of networks, respectively. Figure 6.27 shows the process of optimizing fitness function (represented as Fitness in the figure), EC and FC. The population size and iterations of these two algorithms are both set at 300. The size of the mating pool is 30 and α is set at 0.1 which is consistent with the above. The results of Fitness, EC, and FC over 300 iterations are sampled by an interval of 15, so it does not look so crowded. Points at x-coordinate of zero are the initialization results.

In Fig. 6.27, solid lines denote MeAlign and dotted lines represent Non-LS. The results show that MeAlign always obtains higher fitness function values than Non-LS regardless of the network size. It is proved the effectiveness of the local search module. More specifically, the final EC results of MeAlign are much better than those of Non-LS in the three species pairs. And the FC results of MeAlign in Fig. 6.27b, c also have advantages over Non-LS. All these improvements are attributed to the local search operator. In the local search operator, "contributory" nodes are conserved and they are recombined through the crossover operator to generate new individuals. The topological information and sequence similarities are utilized to find "contributory" nodes. And this is the reason for the effectiveness of the local search module.

In Fig. 6.27a, the final FC result of MeAlign is lower than that of Non-LS. It is a trade-off between EC and FC. At the beginning, as EC increases, FC declines. This also reflects the difficulty of network alignment problems. The expectation is to obtain both higher EC and FC values. It does have the moment that EC and FC increase simultaneously. Regrettably, that one increases and the other declines are more commonly to happen. In Fig. 6.27, the solid lines represent MeAlign. At this time MeAlign is observed separately. The overall trend of MeAlign is that both EC and FC increase as the number of iterations goes up. When EC and FC cannot increase simultaneously, a trade-off between EC and FC is made. The dotted lines represent Non-LS. When Non-LS is observed alone, it is seen that Non-LS gets high FC values but low EC values. The results are attributed to the initialization. In the initialization, α, a weight that controls the relative importance of network topology

Table 6.12 Performance of our memetic algorithm and other methods in some metrics over five species

Species	Metrics	MI-GRAAL	NETAL	MAGNA++	IsoRank	Non-LS	MeAlign
mm-ce (290-2805)	EC	0.7479	**0.7521**	0.1612	0.1860	0.0041	0.4298
	ICS	0.9141	**0.9529**	0.7959	0.0852	0.0204	0.3675
	S3	0.6988	**0.7251**	0.1548	0.0621	0.0034	0.2470
	FC	0.0000	0.0417	0.0219	0.2060	**0.5302**	0.5043
	exact	0.00%	0.00%	0.00%	11.11%	20.93%	**23.40%**
mm-sc (290-5499)	EC	**0.7603**	0.5372	0.2066	0.2107	0.0289	0.6446
	ICS	0.8326	**0.9559**	0.5556	0.0459	0.1346	0.2346
	S3	**0.6595**	0.5242	0.1773	0.0392	0.0244	0.2077
	FC	0.0569	0.0242	0.0063	0.0628	**0.3747**	0.2911
	exact	4.88%	0.00%	0.00%	2.33%	**17.31%**	12.07%
mm-dm (290-7518)	EC	**0.8017**	0.7438	0.1529	0.1116	0.0124	0.3884
	ICS	0.9700	**0.9730**	0.7115	0.0545	0.1579	0.4069
	S3	**0.7823**	0.7287	0.1440	0.0380	0.0116	0.2480
	FC	0.0439	0.0671	0.0253	0.3278	0.5721	**0.5853**
	exact	0.00%	2.78%	0.00%	20.83%	**40.00%**	38.46%
mm-hs (290-9633)	EC	**0.7521**	0.6446	0.1322	0.1612	0.2355	0.4008
	ICS	**0.9333**	0.9176	0.6809	0.0614	0.3098	0.2904
	S3	**0.7137**	0.6094	0.1245	0.0465	0.1545	0.2025
	FC	0.0174	0.0828	0.0403	0.1702	0.6179	**0.6789**
	exact	0.00%	2.94%	0.00%	6.00%	44.74%	**51.28%**
ce-sc (2805-5499)	EC	**0.5066**	0.4496	0.0536	0.0743	0.0093	0.3802
	ICS	0.1984	**0.4308**	0.0404	0.0161	0.0171	0.0888
	S3	0.1663	**0.2821**	0.0236	0.0134	0.0061	0.0776
	FC	0.0483	0.0665	0.0553	0.4840	0.4828	**0.5300**
	exact	1.52%	2.93%	1.83%	34.48%	35.93%	**39.54%**
ce-dm (2805-7518)	EC	**0.5348**	0.5048	0.0374	0.0652	0.0073	0.2554
	ICS	0.3108	**0.7073**	0.0619	0.0234	0.0324	0.1303
	S3	0.2447	**0.4176**	0.0239	0.0175	0.0060	0.0944
	FC	0.0910	0.0421	0.0886	0.5737	0.6774	**0.7588**
	exact	6.07%	1.79%	4.89%	43.75%	54.18%	**61.07%**
ce-hs (2805-9633)	EC	**0.5284**	0.4616	0.0352	0.0612	0.0082	0.2305
	ICS	0.3499	**0.7063**	0.0827	0.0200	0.0298	0.1308
	S3	0.2666	**0.3873**	0.0253	0.0153	0.0065	0.0910
	FC	0.0635	0.0802	0.0653	0.4002	0.6212	**0.6490**
	exact	2.48%	3.40%	3.30%	26.96%	44.23%	**47.07%**

(continued)

Table 6.12 (continued)

Species	Metrics	MI-GRAAL	NETAL	MAGNA++	IsoRank	Non-LS	MeAlign
sc-dm (5499-7518)	EC	0.1582	**0.2078**	0.0119	0.0234	0.0097	0.0967
	ICS	0.2165	**0.2861**	0.0265	0.0338	0.0137	0.1424
	S3	0.1006	**0.1368**	0.0083	0.0140	0.0057	0.0611
	FC	0.0460	0.0547	0.0424	0.5357	0.5289	**0.5909**
	exact	0.93%	2.10%	1.93%	42.57%	40.77%	**45.59%**
sc-hs (5499-7518)	EC	0.1770	**0.2348**	0.0130	0.0232	0.0167	0.1097
	ICS	0.2045	**0.2926**	0.0331	0.0292	0.0301	0.1430
	S3	0.1048	**0.1498**	0.0094	0.0131	0.0109	0.0662
	FC	0.0521	0.0497	0.0309	0.3927	0.4826	**0.5436**
	exact	2.01%	1.40%	0.77%	27.99%	34.87%	**39.92%**
dm-hs (7518-9633)	EC	–	**0.2996**	0.0130	0.0324	0.0082	0.1202
	ICS	–	**0.3003**	0.0167	0.0270	0.0127	0.1061
	S3	–	**0.1765**	0.0073	0.0150	0.0050	0.0597
	FC	–	0.0432	0.0525	0.5597	0.5436	**0.5960**
	exact	–	1.52%	1.83%	42.10%	39.95%	**44.10%**

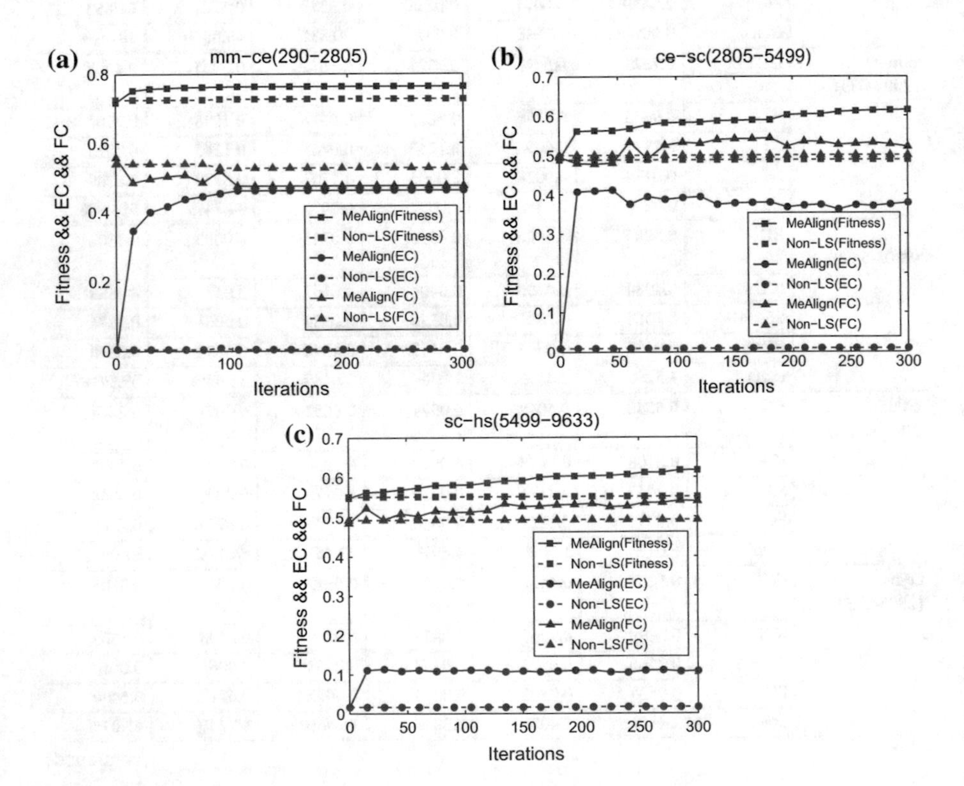

Fig. 6.27 Compare MeAlign with Non-LS algorithm over three species pairs mm-ce, ce-sc and sc-hs. *Solid lines* and *dotted lines* represent MeAlign and Non-LS, respectively. This figure shows the best results of MeAlign and Non-LS in each iteration. The values of Fitness function, EC and FC are sampled by an interval of 15 over 300 iterations

and nodes sequence similarities, is set at 0.1. This is the reason why the FC values are larger than the EC values after the initialization. Besides, the results of Non-LS almost come to a standstill with the increase of the iterations. It reflects the low effectiveness of genetic algorithm to some extent.

In Fig. 6.27b, there are a few points that both EC and FC value decrease simultaneously while the value of the objective function increases. This may be caused by the following reasons. First, PPI data are still incomplete and inaccuracy so far. Second, high sequence similarities do not always consistent with high FC values. This may also reflect both the model and the evaluation indexes can still be improved.

6.4.7.4 Comparison with the Existing Evolutionary Algorithms

MAGNA [53] is the first one to use a genetic algorithm to solve the PPI network alignment problem. In MAGNA, a novel crossover operator is proposed and a random initialization genetic algorithm is adopted to optimize topological metrics like EC. MAGNA++ [58], an extension of MAGNA, changes its optimization model to include both topological similarity and sequence similarities. Different from

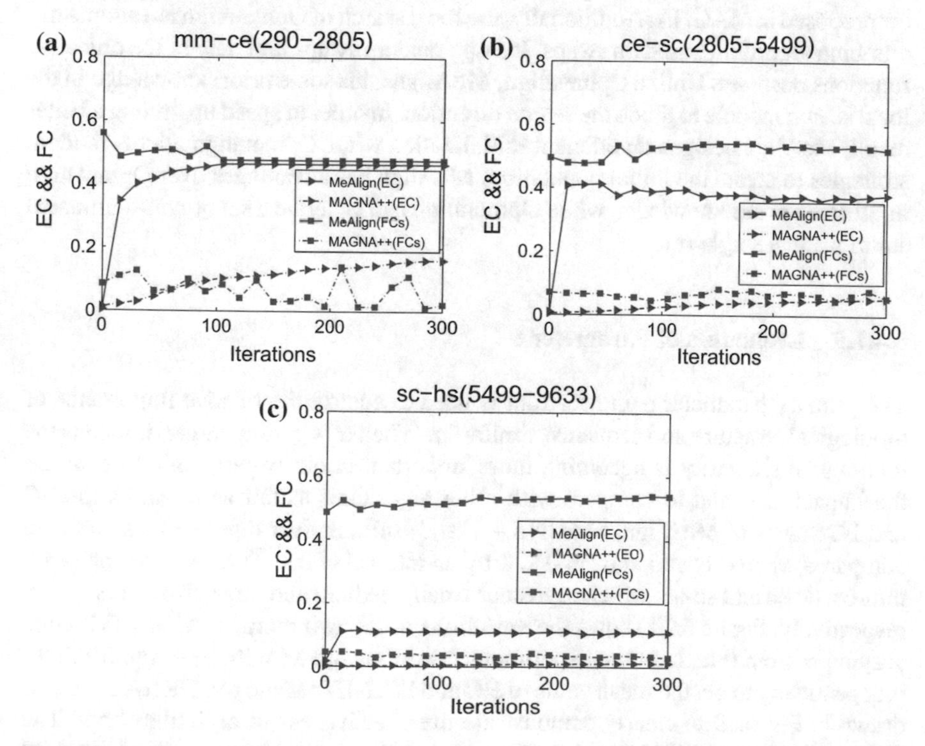

Fig. 6.28 Compare MeAlign with MAGNA++ algorithm over three species pairs mm-ce, ce-sc and sc-hs. *Solid lines* and *dotted lines* represent MeAlign and MAGNA++, respectively. This figure shows the best results of MeAlign and MAGNA++ in each iteration. The values of EC and FC are sampled by an interval of 15 over 300 iterations

MAGNA++, MeAlign designs an efficient initialization and an effective local search module. As is known, a good initialization can help a lot of evolutionary algorithms, especially that NP-hard problem. In addition, MeAlign utilizes priori knowledge in local search module to speed up searching better solutions. Figure 6.28 shows that MeAlign is a more effective algorithm than MAGNA++ in the same iterations. Three species pairs mm-ce, ce-sc, and sc-hs are chosen to represent small, medium, and large network size, respectively. Here the metrics EC and FC are chosen to compare the final result of MeAlign and MAGNA++. ICS and S^3 are not chosen for measuring topological similarity. Because ICS is not a good choice when the size of the two aligned networks differs greatly. In order to more intuitively see the conserved edges of the two algorithm, EC is chosen rather than S^3. Figure 6.28 illustrates that MeAlign is more effective than MAGNA++ in both EC and FC regardless of the network size.

Another recent evolutionary algorithm, denotes as Optnetalign [20], uses a multiobjective memetic algorithm to discover non-dominated alignments. MeAlign is almost completely different from Optnetalign although they are both called memetic algorithms. The crossover module of Optnetalign adopts the Uniform Partially Matched Crossover (UPMX) operator [18] while MeAlign uses the crossover operator proposed in MAGNA. Additionally, the local search of Optnetalign is a simple hill climbing algorithm based on swaps. It stops random swaps until any of the objective functions decrease. Unlike Optnetalign, MeAlign adds some priori knowledge in the local search module to guide the search direction. In order to speed up finding a better result, MeAlign designs an efficient initialization while Optnetalign adopts random strategies to create the initial population. MeAlign has advantages over Optnetalign in utilizing priori knowledge while Optnetalign can generate a set of non-dominated solutions in a single run.

6.4.7.5 Evaluation of Parameter α

The primary parameter α in MeAlign is used to control the relative importance of topological structure and sequence similarity. When α is getting larger, it means the topological similarity is becoming more important in our experiment. To evaluate the impact of α and to compare with other algorithms as fair as possible, the EC and FC values of MeAlign, MAGNA++ and IsoRank over three species pairs are compared when α is varied from 0 to 1 by an interval of 0.1. Three species pairs are mm-ce, ce-sc and sc-hs, which represent small, medium and large size of networks, respectively. Figure 6.29 shows the variations of the two metrics EC and FC while varying α from 0 to 1. In the experiment, MeAlign and MAGNA++ run 30 times independently to get the mean value of EC and FC. NETAL and MI-GRAAL are also drawn in Fig. 6.29 to clearly demonstrate the effectiveness of each algorithm. The EC or FC value of NETAL or MI-GRAAL is a point in each subgraph of Fig. 6.29 and it can be found on the vertical line while α is equal to 1. NETAL is proposed to optimize topological structure only and its result can be regarded as one state when α is equal to 1. MI-GRAAL provides five similarity measures to build a matrix of confidence scores. In the experiments, two topological similarity measures, graphlet degree signature distance, and relative degree difference, are chosen to run MI-GRAAL. In this condition, MI-GRAAL is also shown on the vertical line while α is equal to 1. IsoRank uses two different strategies for extracting the final mapping. One is tantamount to solve a max-weight bipartite matching problem and the other is a heuristic strategy. The former is chosen because it can give a mapping including all nodes of the smaller network. When α is equal to 0, IsoRank doesn't give the results over mm-ce pair, so the corresponding EC and FC values are not shown in Fig. 6.29a, d.

The parameter of two algorithms is not required to be set as the same value just for fairness when the optimization models of these two algorithms are not exactly the same. For example, the topological optimization models of IsoRank and MeAlign are different, so it is not absolutely fair when α is set at a value recommended by IsoRank. However, setting α at the same value can make comparisons as fair as possible in consideration of the meaning of α, a weight controlling the relative importance of topological similarity and sequence similarity. In Fig. 6.29, the EC and FC values of IsoRank, MAGNA++ and MeAlign are compared with each other while α varies from 0 to 1. Figure 6.29 shows the effectiveness of each algorithm on the whole and it can make comparisons as fair as possible.

In Fig. 6.29, the red lines with triangles pointed to the right represent the results of MeAlign. Figure 6.29a-c show the variation of EC on three different sizes of species pairs when α varies from 0 to 1. As a contrast, Fig. 6.29d-f represent the corresponding results of FC. When observing MeAlign alone, it is seen that EC increases and FC decreases simultaneously on the overall trend as α goes up. This is in expectation because the increase of α means the more important of topology. It is found that EC and FC change slowly when α varies between 0.2 and 0.7. This may be related to the reservation strategy used in our local search operator. In the local search module, the neighborhoods of a permutation are defined as set of permutations that keep the position of contributors unchanged. These contributors include nodes with nonzero sequence similarity scores and nodes of conserved edges. For example, when $\alpha = 0.2$ and $\alpha = 0.8$, the reservation strategy in MeAlign retains all contributors in the same way. It does not retain nodes according to α. α only takes effect in the objective function and it controls the general orientation of searching. Unlike α varies from 0.2 to 0.7, values of EC and FC change dramatically when α takes other values. The above is an overall trend of α impacting EC and FC over MeAlign. More carefully, it is found that several "points" do not follow the expectation. For example, in Fig. 6.29d, when α is equal to 0, the FC value of MeAlign over mm-ce is smaller than that of α equaling to 0.1. This may be caused by two reasons. One is that 30 times running is not enough to obtain the final tendency. The other is the case that when considering the topology structure, it promotes to find higher FC in turn.

In Fig. 6.29, the green lines with dimetric marks represent the results of MAGNA++. In the six subfigures of Fig. 6.29, the EC and FC values of our proposed algorithm are better than those of MAGNA++. In other words, no matter what value α takes, MeAlign can get better results than MAGNA++. The blue lines with dot marks show the results of IsoRank in Fig. 6.29. It can be seen that MeAlign can

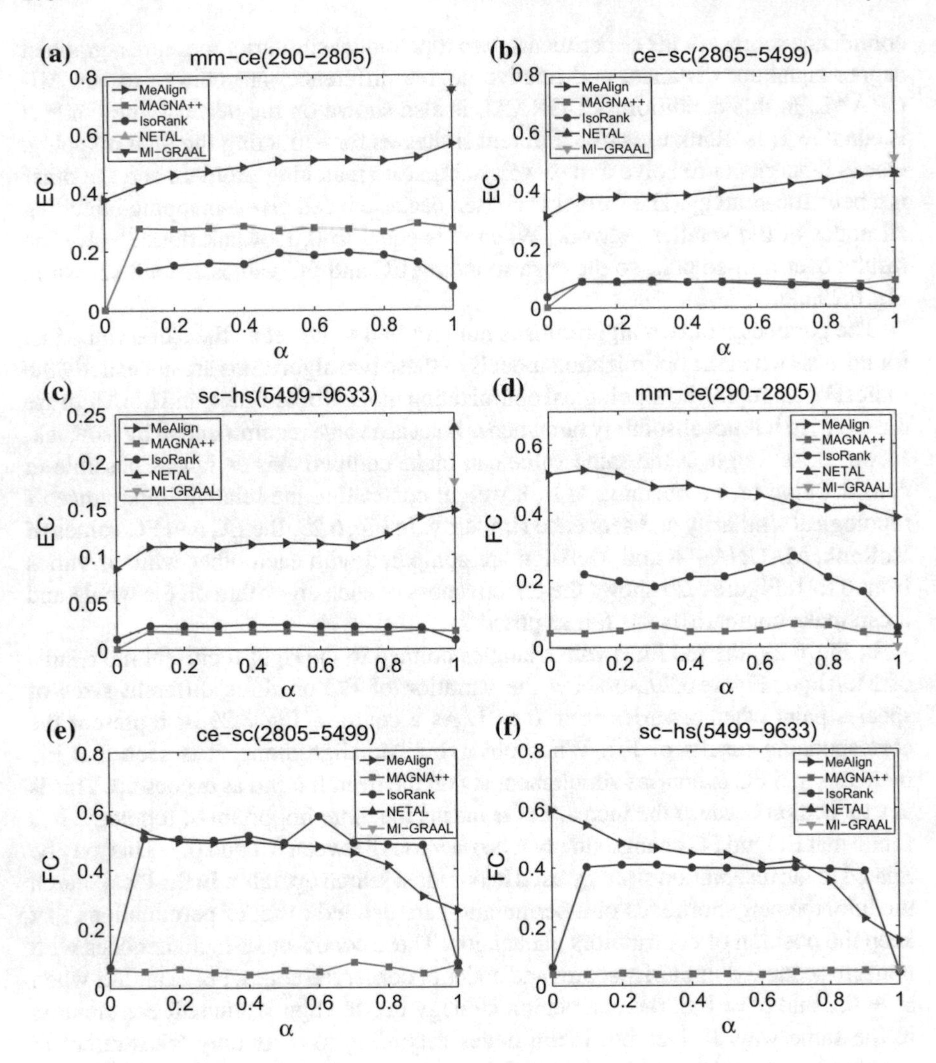

Fig. 6.29 The tendency of EC and FC of MeAlign, MAGNA++ and IsoRank over species pairs mm-ce, ce-sc, and sc-hs while α varies from 0 to 1. For comparison, the results of NETAL and MI-GRAAL can also be found on the vertical line when α is equal to 1. IsoRank does not give the result when α is equal to 0 over mm-ce

obtain better ECs or FCs than those of IsoRank over all values of α in Fig. 6.29a-d. In Fig. 6.29e and f, MeAlign has the advantage over IsoRank when α takes relatively small values. But IsoRank can get better FCs when α adopts bigger values(say 0.8). It can be proved in Fig. 6.29 that MeAlign has the advantage over IsoRank.

A good alignment should ensure that the mapped proteins are evolutionarily related (having higher FC scores) while maximizing the topological similarity. α is chosen at 0.1 after an overall consideration. This is just a recommended value and other values can also be chosen. The guiding intuition in MeAlign is that an alignment obtained in such a way that it keeps a high FC score while optimizing EC value as high as possible should be a good mapping. IsoRank recommends α at 0.6–0.8, and MeAlign recommends 0.1. This seems a big difference in choosing α. In fact, it is caused by the following two reasons. The one is that the topological optimization models of IsoRank and MeAlign are different, so the same α value does not mean the same relative importance of topology and sequence similarity. The other is the reservation strategy used in the local search operator which has been explained above. The different topological optimization models and the different strategies used in the comparing algorithms make α having different impacts on each algorithm.

6.4.8 Conclusions

This section introduces a memetic algorithm MeAlign to solve biological network alignment problem. To obtain a good alignment, an optimization alignment model including both information of topological structure and nodes sequence similarities has been proposed. Besides, a neighborhood-based initialization method and a local search operator are introduced in MeAlign. In the experimental part, MeAlign is tested on five eukaryotic species which are human, fly, yeast, worm and mouse. And the MeAlign is compared against some classical algorithms such as IsoRank, MI-GRAAL, NETAL, and MAGNA++. Moreover, to prove the effectiveness of our initialization and the designed local search operator, comparisons between MeAlign and its variant version without local search has been made. The experimental results have shown that MeAlign outperforms other algorithms and can give a better balance between topology and nodes sequence similarities.

References

1. Adomavicius, G., Kwon, Y.: Improving aggregate recommendation diversity using ranking-based techniques. IEEE Trans. Knowl. Data Eng. **24**(5), 896–911 (2012)
2. Adomavicius, G., Tuzhilin, A.: Toward the next generation of recommender systems: a survey of the state-of-the-art and possible extensions. IEEE Trans. knowl. Data Eng. **17**(6), 734–749 (2005)
3. Aebersold, R., Mann, M.: Mass spectrometry-based proteomics. Nature **422**(6928), 198–207 (2003)
4. Ahuja, R.K., Ergun, Ö., Orlin, J.B., Punnen, A.P.: A survey of very large-scale neighborhood search techniques. Discrete Appl. Math. **123**(1), 75–102 (2002)
5. Aladağ, A.E., Erten, C.: Spinal: scalable protein interaction network alignment. Bioinformatics **29**(7), 917–924 (2013)
6. Albert, R., Jeong, H., Barabási, A.L.: Error and attack tolerance of complex networks. Nature **406**(6794), 378–382 (2000)
7. Blin, G., Sikora, F., Vialette, S.: Querying graphs in protein-protein interactions networks using feedback vertex set. IEEE/ACM Trans. Comput. Biol. Bioinformatics (TCBB) **7**(4), 628–635 (2010)

8. Blondel, V.D., Guillaume, J.L., Lambiotte, R., Lefebvre, E.: Fast unfolding of communities in large networks. J. Stat. Mech.: Theory and Exp. **2008**(10), P10,008 (2008)
9. Bobadilla, J., Ortega, F., Hernando, A., Gutiérrez, A.: Recommender systems survey. Knowl.-Based Syst. **46**, 109–132 (2013)
10. Bogdanov, P., Singh, A.K.: Molecular function prediction using neighborhood features. IEEE/ACM Trans. Comput. Biol. Bioinformatics (TCBB) **7**(2), 208–217 (2010)
11. Brin, S., Page, L.: The anatomy of a large-scale hypertextual web search engine. Comput. Netw. ISDN Syst. **30**(1), 107–117 (1998)
12. Burke, R.: Hybrid recommender systems: survey and experiments. User Model. User-Adap. Inter. **12**(4), 331–370 (2002)
13. Cardona, G., Rossello, F., Valiente, G.: Comparison of tree-child phylogenetic networks. IEEE/ACM Trans. Comput. Biol. Bioinformatics (TCBB) **6**(4), 552–569 (2009)
14. Chen, W., Wang, Y., Yang, S.: Efficient influence maximization in social networks. In: Proceedings of the 15th ACM SIGKDD International Conference on Knowledge Discovery and Data Mining, Paris, France, pp. 199–208. ACM (2009)
15. Chen, X., Ong, Y.S., Lim, M.H., Tan, K.C.: A multi-facet survey on memetic computation. IEEE Trans. Evol. Comput. **15**(5), 591–607 (2011)
16. Chen, Y.C., Zhu, W.Y., Peng, W.C., Lee, W.C., Lee, S.Y.: CIM: community-based influence maximization in social networks. ACM Trans. Intell. Syst. Technol. **5**(2), 25 (2014)
17. Chindelevitch, L., Ma, C.Y., Liao, C.S., Berger, B.: Optimizing a global alignment of protein interaction networks. Bioinformatics **486** (2013)
18. Cicirello, V.A., Smith, S.F.: Modeling ga performance for control parameter optimization. In: GECCO, pp. 235–242 (2000)
19. Clark, C., Kalita, J.: A comparison of algorithms for the pairwise alignment of biological networks. Bioinformatics **30**(16), 2351–2359 (2014)
20. Clark, C., Kalita, J.: A multiobjective memetic algorithm for ppi network alignment. Bioinformatics (2015)
21. Cook, S.A.: The complexity of theorem-proving procedures. In: Proceedings of the Third Annual ACM Symposium on Theory of Computing, pp. 151–158. ACM (1971)
22. Deb, K., Pratap, A., Agarwal, S., Meyarivan, T.: A fast and elitist multiobjective genetic algorithm: Nsga-ii. IEEE Trans. Evol. Comput. **6**(2), 182–197 (2002)
23. Domingos, P., Richardson, M.: Mining the network value of customers. In: Proceedings of the Seventh ACM SIGKDD International Conference on Knowledge Discovery and Data Mining, San Francisco, California, pp. 57–66. ACM (2001)
24. Fields, S.: Song, O.k.: A novel genetic system to detect protein-protein interactions. Nature **340**, 245–246 (1989)
25. Flannick, J., Novak, A., Srinivasan, B.S., McAdams, H.H., Batzoglou, S.: Graemlin: general and robust alignment of multiple large interaction networks. Genome Res. **16**(9), 1169–1181 (2006)
26. Fortunato, S.: Community detection in graphs. Phys. Rep. **486**(3), 75–174 (2010)
27. Gong, M., Cai, Q., Chen, X., Ma, L.: Complex network clustering by multiobjective discrete particle swarm optimization based on decomposition. IEEE Trans. Evol. Comput. **18**(1), 82–97 (2014)
28. Goyal, A., Lu, W., Lakshmanan, L.V.: CELF++: optimizing the greedy algorithm for influence maximization in social networks. In: Proceedings of the 20th International Conference Companion on World Wide Web, Hyderabad, India, pp. 47–48. ACM (2011)
29. Hashemifar, S., Xu, J.: Hubalign: an accurate and efficient method for global alignment of protein-protein interaction networks. Bioinformatics **30**(17), i438–i444 (2014)
30. Herlocker, J.L., Konstan, J.A., Terveen, L.G., Riedl, J.T.: Evaluating collaborative filtering recommender systems. ACM Trans. Inf. Syst. (TOIS) **22**(1), 5–53 (2004)
31. Hu, J., Reinert, K.: Localali: an evolutionary-based local alignment approach to identify functionally conserved modules in multiple networks. Bioinformatics **652** (2014)
32. Kelley, B.P., Yuan, B., Lewitter, F., Sharan, R., Stockwell, B.R., Ideker, T.: Pathblast: a tool for alignment of protein interaction networks. Nucleic Acids Res. **32**(suppl 2), W83–W88 (2004)

33. Kempe, D., Kleinberg, J., Tardos, É.: Maximizing the spread of influence through a social network. In: Proceedings of the Ninth ACM SIGKDD International Conference on Knowledge Discovery and Data Mining, pp. 137–146. ACM (2003)
34. Kerrien, S., Aranda, B., Breuza, L., Bridge, A., Broackes-Carter, F., Chen, C., Duesbury, M., Dumousseau, M., Feuermann, M., Hinz, U., et al.: The intact molecular interaction database in 2012. Nucleic Acids Res. **1088** (2011)
35. Knuth, D.E.: The art of computer programming (1997)
36. Kollias, G., Sathe, M., Mohammadi, S., Grama, A.: A fast approach to global alignment of protein-protein interaction networks. BMC Res. Notes **6**(1), 35 (2013)
37. Koren, Y., Bell, R., Volinsky, C.: Matrix factorization techniques for recommender systems. Computer **42**(8), 30–37 (2009)
38. Koyutürk, M., Subramaniam, S., Grama, A.: Functional coherence of molecular networks in bioinformatics. Springer Science & Business Media (2011)
39. Kuchaiev, O., Milenković, T., Memišević, V., Hayes, W., Pržulj, N.: Topological network alignment uncovers biological function and phylogeny. J. R. Soc. Interface **7**(50), 1341–1354 (2010)
40. Kuchaiev, O., Pržulj, N.: Integrative network alignment reveals large regions of global network similarity in yeast and human. Bioinformatics **27**(10), 1390–1396 (2011)
41. Lee, J.R., Chung, C.W.: A fast approximation for influence maximization in large social networks. In: Proceedings of the Companion Publication of the 23rd International Conference on World Wide Web Companion, Seoul, Korea, pp. 1157–1162. International World Wide Web Conferences Steering Committee (2014)
42. Leskovec, J., Kleinberg, J., Faloutsos, C.: Graph evolution: Densification and shrinking diameters. ACM Trans. Knowl. Dis. Data **1**(1), 2 (2007)
43. Leskovec, J., Krause, A., Guestrin, C., Faloutsos, C., VanBriesen, J., Glance, N.: Cost-effective outbreak detection in networks. In: Proceedings of the 13th ACM SIGKDD International Conference on Knowledge Discovery and Data Mining, San Jose, California, USA, pp. 420–429. ACM (2007)
44. Lü, L., Medo, M., Yeung, C.H., Zhang, Y.C., Zhang, Z.K., Zhou, T.: Recommender systems. Phys. Rep. **519**(1), 1–49 (2012)
45. Lusseau, D., Schneider, K., Boisseau, O.J., Haase, P., Slooten, E., Dawson, S.M.: The bottlenose dolphin community of doubtful sound features a large proportion of long-lasting associations. Behav. Ecol. Sociobiol. **54**(4), 396–405 (2003)
46. Moscato, P., et al.: On evolution, search, optimization, genetic algorithms and martial arts: Towards memetic algorithms. Caltech concurrent computation program, C3P Report **826** (1989)
47. Newman, M.E.: The structure and function of complex networks. SIAM Rev. **45**(2), 167–256 (2003)
48. Neyshabur, B., Khadem, A., Hashemifar, S., Arab, S.S.: Netal: a new graph-based method for global alignment of protein-protein interaction networks. Bioinformatics **29**(13), 1654–1662 (2013)
49. Park, D., Singh, R., Baym, M., Liao, C.S., Berger, B.: Isobase: a database of functionally related proteins across ppi networks. Nucleic Acids Res. **39**(suppl 1), D295–D300 (2011)
50. Patro, R., Kingsford, C.: Global network alignment using multiscale spectral signatures. Bioinformatics **28**(23), 3105–3114 (2012)
51. Rahimkhani, K., Aleahmad, A., Rahgozar, M., Moeini, A.: A fast algorithm for finding most influential people based on the linear threshold model. Expert Syst. Appl. **42**(3), 1353–1361 (2015)
52. Resnick, P., Varian, H.R.: Recommender systems. Commun. ACM **40**(3), 56–58 (1997)
53. Saraph, V., Milenković, T.: Magna: maximizing accuracy in global network alignment. Bioinformatics **30**(20), 2931–2940 (2014)
54. Sharan, R., Suthram, S., Kelley, R.M., Kuhn, T., McCuine, S., Uetz, P., Sittler, T., Karp, R.M., Ideker, T.: Conserved patterns of protein interaction in multiple species. Proc. Nat. Acad. Sci. USA **102**(6), 1974–1979 (2005)

55. Singh, R., Xu, J., Berger, B.: Global alignment of multiple protein interaction networks with application to functional orthology detection. Proc. Nat. Acad. Sci. **105**(35), 12763–12768 (2008)
56. Todor, A., Dobra, A., Kahveci, T.: Probabilistic biological network alignment. IEEE/ACM Trans. Computat. Biol. Bioinform. (TCBB) **10**(1), 109–121 (2013)
57. Uetz, P., Dong, Y.A., Zeretzke, C., Atzler, C., Baiker, A., Berger, B., Rajagopala, S.V., Roupelieva, M., Rose, D., Fossum, E., et al.: Herpesviral protein networks and their interaction with the human proteome. Science **311**(5758), 239–242 (2006)
58. Vijayan, V., Saraph, V., T, M.: Magna++: Maximizing accuracy in global network alignment via both node and edge conservation. Bioinformatics (2015)
59. Wang, Y., Cong, G., Song, G., Xie, K.: Community-based greedy algorithm for mining top-k influential nodes in mobile social networks. In: Proceedings of the 16th ACM SIGKDD International Conference on Knowledge Discovery and Data Mining, Washington, DC, USA, pp. 1039–1048. ACM (2010)
60. Wernicke, S.: Efficient detection of network motifs. IEEE/ACM Trans. Comput. Biol. Bioinform. (TCBB) **3**(4), 347–359 (2006)
61. Xu, R., Wunsch, D.: Survey of clustering algorithms. IEEE Trans. Neural Netw. **16**(3), 645–678 (2005)
62. Zaslavskiy, M., Bach, F., Vert, J.P.: Global alignment of protein-protein interaction networks by graph matching methods. Bioinformatics **25**(12), i259–1267 (2009)
63. Zhang, M., Hurley, N.: Avoiding monotony: improving the diversity of recommendation lists. In: Proceedings of the 2008 ACM conference on Recommender systems, pp. 123–130. ACM (2008)
64. Zhou, T., Kuscsik, Z., Liu, J.G., Medo, M., Wakeling, J.R., Zhang, Y.C.: Solving the apparent diversity-accuracy dilemma of recommender systems. Proc. Nat. Acad. Sci. **107**(10), 4511–4515 (2010)
65. Zhou, T., Ren, J., Medo, M., Zhang, Y.C.: Bipartite network projection and personal recommendation. Phys. Rev. E **76**(4), 046,115 (2007)
66. Zitzler, E., Thiele, L.: Multiobjective optimization using evolutionary algorithms–a comparative case study. In: Eiben, A., Bäck, T., Schoenauer, M., Schwefel, H.P. (eds.) Parallel Problem Solving from Nature–PPSN V, pp. 292–301. Springer, Berlin Heidelberg (1998)

Chapter 7
Concluding Remarks

Abstract This book covers most fundamental network structure analytics topics and computational intelligence methods. In previous chapters, we have reviewed the concepts of complex networks and the emerging topics concerning network structure analytics as well as some basic optimization models of these network structure analytics issues. Besides the addressed topics introduced in previous chapters, there are many other network structure analytics topics, such as network construction, information backbone mining, structure analytics of large-scale networks, etc. These topics can also be formulated as optimization problems and may be well solved by computational intelligence methods. In this chapter, we will give several future research directions that we are working on.

7.1 Future Directions and Challenges

(1) Community detection in large-scale and sparse networks

In this book, we mainly identify the community structures of networks with thousands of nodes and edges. However, in real-world systems, e.g., online communication, interactive online game, and recommendation systems, there are millions or billions of users and interactions. Moreover, most of these systems show the sparsity property. The proposed multiobjective optimization and memetic algorithms cannot find the potential community structure of these systems, and how to identify the potential functionality in these systems is still an open challenge. Recently, the GPU, parallel computation, and distribution techniques have been proposed to accelerate the classical greedy algorithms, and they have widely been used to tackle real large-scale optimization problems. In the near future, how to combine the network-specific multiobjective optimization and memetic algorithms with the GPU-accelerated techniques, parallel computation, and distribution techniques for identifying potential communities in large-scale and sparse systems will be interesting research topics.

(2) Design of network knowledge-based computational intelligence algorithms

With the advance of big data, classical operations in computational intelligence algorithms cannot work well. How to combine the knowledge of potential network structure in complex systems with computational intelligence algorithm is a still

M. Gong et al., *Computational Intelligence for Network Structure Analytics*,
DOI 10.1007/978-981-10-4558-5_7

opening issue. Recently, more and more researchers have adopted network knowledge based techniques (i.e., network compression, network coding, network learning, and networks sampling) and algorithms-based techniques (i.e., parallel computation, distributed computing, GPU, and cloud computing) to solve the structure and behavior issues in big data represented by complex networks.

(3) Structural balance transformation in weighted, directed, and dynamic networks

In Chap. 4, we model the real-world complex systems as unweighted and undirected networks, and solve their structural balance computation and transformation. In an unweighted and undirected network, the individuals connect with each other with the same weight 1. In reality, the weight of each interaction may be different, which results in that the cost of transforming a positive/negative unbalance link between nodes v_a and v_b is different from that of transforming an another unbalanced link. Moreover, there are directed connections among individuals. For instance, in online vote networks, the individual a may support the individual b. However, the individual b may vote the individual c, rather than a. In addition, it is highly likely that multiple relations in complex systems are transformed to each other over time, i.e., dynamic properties. Therefore, how to use exact network models such as weighted, directed, and dynamical networks to express real-world complex systems and how to compute and transform unbalanced factors in these systems are interesting topics.

(4) Robust network construction

Most of existing studies on complex network robustness consider the network attack and the network immunization as two separate single-optimization problems. They try to tune the link topology structure and damage/protect the influential nodes to maximize the attack/defense effects, and overlook the attack/defense cost and the practical constraints. How to use the minimal cost to maximize the attack/defense cost and how to promote the cooperative development between the construction of effective attack ways (attacker) and robust networks (defender) are worthy further being studied.

(5) Design robust interdependent networks

In the real world, the coupling behavior of real systems are more complex than the random coupling model and the classical techniques such as the Degree and PageRank are difficult to identify influential nodes in these systems. How to design robust interdependent networks with different coupling ways is an interesting issue. Moreover, recent studies have found the potential relationship between the inter-similarity and the robustness of coupled networks. How to define the inter-similarity between networks and how to increase the inter-similarity of interdependent networks are necessary to be further studied.

(6) Mining information backbone in large-scale network data

Nowadays, we are facing to make choices from massive and disordered information of real systems, especially for online systems. For instance, there are thousands of movies, millions of books, and billions of web pages in online systems. However, it is difficult for us to select the most suitable one. Moreover, there are many mistaken labels in online systems. How to mine the information backbone of online systems and how to use the identified information backbone to recommend entities to users are worthy to be focused, and their studies can be applied to some practical applications (e.g., recommender systems, online communications, and online vote systems).

(7) Alignment in biological networks

The global alignment in biological networks is to find a good mapping that maximizes the network overlap between the proteins in different species while conserves the biological similarity as far as possible. Generally, the two objectives of high network overlap and high biological similarity are contradictory to some extent. There are two difficulties in modeling the global alignment as an MOP. One of the difficulties is that how to design mathematical criteria to represent the two objectives. The other one is that how to ensure that the designed criteria are contradictory with each other.

The global network alignment is a NP-hard problem, which indicates that the number of local maxima in the optimization of the modeled criteria exponentially grows with the increase of the size of complex networks. In generally, there are thousands of nodes and millions of edges in biological networks. The classical MOEAs are hard to find the optimal mapping solution in reasonable generations. The potential network-specific knowledge can guide MOEAs to search the most interesting regions that obtain the optimal mapping solutions. The search ability of MOEAs can be greatly improved when the network-specific knowledge is incorporated into the genetic operators of MOEAs. The difficulties are that how to transform the potential network structure to network-specific knowledge, and how to combine the transformed network knowledge with the genetic operators of MOEAs.

Finally, this book is suitable for students as a study book or for researchers who have just set their feet in this filed as a reference book. Meanwhile, this book cannot avoid defects like typos, technical drawbacks, etc. It is thus highly expected that readers can help us to find out and correct the errors if any. We also welcome potential students to join us and researchers to communicate with us for useful discussions.

Printed in the United States
By Bookmasters